Using R for Introductory Statistics

Using R for Introductory Statistics

John Verzani

A CRC Press Company
Boca Raton London New York Washington, D.C.

Library of Congress Cataloging-in-Publication Data

Verzani, John.
Using R for introductiory statistics / John Verzani.
p. cm.
Includes index.
ISBN 1-58488-4509 (alk. paper)
1. Statistics—Data processing. 2. R (Computer program language) I. Title

QA276.4.V47 2004
519.5—dc22 2004058244

Direct all inquiries to CRC Press, 2000 N.W. Corporate Blvd., Boca Raton, Florida 33431.

Visit the CRC Press Web site at www.crcpress.com

International Standard Book Number 1-58488-4509
Library of Congress Card Number 2004058244
Printed in the United States of America 2 3 4 5 6 7 8 9 0
Printed on acid-free paper

Contents

1 Data 1

1.1 What is data? 1
  1.1.1 Problems 4
1.2 Some R essentials 4
  1.2.1 Starting R 5
  1.2.2 Using R as a calculator 6
  1.2.3 Assignment 7
  1.2.4 Using c() to enter data 9
  1.2.5 Using functions on a data vector 10
  1.2.6 Creating structured data 13
  1.2.7 Problems 14
1.3 Accessing data by using indices 16
  1.3.1 Assigning values to data vector 17
  1.3.2 Logical values 18
  1.3.3 Missing values 20
  1.3.4 Managing the work environment 21
  1.3.5 Problems 21
1.4 Reading in other sources of data 23
  1.4.1 Using R's built-in libraries and data sets 23
  1.4.2 Using the data sets that accompany this book 26
  1.4.3 Other methods of data entry 27
  1.4.4 Problems 29

2 Univariate data 31

2.1 Categorical data 32
  2.1.1 Tables 32
  2.1.2 Barplots 33
  2.1.3 Pie charts 36
  2.1.4 Dot charts 37

2.1.5 Factors 38
2.1.6 Problems 39
2.2 Numeric data 41
2.2.1 Stem-and-leaf plots 41
2.2.2 Strip charts 42
2.2.3 The center: mean, median, and mode 43
2.2.4 Variation: the variance, standard deviation, and `IQR` 48
2.2.5 Problems 53
2.3 Shape of a distribution 56
2.3.1 Histogram 56
2.3.2 Modes, symmetry, and skew 61
2.3.3 Boxplots 64
2.3.4 Problems 66

3 Bivariate data 69
3.1 Pairs of categorical variables 69
3.1.1 Making two-way tables from summarized data 70
3.1.2 Making two-way tables from unsummarized data 71
3.1.3 Marginal distributions of two-way tables 72
3.1.4 Conditional distributions of two-way tables 73
3.1.5 Graphical summaries of two-way contingency tables 74
3.1.6 Problems 75
3.2 Comparing independent samples 77
3.2.1 Side-by-side boxplots 77
3.2.2 Densityplots 78
3.2.3 Strip charts 79
3.2.4 Quantile-quantile plots 79
3.2.5 Problems 80
3.3 Relationships in numeric data 82
3.3.1 Using scatterplots to investigate relationships 82
3.3.2 The correlation between two variables 86
3.3.3 Problems 89
3.4 Simple linear regression 90
3.4.1 Using the regression model for prediction 92
3.4.2 Finding the regression coefficients using `lm()` 92
3.4.3 Transformations of the data 95
3.4.4 Interacting with a scatterplot 96
3.4.5 Outliers in the regression model 98
3.4.6 Resistant regression lines: `lqs()` and `rlm()` 99
3.4.7 Trend lines 101
3.4.8 Problems 102

4 Multivariate Data 105
4.1 Viewing multivariate data 105

4.1.1 Summarizing categorical data 105
4.1.2 Comparing independent samples 108
4.1.3 Comparing relationships 109
4.1.4 Problems 112
4.2 R basics: data frames and lists 113
4.2.1 Creating a data frame or list 114
4.2.2 Accessing values in a data frame 115
4.2.3 Setting values in a data frame or list 122
4.2.4 Applying functions to a data frame or list 123
4.2.5 Problems 124
4.3 Using model formula with multivariate data 125
4.3.1 Boxplots from a model formula 126
4.3.2 The `plot()` function with model formula 126
4.3.3 Creating contingency tables with `xtabs()` 127
4.3.4 Manipulating data frames: `split()` and `stack()` 129
4.3.5 Problems 130
4.4 Lattice graphics 131
4.4.1 Problems 134
4.5 Types of data in R 135
4.5.1 Factors 136
4.5.2 Coercion of objects 138

5 Describing populations 141
5.1 Populations 141
5.1.1 Discrete random variables 142
5.1.2 Continuous random variables 144
5.1.3 Sampling from a population 146
5.1.4 Sampling distributions 148
5.1.5 Problems 148
5.2 Families of distributions 149
5.2.1 The `d`, `p`, `q`, and `r` functions 149
5.2.2 Binomial, normal, and some other named distributions 150
5.2.3 Popular distributions to describe populations 155
5.2.4 Sampling distributions 157
5.2.5 Problems 158
5.3 The central limit theorem 160
5.3.1 Normal parent population 160
5.3.2 Nonnormal parent population 161
5.3.3 Problems 163

6 Simulation 165
6.1 The normal approximation for the binomial 165
6.2 `for` loops 166
6.3 Simulations related to the central limit theorem 168

6.4 Defining a function 169
6.4.1 Editing a function 169
6.4.2 Function arguments 170
6.4.3 The function body 170
6.5 Investigating distributions 172
6.5.1 Script files and `source()` 173
6.5.2 The geometric distribution 174
6.6 Bootstrap samples 176
6.7 Alternates to for loops 177

7 Confidence intervals....................181
7.1 Confidence interval ideas 181
7.1.1 Finding confidence intervals using simulation 181
7.2 Confidence intervals for a population proportion, p 184
7.2.1 Using `prop.test()` to find confidence intervals 187
7.2.2 Problems 188
7.3 Confidence intervals for the population mean, μ 190
7.3.1 One-sided confidence intervals 193
7.3.2 Problems 194
7.4 Other confidence intervals 197
7.4.1 Confidence interval for σ^2 197
7.4.2 Problems 198
7.5 Confidence intervals for differences 199
7.5.1 Difference of proportions 199
7.5.2 Difference of means 201
7.5.3 Matched samples 204
7.5.4 Problems 205
7.6 Confidence intervals for the median 207
7.6.1 Confidence intervals based on the binomial 207
7.6.2 Confidence intervals based on signed-rank statistic 208
7.6.3 Confidence intervals based on the rank-sum statistic 209
7.6.4 Problems 211

8 Significance tests....................213
8.1 Significance test for a population proportion 218
8.1.1 Using `prop.test()` to compute p-values 220
8.1.2 Problems 221
8.2 Significance test for the mean (t-tests) 222
8.2.1 Problems 225
8.3 Significance tests and confidence intervals 227
8.4 Significance tests for the median 228
8.4.1 The sign test 228
8.4.2 The signed-rank test 230
8.4.3 Problems 232

8.5 Two-sample tests of proportion 233
8.5.1 Problems 235
8.6 Two-sample tests of center 237
8.6.1 Two sample tests of center with normal populations 237
8.6.2 Matched samples 241
8.6.3 The Wilcoxon rank-sum test for equality of center 244
8.6.4 Problems 246

9 Goodness of fit 249
9.1 The chi-squared goodness-of-fit test 249
9.1.1 The multinomial distribution 249
9.1.2 Pearson's χ^2 statistic 250
9.1.3 Problems 254
9.2 The chi-squared test of independence 258
9.2.1 The chi-squared test of homogeneity 262
9.2.2 Problems 264
9.3 Goodness-of-fit tests for continuous distributions 266
9.3.1 Kolmogorov-Smirnov test 266
9.3.2 The Shapiro-Wilk test for normality 270
9.3.3 Finding parameter values using `fitdistr()` 273
9.3.4 Problems 275

10 Linear regression..277
10.1 The simple linear regression model 277
10.1.1 Model formulas for linear models 278
10.1.2 Examples of the linear model 278
10.1.3 Estimating the parameters in simple linear regression 279
10.1.4 Using `lm()` to find the estimates 280
10.1.5 Problems 283
10.2 Statistical inference for simple linear regression 284
10.2.1 Testing the model assumptions 284
10.2.2 Statistical inferences 288
10.2.3 Using `lm()` to find values for a regression model 292
10.2.4 Problems 297
10.3 Multiple linear regression 300
10.3.1 Types of models 301
10.3.2 Fitting the multiple regression model using `lm()` 302
10.3.3 Interpreting the regression parameters 305
10.3.4 Statistical inferences 306
10.3.5 Model selection 307
10.3.6 Problems 310

11 Analysis of variance 313
11.1 One-way ANOVA 313

11.1.1 Using R's model formulas to specify ANOVA models 316
11.1.2 Using `oneway.test()` to perform ANOVA 317
11.1.3 Using `aov()` for ANOVA 317
11.1.4 The nonparametric Kruskal-Wallis test 319
11.1.5 Problems 322
11.2 Using `lm()` for ANOVA 323
11.2.1 Treatment coding for analysis of variance 326
11.2.2 Comparing multiple differences 328
11.2.3 Problems 330
11.3 ANCOVA 332
11.3.1 Problems 334
11.4 Two-way ANOVA 335
11.4.1 Treatment coding for additive two-way ANOVA 336
11.4.2 Testing for row or column effects 336
11.4.3 Testing for interactions 337
11.4.4 Problems 341

12 Two extensions of the linear model 343
12.1 Logistic regression 343
12.1.1 Generalized linear models 346
12.1.2 Fitting the model using `glm()` 346
12.2 Nonlinear models 351
12.2.1 Fitting nonlinear models with `nls()` 352

A Getting, installing, and running R 359
A.1 Installing and starting R 359
A.1.1 Binary installation under Windows 360
A.1.2 Binary installation under Linux 360
A.1.3 Binary installation under Mac OS X 361
A.1.4 Installing from the source code 361
A.1.5 Startup files 361
A.2 Extending R using additional packages 362
A.2.1 Upgrading an existing installation 363

B Graphical user interfaces and R..365
B.1 The Windows GUI 365
B.2 The Mac OS X GUI 367
B.3 Rcdmr 368

C Teaching with R 371

D More on graphics with R..373
D.1 Low- and high-level graphic functions 373
D.1.1 Setting up a plot figure 374
D.1.2 Adding to a figure 378

D.1.3 Printing or saving a figure 381
D.2 Creating new graphics in `R` 381

E Programming in R 387
E.1 Editing functions 387
E.1.1 Using `edit()` 387
E.2 Using functions 388
E.2.1 Function arguments 389
E.2.2 Function body and return values 391
E.2.3 Conditional evaluation 393
E.2.4 Looping 394
E.3 Using files and a better editor 395
E.3.1 Using an external editor 395
E.4 Object-oriented programming with R 396
E.4.1 Method dispatch 397
E.4.2 S3 methods, S4 methods, and the OOP package 399

Index 409

Preface

What is R?

R is a computer language for statistical computing similar to the S language developed at Bell Laboratories. The R software was initially written by Ross Ihaka and Robert Gentleman in the mid 1990s. Since 1997, the R project has been organized by the R Development Core Team. R is open-source software and is part of the GNU project. R is being developed for the Unix, Macintosh, and Windows families of operating systems. The R home page (`http://www.r-project.org`) contains more information about R and instructions for downloading a copy.

R is excellent software to use while first learning statistics. It provides a coherent, flexible system for data analysis that can be extended as needed. The open-source nature of R ensures its availability. R's similarity to S allows you to migrate to the commercially supported S-Plus software if desired. Finally, despite its reputation, R is as suitable for students learning statistics as it is for researchers using statistics.

The purpose of this book

This book started as a set of notes, titled "simpleR," that were written to fill a gap in documentation for using R in an introductory statistics class. The College of Staten Island had been paying a per-seat fee to use a commercial statistics program. The cost of the program precluded widespread installation and curtailed accessibility. It was determined that the students would be better served if they could learn statistics with a software package that taught them good computer skills at the outset, could be installed all over campus and at home with relative ease, and was free to use. However, no suitable materials were available to accompany the class text. Hence, the basis for "simpleR" – a set of notes to accompany an in-class text.

Now, as R gains wider acceptance, for pedagogic, style, and economic reasons, there is an increase, but no abundance, in available documentation. The adoption of R as the statistical software of choice when learning statistics depends on introductory materials. This book aims to serve the needs of students in introductory applied-statistics classes that are based on precalculus skills. An emphasis is put on finding simple-looking solutions, rather than clever ones. Certainly, this material could be covered more quickly (and is in other books such as those by Dalgaard, Fox, and Venables and Ripley). The goal here is to make it as accessible to student-learners as possible.

This book aims to serve a hybrid purpose: to cover both statistical topics and the R software. Though the material stands alone, this book is also intended to be useful as an accompaniment to a standard introductory statistics book.

Description of this book

The pacing and content of this book are a bit different from those in most introductory texts. More time is spent with exploratory data analysis (EDA) than is typical, a chapter on simulation is included, and a unified approach to linear models is given. If this book is being used in a semester-long sequence, keep in mind that the early material is conceptually easier but requires that the student learn more on the computer. The pacing is not as might be expected, as time must be spent learning the software and its idiosyncrasies.

Chapters 1 through 4 take a rather leisurely approach to the material, developing the tools of data manipulation and exploration. The material is broken up so that users who wish only to analyze univariate data can safely avoid the details of data frames, lists, and model formulas covered in Chapter 4. Those wishing to cover all the topics in the book can work straight through these first four chapters.

Chapter 5 covers populations, random samples, sampling distributions, and the central limit theorem. There is no attempt to cover the background probability concepts thoroughly. We go over only what is needed in the sequel to make statistical inference.

Chapter 6 introduces simulation and the basics of defining functions. Since R is a programming language, simulations are a strong selling point for R's use in the classroom.

Traditional topics in statistical inference are covered in chapters 7-11. Chapters 7, 8, and 9 cover confidence intervals, significance tests, and goodness of fit. Chapters 10 and 11 cover linear models. Although this material is broken up into chapters on linear regression and analysis of variance, for the most part we use a common approach to both.

Chapter 12 covers a few extensions to the linear model to illustrate how R is used in a consistent manner with many different statistical models. The necessary background to appreciate the models is left for the reader to find.

The appendices cover some background material and have information on writing functions and producing graphics that goes beyond the scope of the rest of the text.

Typographic conventions

The book uses a few quirky typographic conventions. Variables and commands are typeset with a `data` typeface; functions as `a.function()` (with accompanying parentheses); and arguments to functions as `col=` (with a trailing equal sign). Help-page references have a leading question mark: `?par`. Data sets are typeset like `faithful`. Those that require a package to be loaded prior to usage also have the package name, such as `Animals` (`MASS`). Large blocks of commands are set off with a vertical bar:

```
> hist(rnorm(100))                    # draw histogram
```

Often the commands include a comment, as does the one above. The output is formatted to have 4 digits and 65 characters per column, and the type size is smaller, in order to get more information in a single line. This may cause minor differences if the examples are tried with different settings.

Web accompaniments

The home page for this book is

`http://www.math.csi.cuny.edu/UsingR`

On this page you will find solutions to selected homework problems (a full solutions manual for instructors is available from the publisher), a list of errata, and an accompanying package containing data sets and a few functions used in the text. The `UsingR` package contains data sets collected from various places. Consult the help page of a data set for proper attribution. The package needs to be installed on your computer prior to usage. If your computer has an internet connection, the command

```
> install.packages("UsingR")
```

will fetch the package from CRAN, R's warehouse of add-on packages, and install it. The command `library(UsingR)` will load the package for use.

If for some reason this fails, the package can be retrieved from this book's home page with the commands

```
> where = "http://www.math.csi.cuny.edu/UsingR"
> install.packages("UsingR",contriburl=where)
```

Finally, if that fails, the package can be downloaded from the home page and installed manually as described in Chapter 1.

Access within R to the home page for this book is provided by the function `UsingR()`. By default, it will open a browser window to the UsingR web page. Calling it with `UsingR("errata")` will open to the errata page. The function `getAnswer()` will display a specific answer to a homework problem in a browser window. It is called with a chapter and problem number, as in `getAnswer(chapter=1,problem=4)`.

Using R

The R software is obtained from the Comprehensive R Archive Network (CRAN), which may be reached from the main R web site `http://www.r-project.org`. Some basic details for installation appear in Appendix A and more detail is on the CRAN website. This book was written to reflect the changes introduced by version 2.0.0 of R. R has approximately two new major releases per year (the second number in the version number). Despite the addition of numerous improvements with each new version, the maintainers of R do a very careful job with the upgrades to R. Minor bug fixes appear in maintenance versions (the third number). It is recommended that you upgrade your installation to keep pace with these changes to R, although these new releases may affect some of the details given in this text.

Acknowledgments

The author has many people to thank for this project. First, the numerous contributors to the R software and especially the core members for their seemingly tireless efforts in producing this excellent software. Next, the editorial staff at Chapman Hall/CRC was great. In particular I can't thank an anonymous reviewer enough for his thorough and thoughtful comments and recommendations. As well, thanks go to several people who have contacted me regarding the "simpleR" notes and with corrections to previous printings of this text. Finally, the author wishes to thank his eagle-eyed wife and the rest of his family, for their constant love, help, and support during this project. It couldn't have been done without them.

Chapter 1

Data

1.1 What is data?

When we read the newspaper or watch TV news or read online news sites, we find ourselves inundated with data and its interpretation. Most often the data is presented in a summarized format, leading the reader to draw conclusions. Statistics allow us to summarize data in the familiar terms of counts, proportions, and averages. What is often missing, though, is information telling us how to interpret the statistic. The goal of this book is to learn about data: how to summarize it, how to present it, and how to infer from it when appropriate.

■ **Example 1.1: Tax-cut rhetoric** In spring 2003, while promoting a tax cut, President of the United States George W. Bush said, "Under this plan, 92 million Americans receive an average tax cut of $1,083." Yet the Urban Institute-Brookings Institution Tax Policy Center reports that 80% of Americans would receive less than this, the middle 20% would have an average tax cut of only $256, and almost half of all taxpayers would receive a tax cut of less than $100.

Can this be true? Seemingly not, but it is possible. What is being shown here are various ways of measuring the "center" of some set of numbers: in this case, the center of the amounts people would save under a proposed tax plan. The president uses the familiar mean to find a value of $1,083, yet the median amount is closer to $100. The value of $256 is a trimmed mean. When is it appropriate to use a mean to summarize a center? When is the median or a trimmed mean more appropriate? In this example, we see that the facts can look entirely different based on how we choose to present them. ■

■ **Example 1.2: A public-opinion poll** A news web site runs a daily online poll to record its readers' opinions on a variety of topics. Often, several thousand people "click in" with their opinion. The web site reports the numbers but leaves for discussion the validity of the poll.

What is the difference between this type of poll and a scientific public-opinion poll? The goal of both is to gauge the opinions of some population. What is calculated is the proportion based on a sample from that population. In the news-web site case, the sample is self-selected. With this, no statistical inference can be drawn about a larger population. The results represent only the people who clicked in.

The term *statistical inference* refers to using a probability model to link the data to a wider context. In a scientific poll where the sample is randomly chosen, probability models can be employed to allow us to infer the true opinions of a larger population. In this case, a *statistic* formed from a *sample* is used to estimate an unknown *parameter* of the population. The inference won't be exact, but our intuition is that it is *usually* within some margin of error, which gets smaller as the size of the sample gets larger. ■

■ Example 1.3: Effectiveness of a diet pill

The weight-loss supplement ephedra was popular until its risky side effects became better known. Because of its side effects, ephedra was removed from sale in Canada and the U.S. Its effectiveness is also in question, although in combination with caffeine ephedra is widely thought to work well. The Muscletech company commissioned a number of studies in the year 2001 to see if its ephedra-based product, Hydroxycut, was effective for weight loss. One study found that Hydroxycut users lost 15 pounds of fat mass in 12 weeks, while those taking a placebo (a sugar pill) lost 10.

Even before asking whether the results are statistically significant, a skeptical observer might ask several questions about the trial. We know who funded the trial. Did this fact affect the outcome? Were the groups randomly assigned or chosen to favor the company? Were those in the placebo group aware that they were taking a placebo? Were the researchers aware of who was in the placebo group? Is the difference in weight lost attributable to chance and not the ephedra pill? Is the ephedra pill safe to use?

A randomized experiment is used to measure effectiveness. An idealized one would begin with a group of subjects, or experimental units. These would be randomly allocated into possibly several treatment groups, one being the control group. A treatment is applied to each subject, with those in the control group receiving a placebo. In the example, there are just two groups—those who get a dosage of ephedra and those who get a placebo. After the treatment, observations are made and recorded for further analysis.

The role of the randomization is to avoid *bias*, or a "stacking of the deck." Sometimes, to cut down on variations, the subjects are matched in groups with common characteristics, so that similar treatments would be expected to yield similar results. To ensure that subjects do not improve because they expect they should, a blind experiment may be used. For this, a control group is given a treatment that appears to be the same but is really not. To further eliminate the chance of bias, a *double-blind experiment* is used. In a double-blind experiment, the researchers themselves are unaware of which treatment group a subject is in. This sounds like a lot of work, but it is necessary to try to eliminate the effects of other variables besides the treatment (*confounding variables*) that may affect the results. This is the only way a *cause-and-effect* relationship can be drawn.

Assume for the moment that the industry-sponsored research on ephedra was unbiased. Was the reported difference significant? Not according to a *New York Times* article from June 2003:

> In an internal memorandum accompanying the study, a Muscletech official warned, "None of these results can be deemed significant," adding that "Hydroxycut can't be claimed as superior" to the placebo. To get around that, the official proposed that copy writers simply say, "Lose 15 pounds of fat in 12 weeks with Hydroxycut and exercise!"

How one chooses to compare or present results can have a dramatic effect on what is implied. ■

■ **Example 1.4: The impact of legalized abortion on crime** Does abortion cut down on crime? Steven Levitt, a University of Chicago economist, and John Donohue, a Stanford University law professor, concluded in a paper in the May 2001 *Quarterly Journal of Economics* that legalizing abortion in the United States in 1973 led to the drop in crime seen in the country two decades later. Their data? An analysis of crime rates from 1985 to 1997 correlated against abortion rates of two decades prior; the timing of the decline in crime coincided with the period when children born shortly after *Roe v. Wade* would be reaching their late teenage years. States that were the first to legalize abortion, including New York, Washington, Alaska, and Hawaii, were the first to see a drop in crime, and states with the highest abortion rates had a larger decrease.

Levitt and Donohue may have convinced those who wanted to be convinced, but those who didn't want to be convinced found many flaws in the study. The major problem is that in an observational study such as this one, it is impossible to eliminate confounding variables, despite the ingenuity of the approach. For example, did a higher rate of incarceration lead to a drop in crime? What about a "war on drugs"? In trying to prove a cause and effect with an observational study, we are always open to explanations based on variables that are beyond our control. Remember that it took decades to prove the detrimental effects of smoking on health, despite the results of several observational studies. ■

■ **Example 1.5: What is the maximum heart rate?** A common rule of thumb is that one's maximum heart rate when exercising should be about 220 minus one's age. This is a *linear relationship* between age and maximum heart rate. Although this formula is easy to remember and use, researchers suggest that there are more accurate formulas to use when needed.

The actual relationship between age and maximum heart rate is not exactly linear. It also depends on other factors, such as the type of athlete or the type of activity. However, the ease of understanding linear relationships makes them useful, even when they are not entirely accurate.

The statistical method of fitting a linear relationship to data is called *linear regression*. It can be used in a variety of situations and is one of the most widely used statistical techniques. ■

■ **Example 1.6: Shark populations in decline** Beginning in the 1950s with the advent of large-scale commercial fishing, the populations of several fish species have had a precipitous decline. How can we estimate the size of the original stock given the current population? There were no accurate surveys at the time the fishing rate began to increase. One approach was published in *Nature* by Myers and Worm. They gathered as much data as possible for a species and then fit a nonlinear statistical model to the data. For each species, an estimate can be found for the percentage decline. Then, data for all the surveyed species can be combined to make inferences about the remaining species. It has been estimated, although with controversy, that the populations as of 2003 are 10% of their preindustrial size. ■

1.1.1 Problems

1.1 Find an article in a newspaper or on the internet that shows the results of a poll. Circle any wording indicating how the poll was taken and what results are suggested.

1.2 Find an article in the newspaper or on the internet that shows the results of a clinical trial. Describe the experimental setup. Was there a control group? Was it a scientific study? Was it an observational study? What were the findings?

1.3 Find an article in the newspaper or on the internet that relies on statistics and is not about a survey. Explain what statistics the writer used and how they strengthened the writer's argument.

1.2 Some R essentials

Before we can use the computer to help us look at statistical problems, we need to familiarize ourselves with the way it is used. First we learn some basic concepts for interacting with the computer such as how to store, access, and manipulate data. It is assumed that R is already installed on your computer. For information on installing R please refer to Appendix A.

1.2.1 Starting R

R is started in different ways depending on the platform used and the intent of usage. This book primarily covers using R in an interactive mode. That is, we ask R a question, and it responds with an answer.

To begin in Windows, we click on the R icon on the desktop, or find the program under the `start` menu. A new window pops up with a command-line subwindow. For Linux, R is often started simply by typing "R" at a command prompt. When R is started, a command line and perhaps other things await our usage.

The command line, or console, is where we can interact with R. It looks something like this:

```
R : Copyright 2004, The R Foundation for Statistical Computing
Version 2.0.0  (2004-10-04), ISBN 3-900051-07-0

R is free software and comes with ABSOLUTELY NO WARRANTY.
You are welcome to redistribute it under certain conditions.
Type 'license()' or 'licence()' for distribution details.

R is a collaborative project with many contributors.
Type 'contributors()' for more information and
'citation()' on how to cite R or R packages in publications.

Type 'demo()' for some demos, 'help()' for on-line help, or
'help.start()' for a HTML browser interface to help.
Type 'q()' to quit R.

[Previously saved workspace restored]

>
```

The version number is printed, as is some useful information for new users, including instructions on quitting R from the command line.

When R starts, it searches for any saved work in the current directory. If it finds some, that work will be reloaded and we are reminded that it was done. When we quit R, it will offer to save the current session. Thus we can continue our work from session to session.

The command prompt, >, is the final thing shown. This is where we type commands to be processed by R. This happens when we hit the ENTER key.

Appendix B describes some graphical interfaces available for R. These can make certain tasks in R easier to do. The primary aim of this book, however, is to cover basic usage of the command-line interface.

1.2.2 Using R as a calculator

The simplest usage of R is performing basic arithmetic, as we would do with a calculator. R uses familiar notation for math operations, such as +,-,*, and /. Powers are taken with ^. As usual, we use parentheses to group operations.*

The following example is from an interactive R session. A command is typed at the prompt followed by the ENTER key. Multiple commands per line can be evaluated if separated by a semicolon, ;. The result of the last command is printed in the output.

(We typeset the example with the command prompt showing, but this shouldn't be typed when trying these examples.)

```
> 2 + 2
[1] 4
> 2^2
[1] 4
> (1 - 2) * 3
[1] -3
> 1 - 2 * 3
[1] -5
```

The answer to each "question" is printed starting with a [1]. This notation will make sense once data vectors are explained.

Functions Many mathematical and statistical *functions* are available in R. They are all used in a similar manner. A function has a name, which is typed, followed by a pair of parentheses (required). Arguments are added inside this pair of parentheses as needed.

We show some familiar functions below. (The # is the comment character. All text in the line following this is treated as a comment. In the examples, the comments are provided for comprehension; they should not be typed if you are trying these examples.)

```
> sqrt(2)                        # the square root
[1] 1.414
> sin(pi)                        # the sine function
[1] 1.225e-16                    # this is 0!
> exp(1)                         # this is exp(x) = e^x
[1] 2.718
> log(10)                        # the log base e
[1] 2.303
```

The result of sin(pi) is the very small $1.225 \cdot 10^{-16}$ and not precisely 0, as it should be. Such numeric differences are not uncommon. The logarithm† has base e for its *default behavior*.

---

* The full order of operations is covered in the help page for Syntax.

† This book has a few examples where logarithms and exponentials are important, although for the most part knowledge of these functions is not essential for following the material.

Many functions in R have *extra arguments* that allow us to change the default behavior. For example, to use base 10 for the logarithm, we could use either of the following:

```
> log(10,10)
[1] 1
> log(10, base=10)
[1] 1
```

To understand the first one, `log(10,10)`, we need to know that R expects the base to be the second argument of the function. The second example uses a *named argument*, of the type `base=`, to say explicitly that the base is 10. The first style contains less typing; the second is easier to remember and read. This book will typically use named arguments for clarity.

Warnings and errors When R finds a command it doesn't understand, it will respond with an error message. For example:

```
> squareroot(2)
Error: couldn't find function "squareroot"
> sqrt 2
Error: syntax error
> sqrt(-2)
[1] NaN
Warning message:
NaNs produced in: sqrt(-2)
> sqrt(2                          # the +, like >,  is not typed
+ )
[1] 1.414
```

The first command produced an `Error:` and no output, as R could not find a function with that name. The second command failed to use parentheses around the argument, causing R to issue a syntax error. Parentheses are required to use a function. We typeset function names with parentheses to remind us of that. The third command produced an output, but R guessed it wasn't what we wanted. The output `NaN` means "not a number," in response to the request for a square root of a negative number. Consequently, a `Warning` was printed. The last command shows what happens if R encounters a line that is not complete. The continuation prompt, +, is printed, indicating more input is expected.

1.2.3 Assignment

It is often convenient to name a value so that we can use it later. Doing so is called **assignment**. Assigment is straightforward. We put a name on the left-hand side of the equals sign and the value on the right. Assignment does not produce any printed output.

```
> x = 2                           # assignment is quiet
> x + 3                           # x is now 2
[1] 5
```

```
> pi                               # pi is a built-in constant
[1] 3.142
> e^2                              # e is not
Error: Object "e" not found
> e = exp(1)                       # e is now its familiar value
> e^2
[1] 7.389
```

The variable e is not previously assigned, unlike the built-in constant pi. If we insist though, we can assign it as illustrated.

Assignment with = versus <- Assignment can cause confusion if we are trying to understand the syntax as a mathematical equation. If we write

$$x = 2x + 1$$

as a mathematical equation, we have a single solution: -1. In R, though, the same expression, x = 2*x + 1, is interpreted to *assign* the value of 2*x + 1 to the value of x. This updates the previous value of x. So if x has a value of 2 prior to this line, it leaves with a value of 5.

This type of confusion can be minimized by using the alternate assignment operator <-. The R expression x <- 2*x + 1 then "visually" looks like what it does. In fact, -> also works as assignment only to the right-hand side. Additionally, there is another operator for assignment, <<-. This is useful when we are programming in R.

This book uses the equals sign = for assignment, as it is widely used in other computer-programming languages and shorter to type. Its introduction into R is relatively recent (version 1.4.0).

Acceptable variable names We are free to make variable names out of letters, numbers, and the dot or underline characters. A name starts with a letter or a dot (a leading dot may not be followed by a number). We cannot use mathematical operators, such as +, -, *, and /.[‡] Some examples are

```
> x = 2
> n = 25
> a.really.long.number = 123456789
> AReallySmallNumber = 0.000000001
```

Case is important. Some variable names are naturally used to represent certain types of data. Often n is for a length; x or y stores a data vector; and i and j are for integers and indices. Variables that begin with the dot character are usually reserved for programmers. These conventions are not forced upon us, but consistently using them makes it easier to look back and understand what we've done.

[‡] The help page for make.names() describes this in more detail.

1.2.4 Using c() to enter data

A data set usually contains many observations, not just one. We reference the different observations by an index, as in $x_1, x_2, \ldots, x_n$. We always use n to be the number of observations, unless specified otherwise. For example, the number of whale beachings per year in Texas during the 1990s was

```
74 122 235 111 292 111 211 133 156 79
```

To store this data in R we use a **data vector**. Data vectors can be made with the c() function, which combines its arguments. The whale data can be entered, as follows:

```
> whales = c(74, 122, 235, 111, 292, 111, 211, 133, 156, 79)
```

The values are separated by a comma. Once stored, the values can be printed by typing the variable name

```
> whales
 [1]  74 122 235 111 292 111 211 133 156  79
```

The [1] refers to the first observation. If more than one row is output, then this number refers to the first observation in that row.

The c() function can also combine data vectors. For example:

```
> x = c(74, 122, 235, 111, 292)
> y = c(111, 211, 133, 156, 79)
> c(x,y)
 [1]  74 122 235 111 292 111 211 133 156  79
```

Data vectors have a type One restriction on data vectors is that all the values have the same type. This can be numeric, as in whales, characters strings, as in

```
> simpsons = c("Homer",'Marge',"Bart","Lisa","Maggie")
```

or one of the other types we will encounter. Character strings are made with matching quotes, either double, ", or single, '.

If we mix the type within a data vector, the data will be coerced into a common type, which is usually a character. This can prevent arithmetic operations.

Giving data vectors named entries A data vector can have its entries named. These will show up when it is printed. The names() function is used to retrieve and set values for the names. This is done as follows:

```
> names(simpsons) = c("dad","mom","son","daughter 1","daughter 2")
> names(simpsons)
[1] "dad"          "mom"          "son"          "daughter 1"
[5] "daughter 2"
> simpsons
       dad        mom        son daughter 1 daughter 2
   "Homer"    "Marge"     "Bart"     "Lisa"   "Maggie"
```

When used to assign values for the names, the names() function appears on the other side of the assignment, unlike most functions, as it modifies the attributes of the data vector. The last command shows that a data vector with names is printed with the names above the values.

1.2.5 Using functions on a data vector

Once the data is stored in a variable, we can use functions on it. Most R functions work on vectors exactly as we would want them to. For example, the `sum()` function will add up all the numbers in the data vector, and the `length()` function will return the number of values in the data vector.

```
> sum(whales)                     # total number of beachings
[1] 1524
> length(whales)                  # length of data vector
[1] 10
> sum(whales)/length(whales)      # average no. of beachings
[1] 152.4
> mean(whales)                    # mean function finds average
[1] 152.4
```

We can find the average as the total sum over the length, or we can use the `mean()` function. R has many built-in functions for doing such things. Other useful functions to know about are `sort()`, `min()`, `max()`, `range()`, `diff()`, and `cumsum()`.

```
> sort(whales)                    # the sorted values
 [1]  74  79 111 111 122 133 156 211 235 292
> min(whales)                     # the minimum value
[1] 74
> max(whales)                     # the maximum value
[1] 292
> range(whales)                   # range returns both min and max
[1]  74 292
> diff(whales)                    # diff returns differences
[1]   48  113 -124  181 -181  100  -78   23  -77
> cumsum(whales)                  # a cumulative, or running tally
 [1]   74  196  431  542  834  945 1156 1289 1445 1524
```

Vectorization of functions

Performing arithmetic in R is made much easier by the *vectorization* of functions. That is, most functions will do their operation on each entry of the data vector at the same time. For example, the number of whales that beached in Florida during the 1990s is

89 254 306 292 274 233 294 204 204 90

We enter this in and then look at the sum and differences for the states.

```
> whales.fla = c(89, 254, 306, 292, 274, 233, 294, 204, 204, 90)
> whales + whales.fla
 [1] 163 376 541 403 566 344 505 337 360 169
> whales - whales.fla             # florida usually has more
 [1]  -15 -132  -71 -181   18 -122  -83  -71  -48  -11
> whales - mean(whales)           # difference from average
 [1] -78.4 -30.4  82.6 -41.4 139.6 -41.4  58.6 -19.4   3.6 -73.4
```

The + operator adds up each corresponding entry, and the - operator subtracts each corresponding entry. The last example shows that a single number, in this case `mean(whales)`, can be subtracted from a vector. The result is to subtract the number from each entry in the data vector. This is an example of **data recycling**. R repeats values from one vector so that its length matches the other.

Other arithmetic functions, such as `sin()`, `cos()`, `exp()`, `log()`, `^` and `sqrt()`, are vectorized, as will be seen in the examples throughout this book.

■ **Example 1.7: The variance** A set of numbers has a summary number called the variance, which is related to the average squared distance from the mean. A formula ($\bar{x}$ is the average) is

$$\mathsf{VAR}(x) = \frac{(x_1 - \bar{x})^2 + (x_2 - \bar{x})^2 + \cdots + (x_n - \bar{x})^2}{n-1}.$$

Although the `var()` function will do this work for us in the future, we show how we could do this directly in R using the vectorization of functions. The key here is to find the squared differences and then add up the values.

```
> x = c(2,3,5,7,11)
> xbar = mean(x)
> x - xbar                          # the difference
[1] -3.6 -2.6 -0.6  1.4  5.4
> (x-xbar)^2                        # the squared difference
[1] 12.96  6.76  0.36  1.96 29.16
> sum((x-xbar)^2)                   # sum of squared differences
[1] 51.2
> n = length(x)
> n
[1] 5
> sum((x-xbar)^2)/ (n-1)
[1] 12.8
```

That is, the variance for these numbers is 12.8. ■

Functions are like pets A silly analogy: to remember how to use functions in R, think of them as being like pets. They don't come unless we call them by name (spelled properly). They have a mouth (the parentheses) that likes to be fed (in this case the arguments to the function), and they will complain if they are not fed properly.

Finding help

Using R to do statistics requires knowing a lot of different functions—more than most of us can keep in our head at any given time. Thankfully, R has excellent built-in help facilities. These can be consulted for information about what is returned by a function, for details on additional arguments, and for example usages of the function or data set.

The `help()` function is the primary interface to the help system. For example, `help("mean")` will find help on the `mean()` function. A useful shortcut is the `?`, as in `?mean`, or `?"mean"`. The quoted form is needed for some function names. The help page may show up in the terminal, a separate window, or even in a web browser, depending on your setup.

This works great if we can remember the name of the desired function. If not, there are other ways to search. The function `help.search()` will search each entry in the help system. For example, `help.search("mean")` returns *many* matches of functions that mention the word "mean" in certain parts of their help page. To match just function names, the well-named `apropos()` function will search through the available function names and variables for matches. For example, `apropos("mean")` will return all documented functions and variables with the word "mean" in their names.

If we want to explore the help pages to see what is there, the `help.start()` function will open a web browser to an index of all the available documentation, including the manuals, if installed. All subsequent uses of `help()` or `?` will appear in the browser. A standard installation of R includes a few manuals. These can be viewed in HTML format through `help.start()` or they can be printed out. These manuals may be daunting at first, but they are full of excellent information.

Most help pages have interesting examples. These can be run one-by-one by cutting and pasting into the console, or all at once by using the function `example()`. A typical usage would be `example(mean)`.

Simplifying editing by working smarter not harder

Using the command line in R can involve a fair amount of typing. However, there are ways to reduce the amount of typing necessary.

Using the arrow keys to make editing data easier R's console keeps a history of the commands entered in. The `history()` function shows the last 25. Individually, the commands can be accessed using the up- and down-arrow keys. Repeatedly pushing the up arrow will scroll backward through the history. This can be extremely useful, as we can reuse previous commands. Many times we wish to change only a small part of a previous command, such as when a typo is made. With the arrow commands we can access the previous command then edit it as desired. Table 1.1 has a summary of shortcuts.

Using `data.entry()` or `edit()` to edit data Many ports of R have a primitive spreadsheet interface for editing data. This is available through the `data.entry()` function. For example, `data.entry(x)` will allow us to edit the data vector `x`.

The function does not make a new variable. To use `data.entry()` to make a new variable, we can first create a simple one, as we have done below, and then finish the data entry with the spreadsheet.

```
> x = c(1)                              # 1 will be first entry
```

Table 1.1 Keyboard shortcuts for the command line

↑ (up arrow)	Recalls the previously entered command from the history list; multiple pushes scrolls through the command history list
↓ (down arrow)	Scrolls forward in the history list
← (left arrow)	Moves cursor to the left
→ (right arrow)	Moves cursor to the right
HOME (CTRL-a)	Moves cursor to beginning of current line
END (CTRL-e)	Moves cursor to end of current line

```
> data.entry(x)
```

The `edit()` function also allows us to edit a data vector. We need to assign back its output, otherwise the edits will be lost. For a data vector, the `edit()` command uses a text editor. A spreadsheet interface is used for more complicated data objects.

1.2.6 Creating structured data

Sometimes numbers have some structure or pattern. Take, for example, the integers 1 through 99. To enter these into an R session one by one would be very tedious. Fortunately, R has shortcuts for entering data that is sequential or that is repeated in some way.

Simple sequences A sequence from 1 to 99 by 1's is given by `1:99` in R. The colon operator, `:`, is used to create sequences from a to b by 1's. Some examples:

```
> 1:10
 [1]  1  2  3  4  5  6  7  8  9 10
> rev(1:10)                      # countdown
 [1] 10  9  8  7  6  5  4  3  2  1
> 10:1                           # a > b
 [1] 10  9  8  7  6  5  4  3  2  1
```

Arithmetic sequences An arithmetic sequence is determined by a starting point, a; a step size, h; and a number of points, n. The sequence is

$$a, a+h, a+2h, a+3h, \ldots, a+(n-1)h.$$

These sequences can be created in R directly.

```
> a = 1; h = 4; n=5              # use ; to separate commands
> a + h*(0:(n-1))                # note 0:(n-1) is not 0:n -1
[1]  1  5  9 13 17
```

It is usually desirable to specify either the step size and the starting and ending points or the starting and ending points and the length of the arithmetic sequence. The `seq()` function allows us to do this.

```
> seq(1,9,by=2)                    # odd numbers
[1] 1 3 5 7 9
> seq(1,10,by=2)                   # as 11 > 10 it is not included
[1] 1 3 5 7 9
> seq(1,9,length=5)                # 5 numbers only
[1] 1 3 5 7 9
```

Repeated numbers When a vector of repeated values is desired, the `rep()` function is used. The simplest usage is to repeat its first argument a specified number of times, as in

```
> rep(1,10)
 [1] 1 1 1 1 1 1 1 1 1 1
> rep(1:3,3)
[1] 1 2 3 1 2 3 1 2 3
```

More complicated patterns can be repeated by specifying pairs of equal-sized vectors. In this case, each term of the first is repeated the corresponding number of times in the second.

```
> rep(c("long","short"),c(1,2)) # 1 long and 2 short
[1] "long"  "short" "short"
```

1.2.7 Problems

1.4 Use R as you would a calculator to find numeric answers to the following:

1. $1+2(3+4)$
2. 4^3+3^{2+1}
3. $\sqrt{(4+3)(2+1)}$
4. $\left(\frac{1+2}{3+4}\right)^2$

1.5 Rewrite these R expressions as math expressions, using parentheses to show the order in which R performs the computations:

1. `2 + 3 - 4`
2. `2 + 3 * 4`
3. `2/3/4`
4. `2^3^4`

1.6 Enter the following data into a variable p with `c()`

```
2 3 5 7 11 13 17 19
```

Use `length()` to check its length.

1.7 You recorded your car's mileage at your last eight fill-ups as

```
65311 65624 65908 66219 66499 66821 67145 67447
```

Enter these numbers into the variable `gas`. Use the function `diff()` on the data. What does it give? Interpret what both of these commands return: `mean(gas)` and `mean(diff(gas))`.

1.8 Let our small data set be

```
2 5 4 10 8
```

1. Enter this data into a data vector `x`.
2. Find the square of each number.
3. Subtract 6 from each number.
4. Subtract 9 from each number and then square the answers.

Use the vectorization of functions to do so.

1.9 The asking price of used MINI Coopers varies from seller to seller. An online classifieds listing has these values in thousands:

```
15.9 21.4 19.9 21.9 20.0 16.5 17.9 17.5
```

1. What is the smallest amount? The largest?
2. Find the average amount.
3. Find the differences of the largest and smallest amounts from the mean.

Enter in the data and apply one of R's functions to find answers to the above questions.

1.10 The monthly sales figures of Hummer H2 vehicles in the United States during 2002 were

```
[Jan] 2700 2600 3050 2900 3000 2500 2600 3000 2800
[Oct] 3200 2800 3400
```

(according to a graphic from the *New York Times*). Enter this data into a variable `H2`. Use `cumsum()` to find the cumulative total of sales for 2002. What was the total number sold? Using `diff()`, find the month with the greatest increase from the previous month, and the month with the greatest decrease from the previous month.

1.11 Four successive National Health and Examination surveys showed the average amount of calories consumed by a 20-to-29-year-old male to be 2,450, 2,439, 2,866, and 2,618 (`http://www.cdc.gov`). The percentage of calories from fat was 37%, 36.2%, 34%, and 32.1%. The percentage from carbohydrates was 42.2%, 43.1%, 48.1%, and 50%. Is the average number of fat calories going up or going down? Is this consistent with the fact that over the same time frame (1971 to 2000) the prevalence of obesity in the United States increased from 14.5% to 30.9%?

1.12 Create the following sequences:
1. `"a", "a", "a", "a", "a"`
2. 1, 3, ..., 99 (the odd numbers in $[1, 100]$)
3. 1, 1, 1, 2, 2, 2, 3, 3, 3
4. 1, 1, 1, 2, 2, 3
5. 1, 2, 3, 4, 5, 4, 3, 2, 1

using `:`, `seq()`, or `rep()` as appropriate.

1.13 Store the following data sets into a variable any way you can:
1. 1, 2, 3, 5, 8, 13, 21, 34 (the Fibonacci series)
2. 1, 2, 3, 4, 5, 6, 7, 8, 9, 10 (positive integers)
3. $1/1, 1/2, 1/3, 1/4, 1/5, 1/6, 1/7, 1/8, 1/9, 1/10$ (reciprocals)
4. 1, 8, 27, 64, 125, 216 (the cubes)
5. 1964, 1965, ..., 2003 (some years)
6. 14, 18, 23, 28, 34, 42, 50, 59, 66, 72, 79, 86, 96, 103, 110 (stops on New York's No. 9 subway)
7. 0, 25, 50, 75, 100,..., 975, 1000 (0 to 1000 by 25s)

Use `c()` only when `:` or `seq()` will not work.

1.3 Accessing data by using indices

Using R to access the entries in a data vector is straightforward. Each observation, $x_1, x_2, \ldots, x_n$, is referred to by its index using square brackets, as in `x[1]`, `x[2]`, ..., `x[n]`. Using the indices, we can access and assign to the values stored in the data vector.

We keep track of eBay's Friday stock price in the variable `ebay`. The first two months of data are

`88.8 88.3 90.2 93.5 95.2 94.7 99.2 99.4 101.6`

These are entered as

```
> ebay = c(88.8, 88.3, 90.2, 93.5, 95.2, 94.7, 99.2, 99.4, 101.6)
> length(ebay)
[1] 9
```

The first value is clearly 88.8 and the last 101.6. We can get these directly, as in

```
> ebay[1]
[1] 88.8
> ebay[9]
[1] 101.6
> ebay[length(ebay)]              # in case length isn't known
[1] 101.6
```

Slicing R also allows slicing, or taking more than one entry at a time. If x is the data vector, and vec is a vector of positive indices, then x[vec] is a new vector corresponding to those indices. For the ebay example, the first four entries are for the first month. They can be found by

```
> ebay[1:4]
[1] 88.8 88.3 90.2 93.5
```

The first, fifth, and ninth Fridays can be accessed using c(1,5,9) as the index.

```
> ebay[c(1,5,9)]
[1]  88.8  95.2 101.6
```

Negative indices If the index in x[i] is positive, we can intuit the result. The *i*th value of x is returned if i is between 1 and n. If i is bigger than n, a value of NA is returned, indicating "not available."

However, if i is negative and no less than -n, then a useful convention is employed to return all but the *i*th value of the vector. For example, x[-1] is all of x but the first entry.

```
> ebay[-1]                            # all but the first
[1]  88.3  90.2  93.5  95.2  94.7  99.2  99.4 101.6
> ebay[-(1:4)]                        # all but the 1st - 4th
[1]  95.2  94.7  99.2  99.4 101.6
```

Accessing by names In R, when the data vector has names, then the values can be accessed by their names. This is done by using the names in place of the indices. A simple example follows:

```
> x = 1:3
> names(x) = c("one","two","three") # set the names
> x["one"]
one
  1
```

Parentheses for functions; square brackets for data vectors The usage of parentheses, (), and square brackets, [], can be confusing at first. To add to the confusion, lists will use double square brackets [[ ]]. It helps to remember that R uses parentheses for functions and square brackets for data objects.

1.3.1 Assigning values to data vector

We can assign values to a data vector element by element using indices. The simplest case, x[i] = a, would assign a value of a to the *i*th element of x if i is positive. If i is bigger than the length of the vector x, then x is enlarged. For example, to change the first entry in ebay to 88.0 we could do

```
> ebay[1] = 88.0
> ebay
[1]  88.0  88.3  90.2  93.5  95.2  94.7  99.2  99.4 101.6
```

We can assign more than one value at a time. The case x[vec]<-y will assign to the indices specified by vec the values of y. For example, adding the next month's stock prices to ebay can be done as follows:

```
> ebay[10:13] = c(97.0,99.3,102.0,101.8)
> ebay
 [1]  88.0  88.3  90.2  93.5  95.2  94.7  99.2  99.4 101.6  97.0
[11]  99.3 102.0 101.8
```

Data recycling If y is shorter than the values specified by vec, its values are recycled. That is, y is repeated enough times to fill out the request. If y is too long, then it is truncated.

1.3.2 Logical values

When using R interactively, we naturally imagine that we are having a dialogue with R. We ask a question, and R answers. Our questions may have numeric answers consisting of a single value (e.g., "What is the sum of x?"), a vector of numeric answers (e.g., "What are the differences between x and y?"), or they may by true-or-false answers (e.g., "x is bigger than 2?"). R expressions which involve just values of true or false are called logical expressions. In R the keywords TRUE and FALSE are used to indicate true or false (these can be abbreviated T or F, although it is not recommended as T and F may hold some other value). A question like, "Is x bigger than 2?" is answered for each element of x. For example, "Which values of ebay are more than 100?" is asked with ebay > 100 and answered for each value of x as TRUE and FALSE.

```
> ebay > 100
 [1] FALSE FALSE FALSE FALSE FALSE FALSE FALSE FALSE  TRUE FALSE
[11] FALSE  TRUE  TRUE
```

This output is hard to parse as displayed but is very useful when used as the indices of a data vector. When x is a data vector and vec is a logical vector of the same length as x then, x[vec] returns the values of x for which vec's values are TRUE. These indices can be found with the which() function. For example:

```
> ebay[ ebay > 100 ]              # values bigger than 100
[1] 101.6 102.0 101.8
> which(ebay > 100)               # which indices
[1]  9 12 13
> ebay[c(9,12,13)]                # directly
[1] 101.6 102.0 101.8
```

Some functions are usefully adapted to logical vectors (the logical vector is coerced to a numeric one). In particular, the sum function will add up all the TRUE values as 1 and all the FALSE values as 0. This is exactly what is needed to find counts and proportions.

```
> sum(ebay > 100)                 # number bigger than 100
[1] 3
```

```
> sum(ebay > 100)/length(ebay)  # proportion bigger
[1] 0.2308
```

Table 1.2 summarizes the ways to manipulate a data vector.

Table 1.2 Ways to manipulate a data vector

Suppose x is a data vector of length n = length(x).	
x[1]	the first element of x
x[length(x)]	the last element of x
x[i]	the i th entry if $1 \leq i \leq n$, NA if $i > n$, all but the i th if $-n \leq i \leq -1$, an error if $i < -n$, and an empty vector if $i = 0$
x[c(2,3)]	the second and third entries
x[-c(2,3)]	all but the second and third entries
x[1] = 5	assign a value of 5 to first entry; also x[1] <- 5
x[c(1,4)] = c(2,3)	assign values to first and fourth entries
x[indices] = y	assign to those indices indicated by the values of indices: if y is not long enough, values are recycled; if y is too long, just its initial values are used and a warning is issued
x < 3	vector with length n of TRUE or FALSE depending if x[i] < 3
which(x < 3)	which indices correspond to the TRUE values of x < 3
x[x < 3]	the x values when x < 3 is TRUE. Same as x[which(x < 3)]

Creating logical vectors by conditions

Logical vectors are created directly using c(), or more commonly as a response to some question or *condition*. The **logical operators** are <, <=, >, >=, ==, and !=. The meanings should be clear from their common usage, but equals is == and not simply =. The operator != means not equal. The ! operator will switch values of a logical vector.

Comparisons between logical vectors are also possible: "and" is done with &; "or" is done with |. Each entry is compared and a vector is returned. The longer forms && and || evaluate left to right until a TRUE or FALSE is determined. Unlike the shorter forms, which return a vector of values, the longer forms return a single value for their answer.

To illustrate:

```
> x = 1:5
> x < 5                         # is x less than 5
[1]  TRUE  TRUE  TRUE  TRUE FALSE
> x > 1                         # is x more than 1
[1] FALSE  TRUE  TRUE  TRUE  TRUE
> x > 1 & x < 5                 # is x bigger than 1 and less than 5
[1] FALSE  TRUE  TRUE  TRUE FALSE
> x > 1 && x < 5                # First one is false
[1] FALSE
> x > 1 | x < 5                 # is x bigger than 1 or less than 5
[1] TRUE TRUE TRUE TRUE TRUE
> x > 1 || x < 5                # First one true
[1] TRUE
> x == 3                        # is x equal to 3
[1] FALSE FALSE  TRUE FALSE FALSE
> x != 3                        # is x not equal to 3
[1]  TRUE  TRUE FALSE  TRUE  TRUE
> ! x == 3                      # not ( x equal to 3)
[1]  TRUE  TRUE FALSE  TRUE  TRUE
```

The expression of equality, ==, allows us to compare a data vector with a value. If we wish to use a range of values we can use the `%in%` operator.

```
> x == c(2,4)
[1] FALSE FALSE FALSE  TRUE FALSE
Warning message:
longer object length
        is not a multiple of shorter object length in: x == c(2, 4)
> x %in% c(2,4)
[1] FALSE  TRUE FALSE  TRUE FALSE
```

The last command shows that the second and fourth entries of `x` are in the data vector `c(2,4)`. The first command shows that recycling is applied, then the data vectors are compared element by element. The warning is issued, as this type of implicit usage is often unintended.

For numeric comparisons, the operators == and != do not allow for rounding errors. Consult the help pages `?"<"` and `?all.equal` to see a workaround.

1.3.3 Missing values

Sometimes data is not available. This is different from not possible or null. R uses the value `NA` to indicate this. With operations on a data vector `NA` values are treated as though they can't be found. For example, adding with a value of `NA` returns an `NA`, as the addition cannot be carried out. A natural way to check whether a data value is `NA` would be `x == NA`. However, a value cannot be compared to `NA`, so rather than an answer of `TRUE`, the value `NA` is given. To check whether a value is `NA`, the function `is.na()` is used instead.

For example, the number of O-ring failures for the first six flights of the

United States space shuttle Challenger were (there is no data for the fourth flight):

```
0 1 0 NA 0 0
```

We enter this in using NA as follows:

```
> shuttle = c(0, 1, 0, NA, 0, 0)
> shuttle
[1]  0  1  0 NA  0  0
> shuttle > 0                     # note NA in answer
[1] FALSE  TRUE FALSE    NA FALSE FALSE
> shuttle == NA                   # doesn't work!
[1] NA NA NA NA NA NA
> is.na(shuttle)
[1] FALSE FALSE FALSE  TRUE FALSE FALSE
> mean(shuttle)                   # can't add to get the mean
[1] NA
> mean(shuttle, na.rm=TRUE)       # na.rm means remove NA
[1] 0.2
> mean(shuttle[!is.na(shuttle)]) # hard way
[1] 0.2
```

Many R functions have an argument na.rm=, which can be set to be TRUE in order to remove NAs or FALSE. This is a convenient alternative to using constructs such as x[!is.na(x)].

1.3.4 Managing the work environment

If an R session runs long enough, there typically comes a point when there are more variables defined than can be remembered. The ls() function and the objects() function will list all the objects (variables, functions, etc.) in a given work environment. The browseEnv() function does so using web browser to show the results. The simplest usage is ls(), which shows all the objects that have been defined or loaded into your work environment. To filter the request, the pattern= argument can be used. If this argument is a quoted string then all objects with that string in their names will be listed. More complicated matching patterns are possible.

To trim down the size of the work environment the functions rm() or remove() can be used. These are used by specifying a name of the objects to be removed. The name may or may not be quoted. For example, rm("tmp") or rm(tmp) will remove the variable tmp from the current work environment. Multiple names are possible if separated by commas, or if given in vector form, as quoted strings, to the argument list=.

1.3.5 Problems

1.14 You track your commute times for two weeks (ten days), recording the following times in minutes:

```
17 16 20 24 22 15 21 15 17 22
```

Enter these into R. Use the function `max()` to find the longest commute time, the function `mean()` to find the average, and the function `min()` to find the minimum.

Oops, the 24 was a mistake. It should have been 18. How can you fix this? Do so, and then find the new average.

How many times was your commute 20 minutes or more? What percent of your commutes are less than 18 minutes long?

1.15 According to *The Digital Bits* (`http://www.digitalbits.com`), monthly sales (in 10,000s) of DVD players in 2003 were

```
JAN FEB MAR APR MAY JUN JUL AUG SEP OCT NOV DEC
 79  74 161 127 133 210  99 143 249 249 368 302
```

Enter the data into a data vector `dvd`. By slicing, form two data vectors: one containing the months with 31 days, the other the remaining months. Compare the means of these two data vectors.

1.16 Your cell-phone bill varies from month to month. The monthly amounts in dollars for the last year were

```
46 33 39 37 46 30 48 32 49 35 30 48
```

Enter this data into a variable called `bill`. Use the `sum` function to find the amount you spent last year on the cell phone. What is the smallest amount you spent in a month? What is the largest? How many months was the amount greater than $40? What percentage was this?

1.17 The average salary in major league baseball for the years 1990-1999 are given (in millions) by:

```
0.57 0.89 1.08 1.12 1.18 1.07 1.17 1.38 1.44 1.72
```

Use `diff()` to find the differences from year to year. Are there any years where the amount dropped from the previous year?

The percentage difference is the difference divided by the previous year times 100. This can be found by dividing the output of `diff()` by the first nine numbers (not all ten). After doing this, determine which year has the biggest percentage increase.

1.18 Define `x` and `y` with

```
> x = c(1,3,5,7,9)
> y = c(2,3,5,7,11,13)
```

Try to guess the results of these R commands:

1. `x+1`
2. `y*2`
3. `length(x)` and `length(y)`

4. `x + y` (recycling)
5. `sum(x>5)` and `sum(x[x>5])`
6. `sum(x>5 | x<3)`
7. `y[3]`
8. `y[-3]`
9. `y[x]` (What is `NA`?)
10. `y[y>=7]`

Remember that you access entries in a vector with `[]`.

1.19 Consider the following "inequalities." Can you determine how the comparisons are being done?

```
> "ABCDE" == "ABCDE"
[1] TRUE
> "ABCDE" < "ABCDEF"
[1] TRUE
> "ABCDE" < "abcde"
[1] TRUE
> "ZZZZZ" < "aaaaa"
[1] TRUE
> "11" < "8"
[1] TRUE
```

1.4 Reading in other sources of data

Typing in data sets can be a real chore. It also provides a perfect opportunity for errors to creep into a data set. If the data is already recorded in some format, it's better to be able to read it in. The way this is done depends on how the data is stored, as data sets may be found on web pages, as formatted text files, as spreadsheets, or built in to the R program.

1.4.1 Using R's built-in libraries and data sets

R is designed to have a small code kernel, with additional functionality provided by external packages. A modest example would be the data sets that accompany this book. More importantly, many libraries extend R's base functionality. Many of these come standard with an R installation; others can be downloaded and installed from the Comprehensive R Archive Network (CRAN), `http://www.r-project.org`, as described below and in Appendix A.

Most packages are not loaded by default, as they take up computer memory that may be in short supply. Rather, they are selectively loaded using either the `library()` or `require()` functions. For instance, the package `pkgname` is loaded with `library(pkgname)`. In the Windows and Mac OS X GUIs packages can be loaded using a menu bar item.

In addition to new functions, many packages contain **built-in data sets** to provide examples of the features the package introduces. R comes with a collection of built-in data sets in the datasets package that can be referenced by name. For example, the lynx data set records the number of lynx trappings in Canada for some time period. Typing the data set name will reference the values:

```
> range(lynx)                        # range of values
[1]   39 6991
```

The packages not automatically loaded when R starts need to be loaded, using library(), before their data sets are visible. As of version 2.0.0 of R, data sets in a package may be loaded automatically when the package is. This is the case with the data sets referenced in this text. However, a package need not support this. When this is the case, an extra step of loading the data set using the data() command is needed. For example, to load the survey data set in the MASS package, could be done in this manner:

```
library(MASS)
data(survey)                         # redundant for versions >= 2.0.0
```

To load a data set without the overhead of loading its package the above sequence of commands may be abbreviated by specifying the package name to data(), as in

```
> data(survey, package="MASS")
```

However, this will not load in the help files, if present, for the data set or the rest of the package. In R, most built-in data sets are well documented, in which case we can check what the data set provides, as with ?lynx. See Table 1.3 for more details on data() and library().

Accessing the variables in a data set: $, attach(), and with()

A data set can store a single variable, such as lynx, or several variables, such as the survey data set in the MASS package. Usually, data sets that store several variables are stored as data frames. This format combines many variables in a rectangular grid, like a spreadsheet, where each column is a different variable, and usually each row corresponds to the same subject or experimental unit. This conveniently allows us to have all the data vectors together in one object.

The different variables of a data frame typically have names, but initially these names are not directly accessible as variables. We can access the values by name, but we must also include the name of the data frame. The $ syntax can be used to do this, as in

```
> library(MASS)                      # load package. Includes geyser
> names(geyser)                      # what are variable names of geyser
[1] "waiting"  "duration"            # or ?geyser for more detail
> geyser$waiting                     # access waiting variable in geyser
  [1]  80  71  57  80  75  77  60  86  77  56  81  50  89  54  90
...
```

Table 1.3 `library()` **and** `data()` **usage**

`library()`	list all the installed packages
`library(pkg)`	Load the package pkg. Use `lib.loc=` argument to load package from a non-privileged directory.
`data()`	list all available data sets in loaded packages
`data(package="pkg")`	list all data sets for this package
`data(ds)`	load the data set `ds`
`data(ds,package="pkg")`	load the data set from package
`?ds`	find help on this data set
`update.packages()`	contact CRAN and interactively update installed packages
`install.packages(pkg)`	Install the package named pkg. This gets package from CRAN. Use `lib=` argument to specify a non-privileged directory for installation. The `contriburl=...` allows us to specify other servers to find the package.

Alternately, with a bit more typing, the data can be referred to using index notation as with `geyser[["waiting"]]`. Both these styles use the syntax for a list discussed in Chapter 4.

Having to type the data frame name each time we reference a variable can be cumbersome when multiple references are performed. There are several ways to avoid this.

A convenient method, which requires little typing, is to "attach" a data frame to the current environment with the `attach()` function, so that the column names are visible. Attached data sets are detached with the function `detach()`. As this style works well in interactive sessions, we employ it often in this book. However, as discussed in Chapter 4, this style can be confusing if we plan to change values in the data set or wish to use other variables with the same name as a variable in an attached data set or the same name as a data set.

The function `with()` essentially performs the `attach()` and `detach()` commands at once. A template for its usage is

```
with(data.frame, command)
```

If we desire more than one command, we can give a block of commands by surrounding the commands in curly braces. One caveat: assignment done inside the block using = (or even <-) will be lost.

Beginning in Chapter 3 we will see that many functions in R allow an argument data= to specify a data frame to find the variables in.

Examples of these styles are shown below using the built-in Sitka data set. These illustrate a common task of loading a data set, and finally accessing a variable in the data set. We use names() to show the variable names.

```
> data(Sitka)                    # load data set, optional
> names(Sitka)                   # variable names
[1] "size"  "Time"  "tree"  "treat"
> tree                           # not visible
Error: Object "tree" not found
> length(Sitka$tree)             # length
[1] 395
> with(Sitka,range(tree))        # what is range
[1]  1 79
> attach(Sitka)
> summary(tree)
   Min. 1st Qu.  Median    Mean 3rd Qu.    Max.
      1      20      40      40      60      79
> detach(Sitka)
```

It is a good idea to use detach() to clear out the attached variable, as these variables can be confusing if their names coincide with others we have used.

1.4.2 Using the data sets that accompany this book

The UsingR package contains the data sets that accompany this book. It needs to be installed prior to usage.

Installing an R package

If you have the proper administrative permissions on your computer and the package you wish to install is housed on CRAN, then installing a package on the computer can be done with the command

```
> install.packages(packagename)
```

In the Windows and Mac OS X GUIs a menu bar item allows us to browse and install a package. Once installed, a package can be loaded with library() and the data sets accessed with data(), as with the built-in packages.

If we do not have the necessary administrative permissions we can specify a directory where the packages can be written using the argument lib=. When loading the package with library(), the lib.loc= argument can be used to specify the same directory. In each case, the directory can be specified by a string, such as lib="c:/R/localpackage".

If the package is not on CRAN, but on some other server, the argument contriburl= can be given to specify the server. For example: to install the UsingR package from its home page try these commands:

```
> where = "http://www.math.csi.cuny.edu/UsingR"
> install.packages("UsingR",contriburl=where)
```

If this fails (and it will if the site is not set up properly), download the package file and install it directly. A package is a zip archive under Windows; otherwise it is a tar.gz archive. Once downloaded, the package can be installed from the menu bar in Windows, or from the command line under UNIX. If the package name is `aPackage_0.1.tar.gz`, the latter is done at the command line from a shell (not the R prompt, as these are not R functions) with the command

```
R CMD INSTALL aPackage_0.1.tar.gz
```

1.4.3 Other methods of data entry

What follows is a short description of other methods of data entry. It can be skipped on first reading and referenced as needed.

Cut and paste

Copying and pasting from some other program, such as a web browser, is a very common way to import data. If the data is separated by commas, then wrapping it inside the `c()` function works well. Sometimes, though, a data set doesn't already have commas in it. In this case, using `c()` can be tedious. Use the function `scan()` instead. This function reads in the input until a blank line is entered.

For example, the whale data could have been entered in as

```
> whales = scan()
1: 74 122 235 111 292 111 211 133 156 79
11:
Read 10 items
```

Using `source()` to read in R commands

The function `dump()` can be used to write values of R objects to a text file. For example, `dump("x","somefile.txt")` will write the contents of the variable `x` into the file `somefile.txt`, which is in the current working directory. Find this with `getwd()`. We can dump more than one object per file by specifying a vector of object names. The `source()` function will read in the output of `dump()` to restore the objects, providing a convenient way to transfer data sets from one R session to another.

The function `source()` reads in a file of R commands as though they were typed at the prompt. This allows us to type our commands into a file using a text editor and read them into an R session. There are many advantages to this. For example, the commands can be edited all at once or saved in separate files for future reference.

For the most part, this book uses an interactive style to interface with R, but this is mostly for pedagogic reasons. Once the basic commands are learned, we begin to do more complicated combinations with the commands. At this point using `source()`, or something similar, is much more convenient.

Reading data from formatted data sources

Data can also be found in formatted data files, such as a file of numbers for a single data set, a table of numbers, or a file of comma-separated values (csv). R has ways of reading each of these (and others).

For example, if the Texas whale data were stored in a file called "whale.txt" in this format

```
74 122 235 111 292 111 211 133 156  79
```

then `scan()` could be used to read it in, as in

```
> whale = scan(file="whale.txt")
Read 10 items
```

Options exist that allow some formatting in the data set, such as including a separator, like a comma, (`sep=`), or allowing for comment lines (`comment.char=`).

Tables of data can be read in with the `read.table()` function. For example, if "whale.txt" contained data in this tabular format, with numbers separated by white space,

```
texas  florida
74      89
122    254
...
79      90
```

then the data could be read in as

```
> read.table("whale.txt",header=TRUE)
  texas florida
1    74      89
2   122     254
....
10   79      90
```

The extra argument `header=TRUE` says that a header includes information for the column names. The function `read.csv()` will perform a similar task, only on csv files. Most spreadsheets can export csv files, which is a convenient way to import spreadsheet data.

Both `read.table()` and `read.csv()` return a data frame storing the data.

Specifying the file In the functions `scan()`, `source()`, `read.table()`, and `read.csv()`, the argument `file=` is used to specify the file name. The function `file.choose()` allows us to choose the file interactively, rather than typing it. It is used as follows:

```
> read.table(file=file.choose())
```

We can also specify the file name directly. A file is referred to by its name and sometimes its path. While R is running, it has a working directory to which file names may refer. The working directory is returned by the `getwd()` function and set by the `setwd()` function. If a file is in the working directory, then the file name may simply be quoted.

When a file is not in the working directory, it can be specified with its path. The syntax varies, depending on the operating system. UNIX traditionally uses a forward slash to separate directories, Windows a backward slash. As the backward slash has other uses in UNIX, it must be written with two backward slashes when used to separate directories. Windows users can also use the forward slash.

For example, both `"C:/R/data.txt"` and `"C:\\R\\data.txt"` refer to the same file, `data.txt`, in the `R` directory on the "C" drive.

With a UNIX operating system, we can specify the file as is done at the shell:

```
> source(file="~/R/data.txt")    # tilde expansion works
```

Finding files from the internet R also has the ability to choose files from the internet using the `url()` function. Suppose the webpage `http://www.math.csi.cuny.edu/UsingR/Data/whale.txt` contained data in tabular format. Then the following would read this web page as if it were a local file.

```
> site = "http://www.math.csi.cuny.edu/UsingR/Data/whale.txt"
> read.table(file=url(site), header=TRUE)
```

The `url()` function is used only for clarity, the file will be found without it, as in

```
> read.table(file=site, header=TRUE)
```

1.4.4 Problems

1.20 The built-in data set `islands` contains the size of the world's land masses that exceed 10,000 square miles. Use `sort()` with the argument `decreasing=TRUE` to find the seven largest land masses.

1.21 Load the data set `primes (UsingR)`. This is the set of prime numbers in [1,2003]. How many are there? How many in the range [1,100]? [100,1000]?

1.22 Load the data set `primes (UsingR)`. We wish to find all the twin primes. These are numbers p and $p+2$, where both are prime.

1. Explain what `primes[-1]` returns.
2. If you set `n = length(primes)`, explain what `primes[-n]` returns.
3. Why might `primes[-1] - primes[-n]` give clues as to what the twin primes are?

How many twin primes are there in the data set?

1.23 For the data set `treering`, which contains tree-ring widths in dimensionless units, use an R function to answer the following:

1. How many observations are there?
2. Find the smallest observation.
3. Find the largest observation.
4. How many are bigger than 1.5?

1.24 The data set `mandms (UsingR)` contains the targeted color distribution in a bag of M&Ms as percentages for varies types of packaging. Answer these questions.

1. Which packaging is missing one of the six colors?
2. Which types of packaging have an equal distribution of colors?
3. Which packaging has a single color that is more likely than all the others? What color is this?

1.25 The `times` variable in the data set `nym.2002 (UsingR)` contains the time to finish for several participants in the 2002 New York City Marathon. Answer these questions.

1. How many times are stored in the data set?
2. What was the fastest time in minutes? Convert this into hours and minutes using R.
3. What was the slowest time in minutes? Convert this into hours and minutes using R.

1.26 For the data set `rivers`, which is the longest river? The shortest?

1.27 The data set `uspop` contains decade-by-decade population figures for the United States from 1790 to 1970.

1. Use `names()` and `seq()` to add the year names to the data vector.
2. Use `diff()` to find the inter-decade differences. Which decade had the greatest increase?
3. Explain why you could reasonably expect that the difference will always increase with each decade. Is this the case with the data?

Chapter 2

Univariate data

In statistics, data can initially be considered to be one of three basic types: categorical, discrete numeric, and continuous numeric. Methods for viewing and summarizing data depend on which type it is, so we need to be aware of how each is handled and what we can do with it. In this chapter we look at graphical and numeric summaries of a data set for a single variable. The graphical summaries allow us to grasp qualitative aspects of a data set immediately. Numerical summaries allow us to compare values for the sample with parameters for a population. These comparisons will be the subject of the second half of this text.

When a data set consists of a single variable, it is called a univariate data set. We study univariate data sets in this chapter. When there are two variables in a data set, the data is bivariate, and when there are two or more variables the data set is multivariate.

Categorical data is data that records categories. An example is a survey that records whether a person is for or against a specific proposition. A police force might keep track of the race of the people it pulls over on the highway, or whether a driver was using a cell phone at the time of an accident. The United States census, which takes place every ten years, asks several questions of a categorical nature. In the year 2000, a question regarding race included 15 categories with write-in space for three more answers (respondents could mark themselves as multiracial.) Another example is a doctor's chart, which records patient data. Gender and illness history might be treated as categories.

Let's continue with the medical example. A person's age and weight are numeric quantities. Both are typically discrete numeric quantities usually reported as integers (most people wouldn't say they are 4.673 years old). If the precise values were needed, then they could, in theory, take on a continuum of values. They would then be considered continuous. Why the distinction? We can clearly turn a continuous number into a discrete number by truncation, and into a categorical one by binning (e.g., 40- to 50-year-olds). For some summaries and statistical tests it is important to know whether the data can have ties (two or more data points with the same value). For discrete data there can be ties; for continuous data it is generally not true that there can be ties.

A simple way to remember these is to ask, What is the average value? If it doesn't make sense, then the data is categorical (such as the average of a non-smoker and a smoker); if it makes sense but might not be an answer (such as 18.5 for age when we record only integers), then the data is discrete. Otherwise the data is likely to be continuous.

2.1 Categorical data

Categorical data is summarized by tables or graphically with barplots, dot charts, and pie charts.

2.1.1 Tables

Tables are widely used to summarize data. A bank will use tables to show current interest rates; a newspaper will use tables to show a data set when a graphic isn't warranted; a baseball game is summarized in tables. The main R function for creating tables is, unsurprisingly, `table()`.

In its simplest usage, `table(x)` finds all the unique values in the data vector `x` and then tabulates the frequencies of their occurrence.

For example, if the results of a small survey are "yes," "yes," "no," "yes," and "no," then these can be tabulated as

```
> res = c("Y","Y","N","Y","N")
> table(res)
res
N Y
2 3
```

Such small data sets can, of course, be summarized without the computer. In this next example, we examine a larger, built-in data set, yet there is no more work involved than with a smaller data set.

■ **Example 2.1: Weather in May** The United States National Weather Service collects an enormous amount of data on a daily basis—some categorical, much numeric. It is necessary to summarize the data for consumption by its many users. For example, suppose the day's weather is characterized as "clear," "cloudy," or "partly cloudy." Observations in Central Park for the month of May 2003 are stored in the `central.park.cloud (UsingR)` data set.

```
> library(UsingR)              # need to do once
> central.park.cloud
 [1] partly.cloudy partly.cloudy partly.cloudy clear
```

```
[5] partly.cloudy partly.cloudy clear        cloudy
...
[29] clear         clear         partly.cloudy
Levels: clear partly.cloudy cloudy
```

However, the data is better presented in a tabular format, as in Table 2.1.

Table 2.1 Weather in Central Park for May 2003

clear	partly cloudy	cloudy
11	11	9

The `table()` function will produce this type of output:

```
> table(central.park.cloud)
central.park.cloud
        clear partly.cloudy        cloudy
           11            11             9
```

■

2.1.2 Barplots

Categorical data is also summarized in a graphical manner. Perhaps most commonly, this is done with a barplot (or bar chart). A barplot in its simplest usage arranges the levels of the variable in some order and then represents their frequency with a bar of a height proportional to the frequency.

In R, barplots are made with the `barplot()` function. It uses a *summarized* version of the data, often the result of the `table()` function.* The summarized data can be either frequencies or proportions. The resulting graph will look the same, but the scales on the y-axis will differ.

■ **Example 2.2: A first barplot** Twenty-five students are surveyed about their beer preferences. The categories to choose from are coded as (1) domestic can, (2) domestic bottle, (3) microbrew, and (4) import. The raw data is

```
3 4 1 1 3 4 3 3 1 3 2 1 2 1 2 3 2 3 1 1 1 1 4 3 1
```

Let's make a barplot of both frequencies and proportions. We first use `scan()` instead of `c()`, to read in the data. Then we plot (Figure 2.1) in several ways. The last two graphs have different scales. For barplots, it is most common to use the frequencies.

---

* In version 1.9.0 of R one must coerce the resulting table into a data vector to get the desired plot. This can be done with the command `t(table(x))`

```
> beer = scan()
1: 3 4 1 1 3 4 3 3 1 3 2 1 2 1 2 3 2 3 1 1 1 1 4 3 1
26:
Read 25 items
> barplot(beer)                    # this isn't correct
> barplot(table(beer),             # frequencies
+ xlab="beer", ylab="frequency")
> barplot(table(beer)/length(beer), # proportions
+ xlab="beer", ylab="proportion")
```

The + symbol is the continuation prompt. Just like the usual prompt, >, it isn't typed but is printed by R.

The barplot on the left in Figure 2.1 is not correct, as the data isn't summarized. As well, the `barplot()` function hasn't labeled the x-and y-axes, making it impossible for the reader to identify what is being shown. In the subsequent barplots, the extra arguments `xlab=` and `ylab=` are used to label the x- and y-axes. ■

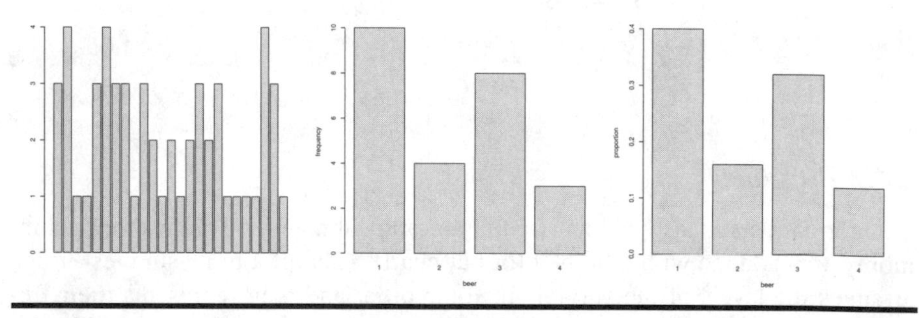

Figure 2.1 Barplot of `beer`: the first needs to be summarized; the second and third show frequencies and proportions

■ **Example 2.3: Misleading barplots** When people view a barplot, they are trying to visualize differences between the data presented. People tend to assume a scale starts at 0. A graphic designer can deliberately mislead the reader by making a graphic non-0 based (and not mentioning having done this). Such misleading barplots appear frequently in the media.

We can see how this is done by looking at a data set on sales figures (Figure 2.2).

```
> sales = c(45,44,46)          # quarterly sales
> names(sales) = c("John","Jack","Suzy") # include names
> barplot(sales, main="Sales", ylab="Thousands") # basic barplot
> barplot(sales, main="Sales", ylab="Thousands",
+ ylim=c(42,46), xpd=FALSE)    # extra arguments to fudge plot
```

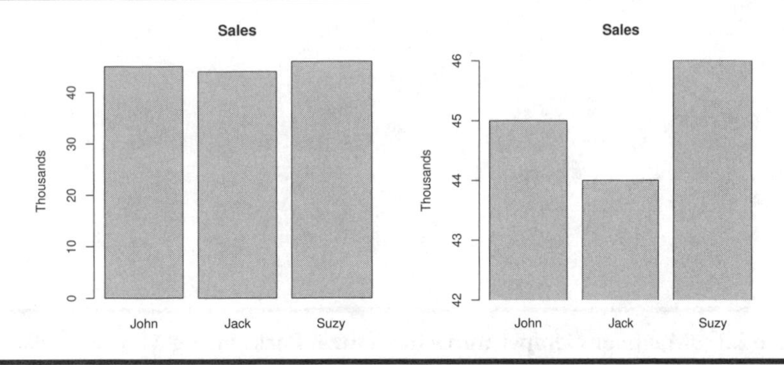

Figure 2.2 Two barplots showing the same data. The right one is misleading, as the *y*-axis does not start at 0.

There are names on the bars because the data vector was given names. The argument `names.arg=` could also be used. We used the argument `ylim=` to set the limits on the *y*-axis and `xpd=` to have R print only within these limits.

The second barplot seems to indicate a much greater difference in sales than the first, despite its representing the same information. As people expect barplots to start at 0, we should always draw them that way. ■

■ **Example 2.4: Time series data shown with barplots** Barplots are often used to illustrate what are technically time series. A time series is a measurement of the same thing taken at several points in time. For example: the high temperature in May 2003 at Central Park is stored in the `MAX` variable of the `central.park (UsingR)` data set. In Figure 2.3 the data is presented using the following commands. We assume the `UsingR` package has already been loaded using the command `library(UsingR)`.

```
> barplot(central.park$MAX,        # used $ notation, not attach()
+ names.arg=1:31,                  # + is continuation prompt
+ xlab="day", ylab="max. temp.")
```

The graph seems familiar, as we regularly see such graphs in the media. Still, it could be criticized for "chart junk" (this term was coined by E. Tufte to refer to excessive amounts of chart ink used to display a relationship). ■

The `barplot()` function is used similarly to the other plotting functions. The basic function provides a graphic that can be adjusted using various arguments. Most of these have names that are consistent from plotting function to plotting function. For the `barplot()` function we showed these arguments: names can be changed with `names.arg=`; axis limits set with `xlim=` or `ylim=`; and the plot region was clipped using `xpd=`. The arguments `horizontal=TRUE` (draws the bars horizontally) and `col=` (sets the bars colors) were not shown.

Although doing so is a bit redundant, some people like to put labels on top of the bars of a barplot. This can be done using the `text()` function. Simply save

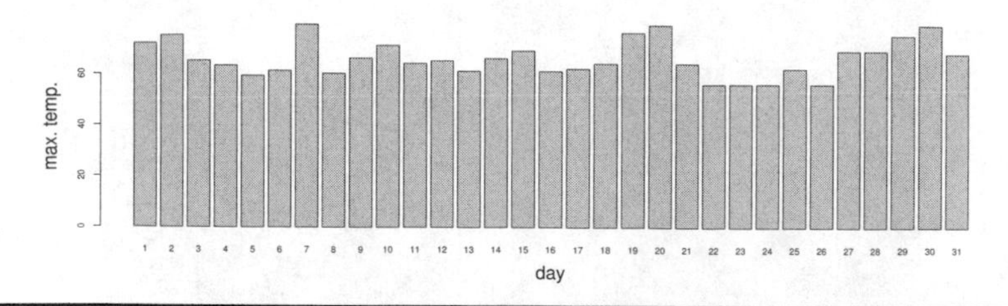

Figure 2.3 Maximum temperatures in Central Park during May 2003. This barplot shows time-series data rather than categorical data.

the output of the `barplot()` function and then call `text()` with postions for the *x* and *y* coordinates; an argument, `labels=`, for the text; and the argument `pos=1` to put the text just below the bar. This example (not shown) illustrates the process:

```
> our.data = c(1,2,2,5); names(our.data)=1:4
> bp = barplot(our.data)
> text(bp, our.data, labels = our.data, pos = 1)
```

The *x*-coordinates are returned by `barplot()`; the *y*-coordinates of the text are the heights of the bars given by the values in the data vector.

2.1.3 Pie charts

The **pie chart** graphic is used to display the relative frequencies or proportions of the levels of a categorical variable. The pie chart represents these as wedges of a circle or pie. Pie charts, like barplots, are widely found in the media. However, unlike barplots, pie charts have lost favor among statisticians, as they don't really do well what they aim to do.

Creating a pie chart is more or less the same as creating a barplot, except that we use the `pie()` function. Similar arguments are used to add names or change the colors.

For example, to produce a pie chart for the sales data in Example 2.3, we use the following commands (Figure 2.4):

```
> sales
John Jack Suzy
  45   44   46
> pie(sales, main="sales")
```

The argument `main=` is used to set a main title. Alternatively, the `title()` function can be used to add a title to a figure. Again, the names attribute of `sales` is used to label the pie chart. Alternatively, the `labels=` argument is available. For the graphic in the text, `col=gray(c(0.7,0.85,0.95))` was used to change the colors.

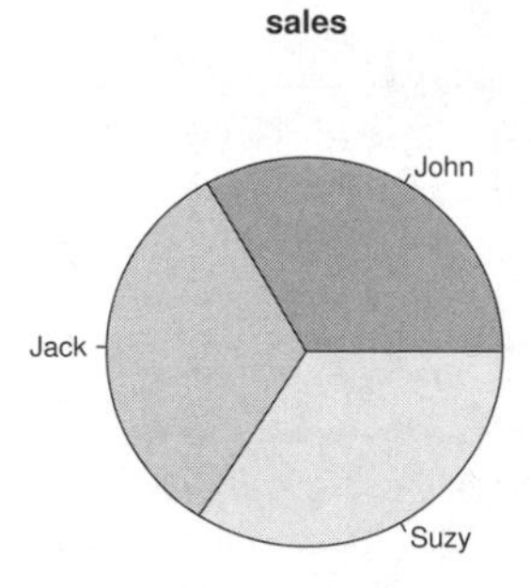

Figure 2.4 An example of a pie chart. Can the sales leader be identified?

Why are pie charts a poor choice? The help page for `pie()`, (`?pie`), gives a clue as to why:

> Pie charts are a very bad way of displaying information. The eye is good at judging linear measures and bad at judging relative areas. A bar chart or dot chart is a preferable way of displaying this type of data.

To illustrate why the pie chart is misleading, look again at Figure 2.4. It is practically impossible to tell who has made the most sales. The pie chart fails at discerning differences. The bar chart example in Figure 2.2 shows that the barplot can be effective in highlighting differences, but using it this way can be misleading.

2.1.4 Dot charts

Using a **dot chart**, also known as a Cleveland dotplot, is one way to highlight differences without being misleading. The default dot chart shows the values of the variables as big dots in a horizontal display over the range of the data. Differences from the maximum and minimum values are very obvious, but to see their absolute values we must look at the scale. The primary arguments to `dotchart()` are the data vector and an optional set of labels specified by the argument `labels=`. Again, if the data vector has names, these will be used by default. Other options exist to change the range of the data (`xlim=`), the colors involved (`color=`, `gcolor=`, `lcolor=`), and the plotting characters (`pch=`, `gpch=`).

```
> dotchart(sales,xlab="Amount of sales")
```

Figure 2.5 shows the differences quite clearly.

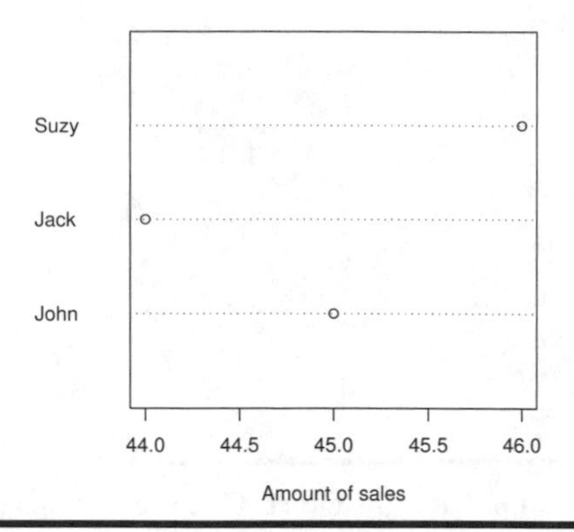

Figure 2.5 The dot chart highlights differences between categories

2.1.5 Factors

Where does the ordering of the categories come from in the examples on creating tables? If we looked at the help page for `table()` we would find that the data should be interpretable as a "factor." R uses factors to store categorical data.

Factors are made with the function `factor()` or the function `as.factor()` and have a specific set of values called `levels()`.

At first glance, factors appear to be similar to data vectors, but they are not. We see below that they are printed differently and do not have numeric values.

```
> 1:5                              # a numeric vector (integer)
[1] 1 2 3 4 5
> factor(1:5)                      # now a factor. Note levels
[1] 1 2 3 4 5
Levels: 1 2 3 4 5
> mean(factor(1:5))                # factors are not numeric
[1] NA
Warning message:
argument is not numeric or logical: returning NA in:
mean.default(factor(1:5))
> letters[1:5]                     # a character vector
[1] "a" "b" "c" "d" "e"
> factor(letters[1:5])             # turned into a factor
[1] a b c d e
Levels: a b c d e
```

The initial order of the levels in a factor is determined by the `sort()` function. In the example with `mean()` an error is returned, as factors are treated as numeric even if their levels are given numeric-looking values. We used `letters` to return a character vector. This built-in variable contains the 26 letters a through z. The capital letters are in LETTERS. Chapter 4 has more on factors.

2.1.6 Problems

2.1 Find an example of a table in the media summarizing a *univariate* variable. Could you construct a potential data set that would have this table?

2.2 Try to find an example in the media of a misleading barplot. Why is it misleading? Do you think it was meant to be?

2.3 Find an example in the media of a pie chart. Does it do a good job of presenting the data?

2.4 Load and attach the data set `central.park (UsingR)`. The `WX` variable contains a list of numbers representing bad weather (e.g., 1 for fog, 3 for thunder, 8 for smoke or haze). `NA` is used when none of the types occurred. Make a table of the data, then make a table with the extra argument `exclude=FALSE`. Why is the second table better?

2.5 Web developers need to know which browsers people use, as they need to support many different platforms. Table 2.2 contains usage percentages based on an analysis of a United States Geological Society web server.

Table 2.2 Web browser statistics

Browser statistics	
Internet Explorer	86%
Gecko-based (Netscape, Mozilla)	4%
Netscape Navigator 4	5%
Opera	1%
unidentified	4%

source `http://www.upsdell.com/BrowserNews/stat.htm`

Make a bar chart, a pie chart, and a dot chart of this data.

2.6 According to the *New York Times*, the top-selling MP3 players for August 2003 are as shown in Table 2.3 with their market shares. Assume the total market share is $22 million.

1. What percent is "other"?
2. Find the dollar amount (not the percentage amount) for each company.
3. Make a bar chart, dot chart, and pie chart of the data, including "other." Which chart shows the relationship best?
4. Comment on how out-of-date this data seems.

Table 2.3 Sales of MP3 players. Total $22 million

MP3 players	
Apple	18%
RCA	15%
Rio	14.4%
iRiver	13.5%
Creative Labs	6.2%

2.7 Make a dot chart of the `mpg` variable in the `mtcars` data set. Specify the argument `labels=` using the command `rownames(mtcars)`, which returns the names of each row of the data frame.

2.8 The data set `npdb (UsingR)` contains information on malpractice awards in the United States. Attach the data set and make a table of the `state` variable. Which state had the most awards? (Using `sort()` on your table is useful here.)

2.9 For the malpractice-award data set `npdb (UsingR)`, the variable `ID` is an identification number unique to a doctor but not traceable back to the doctor. It allows a look at a doctor's malpractice record without sacrificing anonymity.

The commands

```
> table(npdb$ID)
```

create a table of malpractice awards for each of the 6,369 doctors. What does the command `table(table(ID))` do, and why is this interesting?

2.10 The data set `MLBattend (UsingR)` contains attendance information for major league baseball between 1969 and 2000. The following commands will extract just the wins for the New York Yankees, in chronological order.

```
> attach(MLBattend)
> wins[franchise == "NYA"]
 [1]  80  93  82  79  80  89  83  97 100 100  89 103  59  79  91
...
> detach(MLBattend)              # tidy up
```

Add the names `1969:2000` to your variable. Then make a barplot and dot chart showing this data in chronological order.

2.2 Numeric data

For univariate data, we want to understand the distribution of the data. What is the range of the data? What is the central tendency? How spread out are the values? We can answer these questions graphically or numerically. In this section, we'll see how. The familiar mean and standard deviation will be introduced, as will the pth quantile, which extends the idea of a median to measure position in a data set.

2.2.1 Stem-and-leaf plots

If we run across a data set, the first thing we should do is organize the data so that a sense of the values becomes more clear. A useful way to do so for a relatively small data set is with a **stem-and-leaf plot**. This is a way to code a set of numeric values that minimizes writing and gives a fairly clear idea of what the data is, in terms of its range and distribution. For each data point only a single digit is recorded, making a compact display. These digits are the "leaves." The stem is the part of the data value to the left of the leaf.

To illustrate, we have the following data for the number of points scored in a game by each member of a basketball team:

`2 3 16 23 14 12 4 13 2 0 0 0 6 28 31 14 4 8 2 5`

The stem in this case would naturally be the 10s digit. A number like 23 would be written as a 2 for the stem and a 3 for the leaf. The results are tabulated as shown below in the output of `stem()`.

```
> x = scan()
1: 2 3 16 23 14 12 4 13 2 0 0 0 6 28 31 14 4 8 2 5
21:
Read 20 items
> stem(x)
  The decimal point is 1 digit(s) to the right of the |

  0 | 000222344568
  1 | 23446
  2 | 38
  3 | 1
```

The stem is written to the left and the leaf to the right of the `|`. If this isn't clear, look at the values in the row with a stem of 1. They are 12, 13, 14, 14, and 16. For many data sets, the data is faithfully recorded. In some cases, the data values are truncated to accommodate the format.

The stem-and-leaf plot gives us a good understanding of a data set. At a glance we can tell the minimum and maximum values (0 and 31); the shape of the distribution of numbers (mostly in the 0 to 10 range); and, if we want to, the "middle" of the distribution (between 5 and 6).

In practice, stem-and-leaf plots are usually done by hand as data is being collected. This is best done in two passes: a first to get the stem and the leaves, and a second to sort the leaves. We may decide the latter step is not necessary, depending on what we want to know about the numbers.

The `stem()` function As illustrated, stem-and-leaf plots are done in `R` with the `stem()` function. The basic usage is simply `stem(x)`, where `x` is a data vector. If there are too many leaves for a stem, the extra argument `scale=` can be set, as in `stem(x,scale=2)`.

A back-to-back stem and leaf diagram can be used to show two similarly distributed data sets. For the second data set, the left side of the stem is used. There is no built-in `R` function to generate these.

2.2.2 Strip charts

An alternative to a stem-and-leaf plot is a **strip chart** (also referred to as a dot-plot). This graphic plots each value of a data set along a line (or strip). As with a stem-and-leaf plot, we can read off all the data. Unlike the stem-and-leaf plot, however, the strip chart does not take up much vertical space on a page and so is a favorite of some authors. Variants are often found in the media.

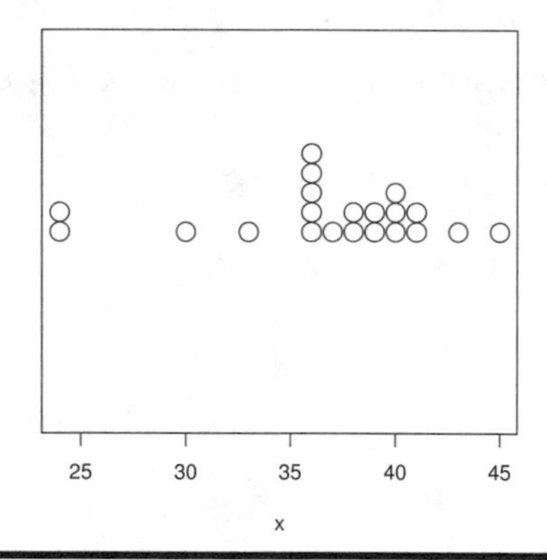

Figure 2.6 An example of a strip chart showing each value in a data set

Strip charts are created in `R` with the `stripchart()` function. The extra argument `method="stack"` will create stacks when there are ties in the data. Otherwise, multiple values will show up as a single plot character.

Figure 2.6 shows the data on heights of 4-year-olds contained in the data set `kid.weights (UsingR)`. The following commands produce the graphic:

```
> attach(kid.weights)
> x = height[48 <= age & age < 60] # four year olds
> stripchart(x,method="stack",xlab="x",pch=1,offset=1,cex=2)
> detach(kid.weights)            # tidy up
```

A lot of extra arguments are needed to make these graphs look right. The argument xlab= adjusts the label on the x-axis, pch=1 uses a different plot character than the default square, cex= changes the size of the plot character, and offset= pushes the points apart.

In this book, we use a visually stripped-down version of this graphic (Figure 2.7) made with the DOTplot() function available in the UsingR package.

```
> DOTplot(x)
```

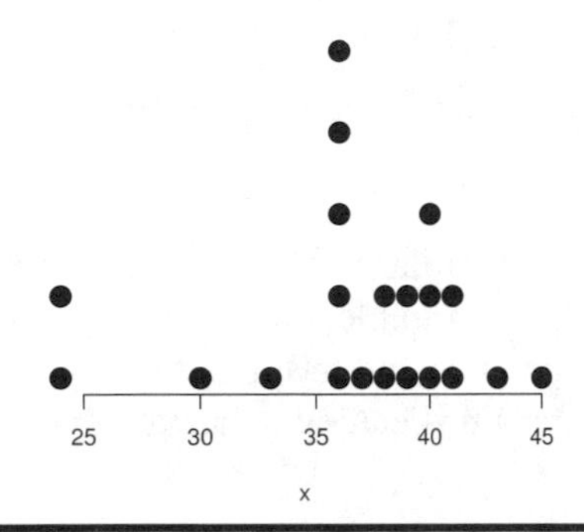

Figure 2.7 Alternative to stripchart(), with fewer lines

2.2.3 The center: mean, median, and mode

Viewing the distribution of a data set with a stem-and-leaf plot or a strip chart can be overwhelming if there is too much data. It is also convenient to have concise, numeric summaries of a data set. Most of these summaries are familiar from everyday usage. Not only will the numeric summaries simplify a description of the data—they also allow us to compare different data sets quantitatively. In this section we cover central tendency; in the next we cover measures of spread.

The most familiar notion of the center of a numeric data set is the average value of the numbers. In statistics, the average of a data set is called the **sample mean** and is denoted by $\bar{x}$.

The sample mean

The sample mean of the numeric data set, $x_1, x_2, \ldots, x_n$, is

$$\bar{x} = \frac{x_1 + x_2 + \cdots + x_n}{n}. \tag{2.1}$$

The `mean()` function will compute the sample mean for a data vector. Additional arguments include `trim=` to perform a trimmed mean and `na.rm=` for removal of missing data.

The mean can be found directly from the formula, as in

```
> x = scan()
1: 2 3 16 23 14 12 4 13 2 0 0 0 6 28 31 14 4 8 2 5
21:
Read 20 items
> sum(x)/length(x)
[1] 9.35
```

Using the `mean()` function is preferable, though:

```
> mean(x)
[1] 9.35
```

The mean is the most familiar notion of center, but there are times when it isn't the best. Consider, for example, the average wealth of the patrons of a bar before and after Microsoft co-founder Bill Gates steps in. Really large values can skew the average, making a misleading measure of center.

The median

A more resistant measure of the center is the **sample median**, which is the "middle" value of a distribution of numbers. Arrange the data from smallest to biggest. When there is an odd number of data points, the median is the middle one; when there is an even number of data points, the median is the average of the two middle ones.

The sample median

The sample median, m, of $x_1, x_2, \ldots, x_n$ is the middle value of the sorted values. Let the *sorted* data be denoted by $x_{(1)} \leq x_{(2)} \leq \cdots \leq x_{(n)}$. Then

$$m = \begin{cases} x_{(k+1)} & n = 2k+1 \text{ (odd)}, \\ \frac{1}{2}\left(x_{(k)} + x_{(k+1)}\right) & n = 2k \text{ (even)}. \end{cases} \tag{2.2}$$

The sample median is found in R with the `median()` function.

For example:

```
> bar = c(50,60,100,75,200)      # bar patrons worth in 1000s
> bar.with.gates = c(bar,50000) # after Bill Gates enters
```

```
> mean(bar)
[1] 97
> mean(bar.with.gates)              # mean is sensitive to large values
[1] 8414
> median(bar)
[1] 75
> median(bar.with.gates)            # median is resistant
[1] 87.5
```

The example shows that a single large value can change the value of the sample mean considerably, whereas the sample median is much less influenced by the large value. Statistics that are not greatly influenced by a few values far from the bulk of the data are called resistant statistics.

Visualizing the mean and median from a graphic

Figure 2.8 shows how to visualize the mean and median from a strip chart (and, similarly, from a stem-and-leaf plot). The strip chart implicitly orders the data from smallest to largest; to find the median we look for the middle point. This is done by counting. When there is an even number, an average is taken of the two middle ones.

The formula for the mean can be interpreted using the physics formula for a center of mass. In this view, the mean is the balancing point of the strip chart when we imagine the points as equal weights on a seesaw.

With this intuition, we can see why a single extremely large or small data point can skew the mean but not the median.

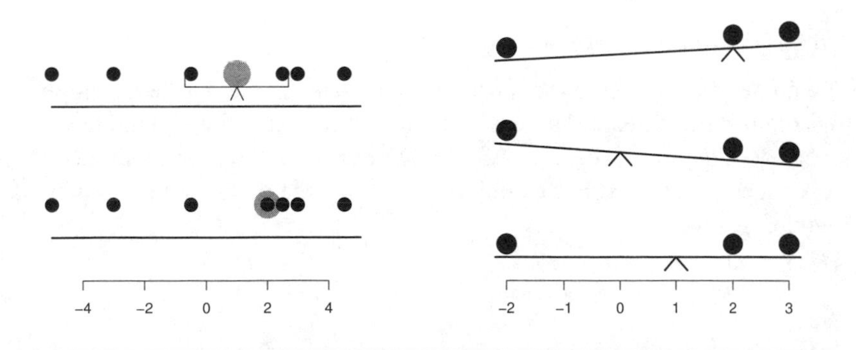

Figure 2.8 The median is the middle point, the mean the balance point

The trimmed mean

A modification of the mean that makes it more resistant is the **trimmed mean**. To compute the trimmed mean, we compute the mean after "trimming" a certain percentage of the smallest and largest values from the data. Consequently, if there are a few values that skew the mean, they won't skew the trimmed mean.

The `mean()` function can be used with the `trim=` argument to compute a trimmed mean. Its value is the proportion to be trimmed from both sides.

■ **Example 2.5: Income distributions are skewed**

The `cfb (UsingR)` data set contains a sampling of the data contained in the Survey of Consumer Finances conducted in the year 2001 by the U.S. Federal Reserve Board. Many of the variables have some values much bigger than the bulk of the data. This is common in income distributions, as some fortunate people accumulate enormous wealth in a lifetime, but few can accumulate enormous debt.

The `INCOME` variable contains yearly income figures by household. For this data, we compare the different measures of center.

```
> income = cfb$INCOME
> mean(income)
[1] 63403
> median(income)
[1] 38033
> mean(income, trim=.2)
[1] 41992
> sum(income <= mean(income))/length(income)*100
[1] 70.5
```

The data is clearly skewed to the right, as the mean is significantly more than the median. The trimmed mean is more in line with the median. The last line shows that 70.5% of the values are less than or equal to the sample mean. ■

The mode and midrange of a data set

The mode of a data set is the most common value in the data set. It applies only to discrete numeric data. There is no built-in function to find the mode, as it is not a very good summary of a data set. However, it can be found using commands we've seen previously. For example, if `x` stores the data, then the mode may be found as follows:

```
> x = c(72,75,84,84,98,94,55, 62)
> which(table(x) == max(table(x)))
84
 5
```

That is, the value of 84, which is the fifth, after sorting, of `x`. Alternately, the function `which.max()`, which determines the position of the max in a data vector, finds this value with `which.max(table(x))`.

The midrange is a natural measure of center—the middle of the range. It can be found using `mean(range(x))`. For some data sets it is close to the mean, but not when there are outliers. As it is even more sensitive to these than the mean, it isn't widely used to summarize the center.

Summation notation

The definition of the mean involves a summation:

$$\bar{x} = \frac{x_1 + x_2 + \cdots + x_n}{n}.$$

In statistics, this is usually written in a more compact form using **summation notation**. The above sum is rewritten as

$$\bar{x} = \frac{1}{n}\sum_{i=1}^{n} x_i.$$

The symbol $\sum$, the Greek capital *sigma*, is used to indicate a sum. The $i = 1$ on the bottom and n on top indicate that we should include x_i for $i = 1, 2, \ldots, n$, that is $x_1, x_2, \ldots, x_n$. Sometimes the indices are explicitly indicated, as in

$$\frac{1}{n}\sum_{i \text{ in } \{1,2,\ldots,n\}} x_i.$$

When the variable that is being summed over is not in doubt, the summation notation is often shortened. For example,

$$\bar{x} = \frac{1}{n}\sum x_i.$$

Notationally, this is how summations are handled in R using the `sum()` function. If `x` is a data vector, then `sum(x)` adds up `x[1] + x[2] + ... + x[n]`.

The summation notation can be confusing at first but offers the advantages of being more compact to write and easier to manipulate algebraically. It also forces our attention on the operation of addition.

■ **Example 2.6: Another formula for the mean** We can use the summation formula to present another useful formula for the mean. Let $\mathsf{Range}(x)$ be all the values of the data set. When we add $x_1 + x_2 + \cdots + x_n$, if there are ties in the data, it is natural to group the same numbers first and then add. For example, if there are four 5's, we would just add a single value of $4 \cdot 5$, or 20. Let $n_k = \#\{i : x_i = k\}$; that is, the number of data points equal to k. We can then write the sample mean as

$$\bar{x} = \frac{1}{n}\sum x_i = \frac{1}{n}\sum_{k \text{ in } \mathsf{Range}(x)} k \cdot n_k = \sum_{k \text{ in } \mathsf{Range}(x)} k \cdot p_k.$$

Here $p_k = n_k/n$ is the proportion of times the data is k.

The last sum is a **weighted average**. The weights are the p_k—nonnegative numbers that add to 1. ■

2.2.4 Variation: the variance, standard deviation, and `IQR`

The center of a distribution of numbers may not adequately describe the entire distribution. For example, consider the data in Table 2.4 on test scores for two different tests presented in a back-to-back stem-and-leaf plot.

Table 2.4 Two test results

first test	stem	second test
	4	07
	5	
	6	
75	7	
87520	8	260
	9	
	10	00

The means are about the same, but clearly there is more variation in the second test—two students aced it with 100s, but two failed miserably. In the first test, the students all did roughly the same. In short, there is more "spread," or variation, in the second test.

The sample range

There are many ways to measure spread. The simplest is the **range** of the data. Sometimes the range refers to the distance between the smallest and largest values, and other times it refers to these two values as a pair. The function `range()` returns the smallest and largest values, as in `range(x)`. The distance between the two is computed by `diff(range(x))`.

The term *distance* between two points x and y refers to the value $|x-y|$, which is nonnegative. The *difference* between x and y is $x-y$, which may be negative. We also call this the *deviation*.

Sample variance

Using the idea of a center, we can think of variation in terms of deviations from the center. Using the mean for the center, the variation of a single data point can be assessed using the value $x_i - \bar{x}$. A sum of all these differences will give a sense of the total variation. Just adding these values gives a value of 0, as terms cancel. A remedy is to add the squared deviations $(x_i - \bar{x})^2$.

If there is a lot of spread in the data, then this sum will be relatively large; if there is not much spread, it will be relatively small. Of course, this sum can be large because n is large, not just because there is much spread. We take care of that by dividing our sum by a scale factor. If we divided by n we would have the "average squared deviation." It is conventional, though, to divide by $n - 1$, producing the **sample variance**:

$$s^2 = \frac{1}{n-1} \sum_{i=1}^{n} (x_i - \bar{x})^2.$$

We will see that many of the statistics we consider can be analyzed this way: one piece that intuitively makes sense and a divisor that allows us to compare the statistic among different data sets.

The **sample standard deviation** is the square root of the variance. It has the advantage of having the same units as the mean. However, the interpretation remains: large values indicate more spread.

The sample variance and standard deviation

For a numeric data set $x_1, x_2, \ldots, x_n$, the sample variance is defined by

$$s^2 = \frac{1}{n-1} \sum_{i=1}^{n} (x_i - \bar{x})^2. \tag{2.3}$$

The sample standard deviation is the square root of the sample variance:

$$s = \sqrt{\frac{1}{n-1} \sum_{i=1}^{n} (x_i - \bar{x})^2}. \tag{2.4}$$

The sample variance is computed in R using the `var()` function, the sample standard deviation with the `sd()` function.

To illustrate on the test-scores data:

```
> test.scores = c(80,85,75,77,87,82,88)
> test.scores.b = c(100,90,50,57,82,100,86)
> mean(test.scores)
[1] 82
> mean(test.scores.b)                # means are similar
```

```
[1] 80.71
> n = length(test.scores)
# compute directly
> (1/(n-1)) * sum( (test.scores - mean(test.scores))^2 )
[1] 24.67
> var(test.scores)                  # built-in var function
[1] 24.67
> var(test.scores.b)                # larger, as anticipated
[1] 394.2
> sd(test.scores)
[1] 4.967
```

Quantiles, quintiles, percentiles, and more

The standard deviation, like the mean, can be skewed when an exceptionally large or small value is in the data. More resistant alternatives are available. A conceptually simple one (the IQR) is to take the range of the middle 50% of the data. That is, trim off 25% of the data from the left and right, and then take the range of what is remaining.

To be precise, we need to generalize the concept of the median. The median splits the data in half—half smaller than the median and half bigger. The quantiles generalize this. The p**th quantile** is at position $1+p(n-1)$ in the sorted data. When this is not an integer, a weighted average is used. † This value essentially splits the data so $100p\%$ is smaller and $100(1-p)\%$ is larger. Here p ranges from 0 to 1. The median then is the 0.5 quantile.

The **percentiles** do the same thing, except that a scale of 0 to 100 is used, instead of 0 to 1. The term **quartiles** refers to the 0, 25, 50, 75, and 100 percentiles, and the term **quintiles** refers to the 0, 20, 40, 60, 80, and 100 percentiles.

The `quantile()` function returns the quantiles. This function is called with the data vector and a value (or values) for p. We illustrate on a very simple data set, for which the answers are easily guessed.

```
> x = 0:5                           # 0,1,2,3,4,5
> length(x)
[1] 6
> sum(sort(x)[3:4])/2               # the median the hard way
[1] 2.5
> median(x)                         # easy way. Clearly the middle
[1] 2.5
> quantile(x,.25)
 25%
1.25
> quantile(x,c(0.25,0.5,0.75))  # more than 1 at a time
```

† There are other definitions used for the pth quantile implemented in the `quantile()` function. These alternatives are specified with the `type=` argument. The default is type `7`. See `?quantile` for the details.

```
 25%  50%  75%
1.25 2.50 3.75
> quantile(x)                    # default gives quartiles
  0%  25%  50%  75% 100%
0.00 1.25 2.50 3.75 5.00
```

■ **Example 2.7: Executive pay** The `exec.pay (UsingR)` data set contains compensation to CEOs of 199 U.S. companies in the year 2000 in units of $10,000. The data is not symmetrically distributed, as a stem-and-leaf plot will show. Let's use the `quantile()` function to look at the data:

```
> sum(exec.pay > 100)/length(exec.pay) # proportion more
[1] 0.09045                      # 9% make more than 1 million
> quantile(exec.pay,0.9)         # 914,000 dollars is 90 percentile
 90%
91.4
> quantile(exec.pay,0.99)        # 9 million is top 1 percentile
  99%
906.6
> sum(exec.pay <= 10)/length(exec.pay)
[1] 0.1457                       # 14 percent make 100,000 or less
> quantile(exec.pay,.10)         # the 10 percentile is 90,000
10%
  9
```

■

Quantiles versus proportions For a data vector x we can ask two related but inverse questions : what proportion of the data is less than or equal to a specified value? Or for a specified proportion, what value has this proportion of the data less than or equal? The latter question is answered by the quantile function.

The inter-quartile range

Returning to the idea of the middle 50% of the data, this would be the distance between the 75th percentile and the 25th percentile. This is known as the **inter-quartile range** and is found in R with the `IQR()` function.

For the executive pay data the IQR is

```
> IQR(exec.pay)
[1] 27.5
```

Whereas, for comparison, the standard deviation is

```
> sd(exec.pay)
[1] 207.0
```

This is much bigger, as the largest values of `exec.pay` are much larger than the others and skew the results.

z-scores

The z-score of a value is the number of standard deviations the value is from the sample mean of the data set. That is,

$$z\text{-score} = \frac{x_i - \bar{x}}{s}.$$

As with the quantiles, z-scores give a sense of the size a value has within a set of data. In R the collection of z scores for a data set are returned by the `scale()` function. The set of z-scores will have sample mean of 0 and standard deviation 1, allowing for comparisons among samples with different senses of scale.

Numeric summaries of the data

A short description of a distribution of numbers could be made with the range, the mean, and the standard deviation. It might also make sense to summarize the data with the range, the quartiles, and the mean. In R, the `summary()` function does just this for a numeric data vector.

For the executive-pay data set we have

```
> summary(exec.pay)
   Min. 1st Qu.  Median    Mean 3rd Qu.    Max.
    0.0    14.0    27.0    59.9    41.5  2510.0
```

There is a large difference between the mean and the median. We would want to be careful using the mean to describe the center of this data set. In fact, the mean is actually the 84th percentile:

```
> sum(exec.pay <= mean(exec.pay))/length(exec.pay)
[1] 0.8392
```

That is, only 16% make more than the average.

In the sequel we will see that the `summary()` function returns reasonable numeric summaries for other types of objects in R.

Hinges and the five-number summary

There is a historically popular set of alternatives to the quartiles called the hinges, which are somewhat easier to compute by hand. Quickly put, the lower hinge is the median of the lower half of the data, and the upper hinge the median of the upper half of the data. In Figure 2.9, when $n = 6$, the upper and lower halves include three ($n/2$) data points; when $n = 7$, there are still three ($(n-1)/2$) points in each. The difference is that when n is odd, the median is removed from the data when considering the upper and lower halves.

The hinges are returned as part of the five-number summary, which is output by the `fivenum()` function.

The lower and upper hinges can be different from the quartiles Q_1 and Q_3. For example, with $n = 6$, the first quartile is at position $1 + (1/4)(6-1) = 2.25$. That is, a quarter of the way between the second and third data points after sorting.

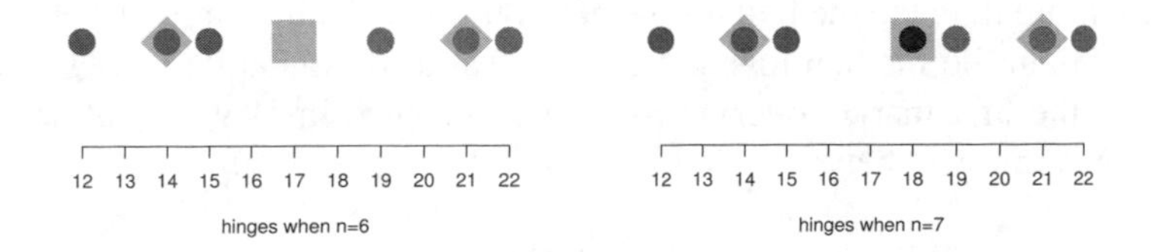

Figure 2.9 Hinges, marked with diamonds, are medians of left and right halves of the data. The left and right halves of data consist of $n/2$ points when n is even, and $(n-1)/2$ points when n is odd.

This is different from the lower hinge, which from Figure 2.9 is seen to be the second data point after sorting.

The IQR is the difference between the third and first quartiles. The **H-spread** is used for the difference of the upper and lower hinges. Some books call the IQR the Q-spread; others refer to the H-spread as the IQR.

2.2.5 Problems

2.11 Read this stem-and-leaf plot. First find the median by hand. Then enter in the data and find the median using `median()`.

```
  The decimal point is 1 digit(s) to the right of the |

   8 | 028
   9 | 115578
  10 | 1669
  11 | 01
```

2.12 Figure 2.10 contains a strip chart of a data set. Estimate the median, mean, and 10% trimmed mean. Enter in the data as accurately as you can and then check your estimates using the appropriate function in R.

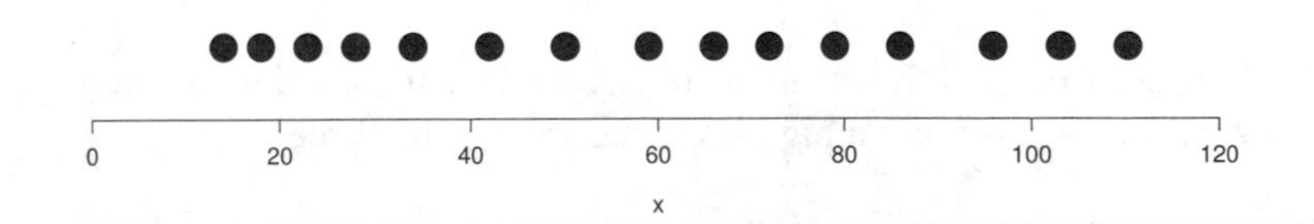

Figure 2.10 Strip chart of a data set

2.13 Can you copyedit this paragraph from the August 16, 2003 *New York Times*?

The median sales price, which increased to $575,000, almost 12 per-

> cent more than the median for the previous quarter and almost 13 percent more than the median for the period a year ago, was at its highest level since the first market overview report was issued in 1989. (The median price is midway between the highest and lowest prices.)

2.14 In real estate articles the median is often used to describe the center, as opposed to the mean. To see why, consider this example from the August 16, 2003 *New York Times* on apartment prices:

> The average and the median sales prices of cooperative apartments were at record highs, with the average up almost 9 percent to $775,052 from the first quarter this year, and the median price at $479,000, also an increase of almost 9 percent.

Explain how using the median might affect the reader's sense of the center.

2.15 The data set `pi2000 (UsingR)` contains the first 2,000 digits of π. What is the percentage of digits that are 3 or less? What percentage of the digits are 5 or more?

2.16 The data set `rivers` contains the lengths (in miles) of 141 major rivers in North America.

1. What proportion are less than 500 miles long?
2. What proportion are less than the mean length?
3. What is the 0.75 quantile?

2.17 The `time` variable in the `nym.2002 (UsingR)` data set contains the time to finish the 2002 New York City marathon for a random sample of the finishers.

1. What percent ran the race in under 3 hours?
2. What is the time cutoff for the top 10%? The top 25%?
3. What time cuts off the bottom 10%?

Do you expect this data set to be symmetrically distributed?

2.18 Compare values of the mean, median, and 25% trimmed mean on the built-in `rivers` data set. Is there a big difference among the three?

2.19 The built-in data set `islands` contains the size of the world's land masses that exceed 10,000 square miles. Make a stem-and-leaf plot, then compare the mean, median, and 25% trimmed mean. Are they similar?

2.20 The data set `OBP (UsingR)` contains the on-base percentages for the 2002 major league baseball season. The value labeled `bondsba01` contains this value for Barry Bonds. What is his z-score?

2.21 For the `rivers` data set, use the `scale()` function to find the z-scores. Verify that the z-scores have sample mean 0 and sample standard deviation 1.

2.22 The median absolute deviation is defined as

$$\mathsf{mad}(x) = 1.4826 \cdot \mathsf{median}(|x_i - \mathsf{median}(x)|). \tag{2.5}$$

This is a resistant measure of spread and is implemented in the `mad()` function. Explain in words what it measures. Compare the values of the sample standard deviation, IQR, and median absolute deviation for the `exec.pay (UsingR)` data set.

2.23 The data set `npdb (UsingR)` contains malpractice-award information. The variable `amount` is the size of malpractice awards in dollars. Find the mean and median award amount. What percentile is the mean? Can you explain why this might be the case?

2.24 The data set `cabinet (UsingR)` contains information on the amount each member of President George W. Bush's cabinet saved due to the passing of a tax bill in 2003. This information is stored in the variable `est.tax.savings`. Compare the median and the mean. Explain the difference.

2.25 We may prefer the standard deviation to measure spread over the variance as the units are the same as the mean. Some disciplines, such as ecology, prefer to have a unitless measurement of spread. The **coefficient of variation** is defined as the standard deviation divided by the mean.

One advantage is that the coefficient of variation matches our intuition of spread. For example, the numbers 1, 2, 3, 4 and 1001, 1002, 1003, 1004 have the same standard deviation but much different coefficient of variations. Somehow, we mentally think of the latter set of numbers as closer together.

For the `rivers` and `pi2000 (UsingR)` data sets, find the coefficient of variation.

2.26 A **lag plot** of a data vector plots successive values of the data against each other. By using a lag plot, we can tell whether future values depend on previous values: if not, the graph is scattered; if so, there is often a pattern.

Making a lag plot (with lag 1) is quickly done with the indexing notation of negative numbers. For example, these commands produce a lag plot[‡] of `x`:

```
> n = length(x)
> plot(x[-n],x[-1])
```

(The `plot()` function plots pairs of points when called with two data vectors.) Look at the lag plots of the following data sets:

‡ This is better implemented in the `lag.plot()` function from the `ts` package.

1. `x = rnorm(100)` (random data)
2. `x = sin(1:100)` (structured data, but see `plot(x)`)

Comment on any patterns you see.

2.27 Verify that the following are true for the summation notation:

$$\sum_i ax_i = a\sum_i x_i, \quad \sum_i (x_i + y_i) = (\sum_i x_i) + (\sum_i y_i).$$

2.28 Show that for any data set

$$\sum_i (x_i - \bar{x}) = 0.$$

2.29 The sample variance definition, Equation (2.3), has a nice interpretation, but the following formula is easier to compute by hand:

$$s^2 = \frac{n}{n-1}\left[(\bar{x^2}) - (\bar{x})^2\right].$$

The term $\bar{x^2}$ means to square the data values, then find the sample average, whereas $(\bar{x})^2$ finds the sample average, then squares the answer. Show that the equivalence follows from the definition.

2.3 Shape of a distribution

The stem-and-leaf plot tells us more about the data at a glance than a few numeric summaries, although not as precisely. However, when a data set is large, it tells us too much about the data. Other graphical summaries are presented here that work for larger data sets too. These include the histogram, which at first glance looks like a barplot, and the boxplot, which is a graphical representation of the five-number summary.

In addition to learning these graphical displays, we will develop a vocabulary to describe the shape of a distribution. Concepts will include the notion of modes or peaks of a distribution, the symmetry or skew of a distribution, and the length of the tails of a distribution.

2.3.1 Histogram

A histogram is a visual representation of the distribution of a data set. At a glance, the viewer should be able to see where there is a relatively large amount of data, and where there is very little. Figure 2.11 is a histogram of the `waiting` variable from the data set `faithful`, recording the waiting time between eruptions of Old Faithful. The histogram is created with the `hist()` function. Its simplest usage is just `hist(x)`, but many alternatives exist. This histogram has two distinct peaks or modes.

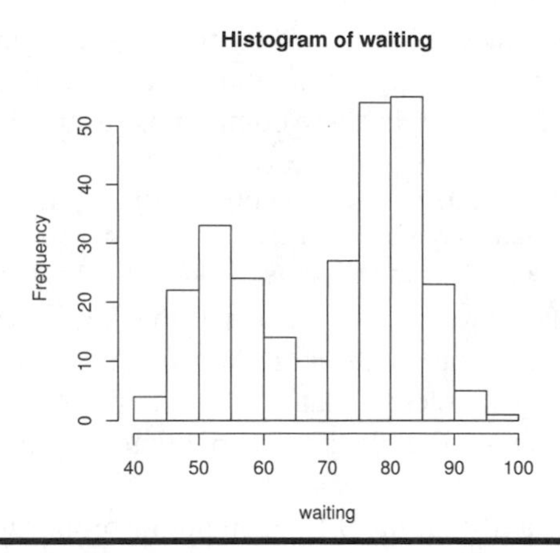

Figure 2.11 Histogram of `waiting` variable in `faithful` data set

Figure 2.11 was created with these commands:

```
> attach(faithful)
> hist(waiting)
```

The graphic is similar, but not identical, to a barplot. The histogram also uses bars to indicate frequency or proportion, but for an interval not a category. The construction is as follows. First, a contiguous collection of disjoint intervals, called bins, covering all the data points is chosen. "Disjoint" means no overlap, so the intervals look like $(a,b]$ or $[a,b)$. That is, the first interval contains all the values from a to b including b but not a, and the second all the values including a but not b. Next, the number of data points, or frequency, in each of these intervals is counted. Finally, a bar is drawn above the interval so that the *area* of the bar is proportional to the frequency. If the intervals defining the bins all have the same length, then the *height* of the bar is proportional to the frequency.

Finding the mean and median from a histogram As described for the strip chart, the mean is a balance point. From a histogram the mean can be estimated from the balance point of the graph, were the figure constructed from some uniform material. The median, on the other hand, should visually separate the area into two equal-area pieces.

Creating histograms in R with `hist()`

When constructing a histogram, we make a decision as to which bins to use and how high to draw the bars, as they need be only in the correct proportion. R has a few built-in choices for the bin selection. Two common choices for the height of the bars are either the frequency or total count, or the proportion of the whole. In the latter case, the total area covered by the bars will be 1, a desirable feature when probability models are considered.

For `hist()`, the bin size is controlled by the `breaks=` argument. This can be specified by the name of an algorithm, the number of breaks desired, or the location of the breaks. For example, these commands would all make histograms:

```
> hist(waiting)                     # use defaults
> hist(waiting,breaks=10)           # suggest 10 breaks
> hist(waiting,breaks=seq(43,108,length=10)) # use these breaks
> hist(waiting,breaks="scott")      # use "Scott" algorithm
```

If these graphs are made, we will be surprised that the second histogram has more than ten bins, despite our suggestion. We directly specify the breaks as a vector of cut points to get exactly what is wanted. The "Sturges" algorithm is the default; "Scott" is an alternative, as is "Friedman-Diaconis," which may be abbreviated as `FD`.

The choice to draw a histogram of frequencies or proportions is made by the argument `probability=`. By default, this is `FALSE` and frequencies are drawn. Setting it to `TRUE` will create histograms where the total area is 1. For example, the commands

```
> hist(waiting)
> hist(waiting,prob=T)              # shortened probability = TRUE
```

will create identical-looking graphs, but the y-axes will differ. We used `prob=T` to shorten the typing of `probability=TRUE`. Although `T` can usually be used as a substitute for `TRUE`, there is no guarantee it will work, as we can assign new values to a variable named `T`.

By default, R uses intervals of the type $(a,b]$. If we want the left-most interval to be of the type $[a,b]$ (i.e., include a), we use the argument `include.lowest=TRUE`.

■ **Example 2.8: Baseball's on-base percentage** Statistical summaries are very much a part of baseball. A common statistic is the "on-base percentage" (OBP), which indicates how successful a player is as a batter. This "percentage" is usually given as a "proportion," or a number between 0 and 1. The data set `OBP` (`UsingR`) contains the OBP for the year 2002, according to the Sam Lahman baseball database (`http://www.baseball1.com`).

This command will produce the histogram in Figure 2.12.

```
> hist(OBP,breaks="Scott",prob=TRUE,col=gray(0.9))
```

The distribution has a single peak and is fairly symmetric, except for the one outlier on the right side of the distribution. The outlier is Barry Bonds, who had a tremendous season in 2002.

The arguments to `hist()` are good ones, but not the default. They are those from the `truehist()` function in the `MASS` package, which may be used as an alternate to `hist()`. ■

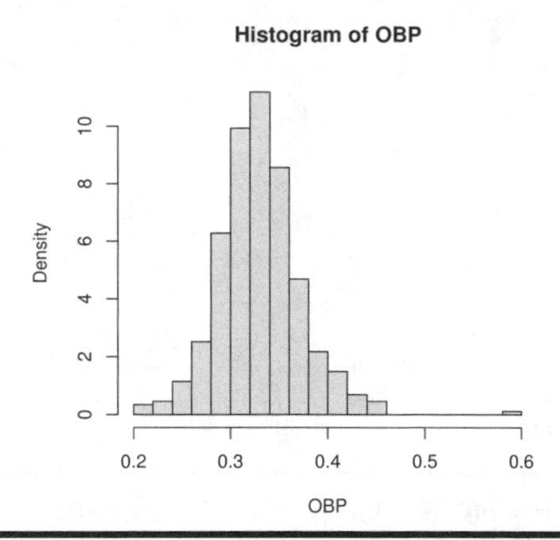

Figure 2.12 Histogram of on-base percentage for the 2002 MLB season

Adding a title to a histogram or other graphic The `hist()` function adds a default title to a histogram. This can be changed with the `main=` argument. This argument is common to many of the plotting functions we will encounter. For example, this command produces a histogram with a custom title:

```
> hist(OBP, main = "My histogram of the OBP dataset")
```

Setting `main=` to an empty string or `NULL` will print no title. In this case, one can be added at a later point with the `title()` function. In addition, this function can be used to label the x- and y-axes with the arguments `xlab=` and `ylab=`.

Density estimates and frequency polygons

In many statistics books, a **frequency polygon** is presented in addition to a histogram. Figure 2.13 displays such a frequency polygon for the `waiting` variable. To draw a frequency polygon, we select the bins (all the same size) and find the frequencies, as we would for the histogram. Rather than draw a bar, though, we draw a point at the midpoint of the bin with height given by the frequency, then connect these points with straight lines to form a polygon.

Creating a frequency polygon The commands to create the frequency polygon in Figure 2.13 are:

```
> bins = seq(42, 109, by=10)
> freqs <- table(cut(waiting, bins))
> y.pts = c(0, freqs, 0)
> x.pts = seq(37,107,by=10)
> plot(x.pts,y.pts,type="l")    # connect points with lines
> rug(waiting)                  # show values
```

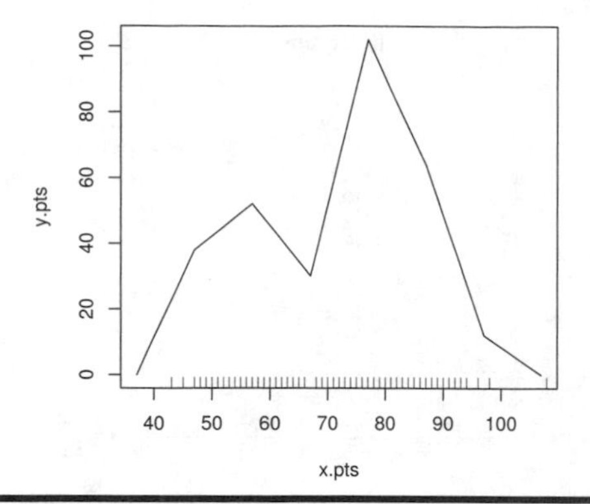

Figure 2.13 Frequency polygon for `waiting` variable of the `faithful` data set

The `plot()` function is used to plot points. It will be discussed more thoroughly in the next chapter. The `type="l"` argument to `plot()` is used to draw line segments between the points instead of plotting the points. The `rug()` function is used to display the data points using hash marks along the x-axis. This example shows how we can use the `cut()` function and the `table()` function to turn continuous numeric data into discrete numeric data, or even categorical data. The output of `cut()` is simply the bin that the data point is in, where bins are specified with a vector of endpoints. For example, if this vector is `c(1,3,5)` then the bins are $(1,3],(3,5]$. The left-most endpoint is not included by default; if the extra argument `include.lowest=TRUE` is given, it will be included. (We could also use the output from `hist()` to do most of this.)

The frequency polygon is used to tie in the histogram with the notion of a probability density, which will be discussed when probabilities are discussed in Chapter 5. However, it is more desirable to estimate the density directly, as the frequency polygon, like the histogram, is very dependent on the choice of bins.

Estimating the density The `density()` function will find a **density estimate** from the data. To use it, we give it the data vector and, optionally, an argument as to what algorithm to use. The result can be viewed with either the `plot()` function or the `lines()` function. A new graphic showing the densityplot is produced by the command `plot(density(x))`. The example uses `lines()` to add to the existing graphic.

```
> attach(faithful)
> hist(waiting, breaks="scott", prob=TRUE, main="",ylab="")
> lines(density(waiting))          # add to histogram
> detach(waiting)                  # tidy up
```

In Figure 2.14, the density estimate clearly shows the two peaks in this data set. It is layered on top of the histogram plotted with total area 1 (from `prob=TRUE`).

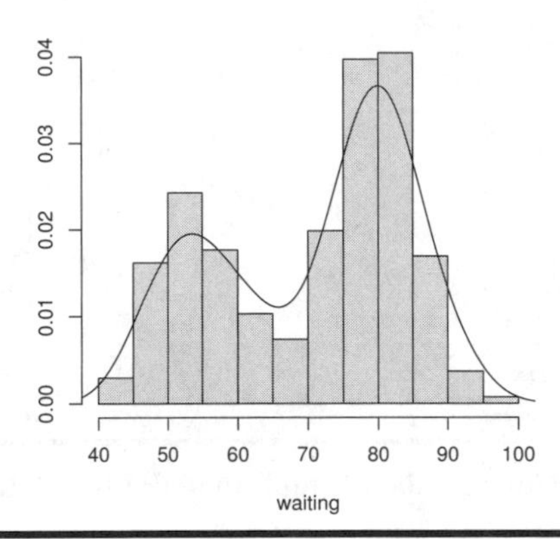

Figure 2.14 Histogram of `waiting` with density estimate

2.3.2 Modes, symmetry, and skew

Using the histogram and density estimate of a univariate data set, we can broadly classify the distribution according to the number of peaks, the symmetry, and the size of the tails. These attributes are essential to know when we want to make statistical inferences about the data.

Modes

A **mode of a distribution** is a peak, or a local maximum, in its density (found using the density estimate). A data set can be characterized by its number of modes. A **unimodal distribution** has a single mode—it occurs at "the mode." The mode is sometimes used to represent the center of a distribution. Distributions with two modes are termed bimodal distributions; those with two or more modes are multimodal distributions.

For example, the `waiting` data set shown in Figure 2.14 is bimodal. The data set `galaxies` (`MASS`) shown in Figure 2.15 is an example of a multimodal data set. In the same figure, we see that the `OBP` data set could be considered unimodal if the Barry Bonds outlier is removed from the data.

Symmetry

A univariate data set has a **symmetric distribution** if it spreads out in a similar way to the left and right of some central point. That is, the histogram or density estimate should have two sides that are nearly mirror images of one another. The `OBP` data set (Figure 2.15) is an example of a symmetric data set if once again the Barry Bonds outlier is removed. The `waiting` data set in Figure 2.14 is *not* symmetric.

Another type of a symmetric data set is the "well-shaped" distribution. These

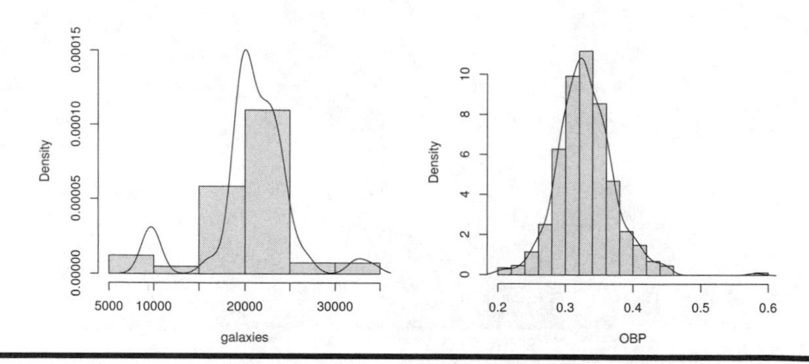

Figure 2.15 Galaxies data is multimodal; OBP data is unimodal

distributions have very little in the middle and a lot at the ends. Surprisingly, these distribution show up in unlikely places. Economic data can show this shape—e.g., the vanishing middle class—as can grade distributions. A more theoretical example is the location of the last tie in a game of chance. Imagine a coin is tossed 100 times, and a running count of heads and tails is kept. After 100 tosses, the number of heads may be more than the number of tails, the same, or less. The last tie is defined as the last toss on which there were the same number of heads as tails. This is a number in the range of 0 to 100. A simulation of this was done 200 times. The results are stored in the data set `last.tie` (`UsingR`). A histogram of the data is shown in Figure 2.16.

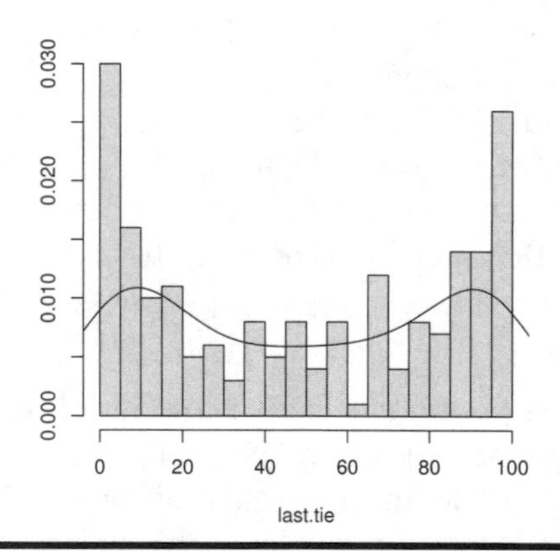

Figure 2.16 An example of a symmetric, well-shaped distribution. This graphic shows 200 simulations of 100 coin tosses. For each simulation, the location of the last time there are an equal number of heads and tails is recorded.

Tails of a distribution and skew

The **tails of a distribution** are the very large and very small values of the distribution. They give the shape of the histogram on the far left and right—hence the name. Many inferences about a distribution are affected by its tails. A distribution is called a **long-tailed distribution** if the data set contains values far from the body of the data. This is made precise after the normal distribution is introduced as a reference. Long tails are also know as "fat tails." Alternatively, a distribution is called a **short-tailed distribution** if there are no values far from the body.

A distribution is a **skewed distribution** if one tail is significantly fatter or longer than the other. A distribution with a longer left tail is termed **skewed left**; a distribution with a longer right tail is termed **skewed right**.

We've seen how very large or very small values in a data set can skew the mean. We will call a data point that doesn't fit the pattern set by the majority of the data an **outlier**. Outliers may be the result of an underlying distribution with a long tail or a mixture of distributions, or they may indicate mistakes of some sort in the data.

■ **Example 2.9: Asset distributions are long-tailed** The distributions of assets, like incomes, are typically skewed right. For example, the amount of equity a household has in vehicles (cars, boats, etc.) is contained in the `VEHIC` variable of the `cfb (UsingR)` data set. Figure 2.17 shows the long-tailed distribution. The `summary()` function shows a significant difference between the median and mean as expected in these situations.

```
> attach(cfb)                        # it is a data frame
> summary(VEHIC)
   Min. 1st Qu.  Median    Mean 3rd Qu.    Max.
      0    3880   11000   15400   21300  188000
> hist(VEHIC,breaks="Scott",prob=TRUE)
> lines(density(VEHIC))
> detach(cfb)
```

■

Measures of center for symmetric data When a data set is symmetric and not too long tailed, then the mean, trimmed mean, and median are approximately the same. In this case, the more familiar mean is usually used to measure center.
Measuring the center for long-tailed distributions If a distribution has very long tails, the mean may be a poor indicator of the center, as values far from the mean may have a significant effect on the mean. In this case, a trimmed mean or median is preferred if the data is symmetric, and a median is preferred if the data is skewed.

For similar reasons, the IQR is preferred to the standard deviation when summarizing spread in a long-tailed distribution.

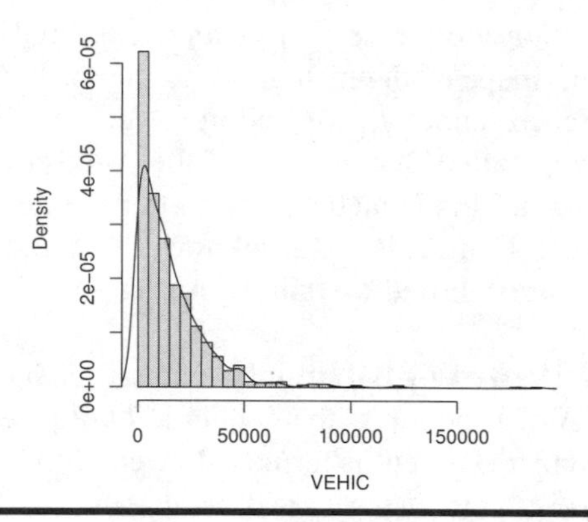

Figure 2.17 Amount of equity in vehicles

2.3.3 Boxplots

A histogram with a density is a good plot for visually finding the center, the spread, the tails, and the shape of a distribution. However, it doesn't work so well to compare distributions, as histograms are hard to read when overlapped and take up too much space when stacked on top of each other. We will use layered densityplots in the sequel instead. But this too, works well only for a handful of data sets at once. A clever diagram for presenting just enough information to see the center, spread, skew, and length of tails in a data set is the **boxplot** or box-and-whisker plot. This graphic allows us to compare many distributions in one figure.

A boxplot graphically displays the five-number summary, which contains the minimum, the lower hinge, the median, the upper hinge, and the maximum. (The hinges give essentially the same information as the quartiles.) The choice of hinges over the quartiles was made by John Tukey, who invented the boxplot.

To show spread, a box is drawn with side length stretching between the two hinges. This length is basically the IQR. The center is illustrated by marking the median with a line through the box. The range is shown with whiskers. In the simplest case, these are drawn extending from the box to the minimum and maximum data values. Another convention is to make the length of the whiskers no longer than 1.5 times the length of the box. Data values that aren't contained in this range are marked separately with points.

Symmetry of the distribution is reflected in symmetry of the boxplot in both the location of the median within the box and the lengths of the two whiskers.

■ **Example 2.10: All-time gross movie sales** Figure 2.18 shows a boxplot of the `Gross` variable in the data set `alltime.movies (UsingR)`. This records the gross domestic (U.S.) ticket sales for the top 79 movies of all time. The minimum, lower hinge, median, upper hinge, and maximum are marked. In addition, the upper whisker extends from the upper hinge to the largest data point that is less than 1.5 times the H-spread plus the upper hinge. Points larger than this are marked separately, including the one corresponding to the maximum. This boxplot shows a data set that is skewed right. It has a long right tail and short left tail. ■

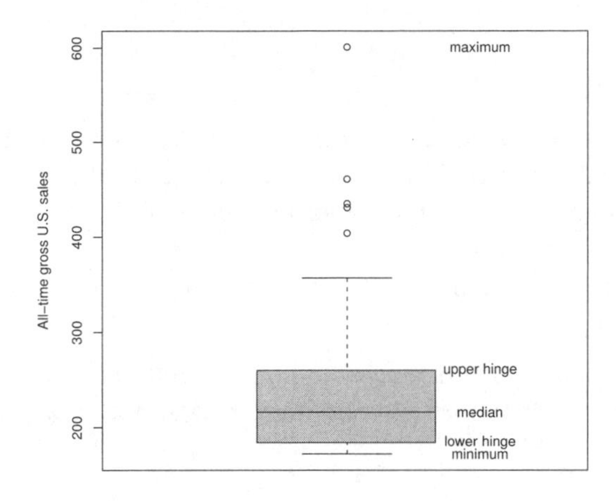

Figure 2.18 Boxplot of all-time gross movie revenues in United States

Making boxplots in R Boxplots are drawn in R with the `boxplot()` function. Figure 2.18 was made with these commands:

```
> attach(alltime.movies)
> boxplot(Gross,ylab="All-time gross sales")
> f= fivenum(Gross)
> text(rep(1.3,5),f,labels=c("minimum","lower hinge",
+    "median","upper hinge","maximum"))
```

The `text()` function places the values of `labels=` on the graphic as specified. Common arguments for `boxplot()` are `col=` to set a color for the box, `horizontal=TRUE` to change the orientation of the boxplot, and `notch=TRUE` to add a notch to the waist of the box where the median is marked.

Getting the outliers If we are curious as to what the top five movies are that are marked separately, we can use the `fivenum()` function to find out. (First we get the names using the `rownames()` function.)

```
> f = fivenum(Gross)
> the.names = rownames(alltime.movies)
```

```
> the.names[Gross > f[4] + 1.5*(f[4]-f[2])]
[1] "Titanic                              "
[2] "Star Wars                            "
[3] "E.T.                                 "
[4] "Star Wars: The Phantom Menace        "
[5] "Spider-Man                           "
> detach(alltime.movies)        # tidy up
```

Alternately, this information is available in the (invisible) output of the `boxplot()` function in the list element named "`out`".

2.3.4 Problems

2.30 For the data sets `bumpers (UsingR)`, `firstchi (UsingR)`, and `math (UsingR)`, make histograms. Try to predict the mean, median, and standard deviation. Check your guesses with the appropriate `R` commands.

2.31 We can generate random data with the "r" functions. For example,

```
> x = rnorm(100)
```

produces 100 random numbers with a normal distribution. Create two different histograms for two different times of defining `x` as above. Do you get the same histogram?

2.32 Fit a density estimate to the data set `pi2000 (UsingR)`. Compare with the appropriate histogram. Why might you want to add an argument like `breaks = 0:10-.5` to `hist()`?

2.33 The data set `normtemp (UsingR)` contains body measurements for 130 healthy, randomly selected individuals. The variable `temperature` contains normal body temperature. Make a histogram. Estimate the sample mean body temperature, and then check using `mean()`.

2.34 The data set `DDT (MASS)` contains independent measurements of the pesticide DDT on kale. Make a histogram and a boxplot of the data. From these, estimate the mean and standard deviation. Check your answers with the appropriate functions.

2.35 There are several built-in data sets on the 50 United States. For instance, `state.area(,)` showing the area of each U.S. state, and `state.abb(,)` showing a common abbreviation. First, use `state.abb` to give names to the `state.area` variable, then find the percent of states with area less than New Jersey (`NJ`). What percent have area less than New York (`NY`)? Make a histogram of all the data. Can you identify the outlier?

2.36 The `time` variable of the `nym.2002 (UsingR)` data set contains the time to finish the 2002 New York City marathon for a random sample of runners. Make a histogram and describe the shape. Can you explain why the shape is as it is?

2.37 The `lawsuits (UsingR)` data set contains simulated data on the settlement amounts of 250 common fund class actions in $10,000s. Look at the differences between the mean and the median. Explain why some would say the average is too high and others would say the average is the wrong way to summarize the data.

2.38 The data set `babyboom (UsingR)` contains data on the births of 44 children in a one-day period at a Brisbane, Australia, hospital. Make a histogram of the `wt` variable, which records birth weight. Is it symmetric or skewed?

The variable `running.time` records the time after midnight of each birth. The command `diff(running.time)` records the differences or inter-arrival times. Make a histogram of this data. What is the general shape? Is it uniform?

2.39 The data set `hall.fame (UsingR)` contains baseball statistics for several baseball players. Make histograms of the following variables and describe their shapes: `HR`, `BA`, and `OBP`.

2.40 Find a graphic in the newspaper or on the web. Try to use `R` to produce a similar figure.

2.41 Why are the boxplot whiskers chosen with the factor of 1.5? Why not some other factor? You can see the results of other choices by setting the `range=` argument. Use `x = rnorm(1000)` for your data. Try values of 0.5, 1, 1.5, and 2 to see which shows the tails of the distribution best. (This random sample should not have a heavy tail or a light tail, meaning it will usually have a handful of points beyond the whiskers in a sample of this size.)

2.42 The data set `cfb (UsingR)` contains a sampling of the data from a survey of consumer finances. For the variables `AGE`, `EDUC`, `NETWORTH`, and `log(SAVING + 1)`, describe their distribution using the concepts of modes, symmetry, and tails. Can you convince yourself that these distributions should have the shape they do? Why?

2.43 The `brightness (UsingR)` data set contains the brightness for 966 stars in a sector of the sky. It comes from the Hipparcos catalog. Make a histogram of the data. Describe the shape of the distribution.

2.44 It can be illuminating to view two different graphics of the same data set at once. A simple way to stack graphics is to specify that a figure will contain two graphics by using the command

```
> par(mfrow=c(2,1))                # 2 rows, 1 column for graphic figures
```

Then, if x is the data set, the commands

```
> hist(x)
> boxplot(x, horizontal=TRUE)
```

will produce stacked graphics. (The graphics device will remain divided until you change it back with a command such as `par(mfrow=c(1,1))` or close the device.)

For the data set `lawsuits (UsingR)`, make stacked graphics of `lawsuits` and `log(lawsuits)`. Could you have guessed where the middle 50% of the data would have been without the help of the boxplot?

2.45 Sometimes a data set is so skewed that it can help if we transform the data prior to looking at it. A common transformation for long-tailed data sets is to take the logarithm of the data. For example, the `exec.pay (UsingR)` data set is highly skewed. Look at histograms before and after taking a logarithmic transform. Which is better at showing the data and why? (You can transform with the command `log(1 + exec.pay,10)`.) Find the median and the mean for the transformed data. How do they correspond to the median and mean of the untransformed data?

2.46 The skew of a data set is sometimes defined as

$$\sum_i \left[\frac{(x_i - \bar{x})}{\mathsf{SD}(x)}\right]^3.$$

Explain why this might make sense as a measurement of skew. Find the skew for the `pi2000 (UsingR)` data set and the `exec.pay (UsingR)` data sets.

Chapter 3

Bivariate data

This chapter looks at data contained in two variables (bivariate data). With univariate data, we summarized a data set with measures of center and spread and the shape of a distribution with words such as "symmetric" and "long-tailed." With bivariate data we can ask additional questions about the relationship between the two variables.

Take, for instance, data on test scores. If two classes take the same test, the students' scores will be two samples that should have similarly shaped distributions but will be otherwise unrelated as pairs of data. However, if we focus on two exams for the same group of students, the scores should be related. For example, a better student would be expected to do better on both exams. Consequently, in addition to the characterization of data as categorical or numeric, we will also need to know when the data is paired off in some way.

3.1 Pairs of categorical variables

Bivariate, categorical data is often presented in the form of a (two-way) contingency table. The table is found by counting the occurrences of each possible pair of levels and placing the frequencies in a rectangular grid. Such tables allow us to focus on the relationships by comparing the rows or columns. Later, statistical tests will be developed to determine whether the distribution for a given variable depends on the other variable.

Our data may come in a summarized or unsummarized format. The data entry is different for each.

3.1.1 Making two-way tables from summarized data

If the data already appears in tabular format and we wish to analyze it inside R, how is the data keyed in? Data vectors were created using the c() function. One simple way to make a table is to combine data vectors together as rows (with rbind()) or as columns (with cbind()).

To illustrate: an informal survey of seat-belt usage in California examined the relationship between a parent's use of a seat belt and a child's. The data appears in Table 3.1. A quick glance at the table shows a definite relationship between the two variables: the child's being buckled is greatly determined by the parent's.

Table 3.1 Seat-belt usage in California

	Child	
Parent	buckled	unbuckled
buckled	56	8
unbuckled	2	16

We can enter these numbers into R in several ways.

Creating the table as a combination of the row (or column) vectors is done as follows:

```
> rbind(c(56,8),c(2,16))          # combine rows
     [,1] [,2]
[1,]   56    8
[2,]    2   16
> cbind(c(56,2),c(8,16))          # bind as columns
     [,1] [,2]
[1,]   56    8
[2,]    2   16
```

Combining rows (or columns) of numeric vectors results in a matrix—a rectangular collection of numbers. We can also make a matrix directly using the matrix() function. To enter in the numbers we need only specify the correct size. In this case we have two rows. The data entry would look like this:

```
> x = matrix(c(56,2,8,16),nrow=2)
> x
     [,1] [,2]
[1,]   56    8
[2,]    2   16
```

The data is filled in column by column. Set byrow=TRUE to do this row by row.

Alternately, we may enter in the data using the edit() function. This will open a spreadsheet (if available) when called on a matrix. Thus the commands

```
> x = matrix(1)                   # need to initialize x
> x = edit(x)                     # will edit matrix with spreadsheet
```

will open the spreadsheet and store the answer into x when done. The 1 will be the first entry. We can edit this as needed.

Giving names to a matrix It isn't necessary, but it is nice to give the matrix row and column names. The `rownames()` and `colnames()` functions will do so. As with the `names()` function, these are used in a slightly different manner. As they modify the attributes of the matrix, the functions appear on the left side of the assignment.

```
> rownames(x) = c("buckled","unbuckled")
> colnames(x) = c("buckled","unbuckled")
> x
          buckled unbuckled
buckled        56         8
unbuckled       2        16
```

The `dimnames()` function can set both at once and allows us to specify variable names. A list is used to specify these, as made by `list()`. Lists are discussed further in Chapter 4. For this usage, the variable name and values are given in `name=value` format. The row variable comes first, then the column.

```
> tmp = c("buckled","unbuckled") # less typing
> dimnames(x) = list(parent=tmp,child=tmp) # uses a named list
> x
           child
parent      buckled unbuckled
    buckled      56         8
  unbuckled       2        16
```

If the matrix is made with `rbind()`, then names for the row vectors can be specified in `name=value` format. Furthermore, column names will come from the vectors if present.

```
> x = c(56,8); names(x) = c("buckled","unbuckled")
> y = c(2,16)
> rbind(buckled=x, unbuckled=y)  # names rows, columns come from x
          buckled unbuckled
  buckled      56         8
unbuckled       2        16
```

3.1.2 Making two-way tables from unsummarized data

With unsummarized data, two-way tables are made with the `table()` function, just as in the univariate case. If the two data vectors are x and y, then the command `table(x,y)` will create the table.

■ **Example 3.1: Is past performance an indicator of future performance?**

A common belief is that an A student in one class will be an A student in the next. Is this so? The data set `grades (UsingR)` contains the grades students received in a math class and their grades in a previous math class.

```
> library(UsingR)                      # once per session
> grades
   prev grade
1    B+    B+
2    A-    A-
3    B+    A-
...
122  B     B
> attach(grades)
> table(prev,grade)                    # also table(grades) works
      grade
prev    A   A-  B+  B   B-  C+  C   D   F
   A   15   3   1   4   0   0   3   2   0
   A-   3   1   1   0   0   0   0   0   0
   B+   0   2   2   1   2   0   0   1   1
   B    0   1   1   4   3   1   3   0   2
   B-   0   1   0   2   0   0   1   0   0
   C+   1   1   0   0   0   0   1   0   0
   C    1   0   0   1   1   3   5   9   7
   D    0   0   0   1   0   0   4   3   1
   F    1   0   0   1   1   1   3   4  11
```

A quick glance at the table indicates that the current grade relates quite a bit to the previous grade. Of those students whose previous grade was an A, fifteen got an A in the next class; only three of the students whose previous grade was a B or worse received an A in the next class. ■

3.1.3 Marginal distributions of two-way tables

A two-way table involves two variables. The distribution of each variable separately is called the **marginal distribution**. The marginal distributions can be found from the table by summing down the rows or columns. The `sum()` function won't work, as it will add all the values. Rather, we need to apply the `sum()` function to just the rows or just the columns. This is done with the function `apply()`. The command `apply(x,1,sum)` will sum the rows, and `apply(x,2,sum)` will sum the columns. The `margin.table()` function conveniently implements this. Just remember that 1 is for rows and 2 is for columns.

For the seat-belt data stored in x we have:

```
> x
            child
parent      buckled unbuckled
   buckled       56         8
 unbuckled        2        16
> margin.table(x,1)                    # row sum is for parents
[1] 64 18
> margin.table(x,2)                    # column sum for kids
[1] 58 24
```

The two marginal distributions are similar: the majority in each case wore seat belts.

Alternatively, the function `addmargins()` will return the marginal distributions by extending the table. For example:

```
> addmargins(x)
           child
parent      buckled unbuckled Sum
  buckled        56         8  64
  unbuckled       2        16  18
  Sum            58        24  82
```

Looking at the marginal distributions of the `grade` data also shows two similar distributions:

```
> margin.table(table(prev,grade),1) # previous. Also table(prev)
prev
 A  A-  B+  B  B-  C+  C  D  F
 28  5   9  15  4   3 27  9 22
> margin.table(table(prev,grade),2) # current
grade
 A  A-  B+  B  B-  C+  C  D  F
 21  9   5  14  7   5 20 19 22
```

The grade distributions, surprisingly, are somewhat "well-shaped."

3.1.4 Conditional distributions of two-way tables

We may be interested in comparing the various rows of a two-way table. For example, is there a difference in the grade a student gets if her previous grade is a B or a C? Or does the fact that a parent wears a seat belt affect the chance a child does? These questions are answered by comparing the rows or columns in a two-way table. It is usually much easier to compare proportions or percentages and not the absolute counts.

For example, to answer the question of whether a parent wearing a seat belt changes the chance a child does, we might want to consider Table 3.2.

Table 3.2 Proportions of children with seat belt on

	Child	
Parent	buckled	unbuckled
buckled	0.875	0.125
unbuckled	0.1111	0.8889

From this table, the proportions clearly show that 87.5% of children wear seat belts when their parents do, but only 11% do when their parents don't. In this example, the rows add to 1 but the columns need not, as the rows were divided by the row sums.

For a given row or column, calculating these proportions is done with a command such as `x/sum(x)`. But this needs to be applied to each row or column. This can be done with `apply()`, as described before, or with the convenient function `prop.table()`. Again, we specify whether we want the conditional rows or columns with a 1 or a 2.

For example, to find out how a previous grade affects a current one, we want to look at the proportions of the rows.

```
> options("digits"=1)            # to fit on the page
> prop.table(table(prev,grade),1)
     grade
prev    A    A-   B+   B    B-   C+   C    D    F
   A   0.54 0.11 0.04 0.14 0.00 0.00 0.11 0.07 0.00
   ...
   C   0.04 0.00 0.00 0.04 0.04 0.11 0.19 0.33 0.26
   D   0.00 0.00 0.00 0.11 0.00 0.00 0.44 0.33 0.11
   F   0.05 0.00 0.00 0.05 0.05 0.05 0.14 0.18 0.50
> options("digits"=4)            # set back to original
> detach(grades)                 # tidy up
```

From comparing the rows, it is apparent that the previous grade has a big influence on the current grade.

The `options()` function is used to set the number of digits that are displayed in the output of decimal numbers. It was set to 1 to make the table print without breaking in the space provided.

3.1.5 Graphical summaries of two-way contingency tables

Barplots can be used effectively to show the data in a two-way table. To do this, one variable is chosen to form the categories for the barplot. Then, either the bars for each level of the category are segmented, to indicate the proportions of the other variable, or separate bars are plotted side by side.

The `barplot()` function will plot segmented barplots when its first argument is a two-way table. Levels of the columns will form the categories, and the sub-bars or segments will be proportioned by the values in each column. Segmented bar graphs are the default; use `beside=TRUE` to get side-by-side bars.

If `x` stores the seat-belt data, we have:

```
> barplot(x, xlab="Parent", main="Child seat-belt usage")
> barplot(x, xlab="Parent", main="Child seat-belt usage",beside=TRUE)
```

We can add a legend to the barplot with the argument `legend.text=TRUE`, or by specifying a vector of names for `legend.text=`. For example, try

```
> barplot(x,main="Child seat belt usage",legend.text=TRUE)
```

For the seat-belt data, if we wanted the parents' distribution (the rows) to be

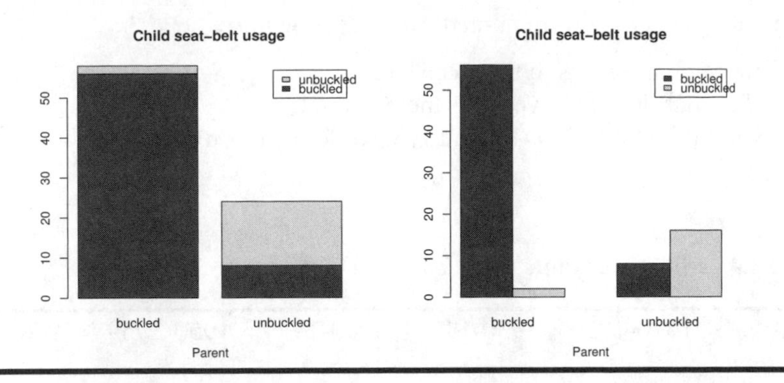

Figure 3.1 Segmented and side-by-side barplots showing distribution of child's seat-belt usage depending on whether parent is buckled or unbuckled

the primary distribution, then we need to flip the table around. This is done with the transpose function, `t()`, as in `barplot(t(x))`.

Sometimes a relationship is better presented as proportions than counts. To do this, we apply `prop.table()` prior to the barplot.

3.1.6 Problems

3.1 Find an example of a two-way contingency table in the media. Identify the two variables and summarize the data that is presented.

3.2 *Wired* magazine announced that as of July 2003 the percentage of all e-mail that is spam (junk e-mail) is above 50% and climbing. A user may get over 100 e-mail messages a day, making spam a time-consuming and expensive reality. Table 3.3 lists the amount of spam in commercial e-mail and the total amount of commercial e-mail by year with some predicted amounts. Enter in the data and then recreate the table. Make a segmented barplot showing the amount of spam and the total amount of e-mail.

Table 3.3 Volume of spam in commercial e-mail (in billions)

	2000	2001	2002	2003	2004	2005
spam	50	110	225	315	390	450
total	125	210	375	475	590	700

Source: *Wired* magazine September 2003

3.3 The data set `coins (UsingR)` contains the number of coins in a change bin and the years they were minted. Do the following:

1. How much money is in the change bin?
2. Make a barplot of the years. Is there a trend?
3. Try to fill in Table 3.4. (Use `cut()`, but look at `?cut` and its arguments.)

Table 3.4 Fill in this table using `coins`

Year	1920-1929	1930-1939	1940-1949	1950-1959	1960-1969
Amount	3			2	
Year	**1970-1979**	**1980-1989**	**1990-1999**	**2000-2009**	
Amount				88	

3.4 The data set `dvdsales (UsingR)` contains monthly sales of DVD players from their initial time on the market through May 2004. Make side-by-side barplots of monthly sales by year. (The data needs to be transposed using `t()`; otherwise the barplots will be broken up by month.)

3.5 The `florida (UsingR)` data set contains county-by-county tallies of the votes cast in the 2000 United States presidential election for the state of Florida. The main candidates were George Bush and Al Gore. Make a segmented barplot of the proportion of Bush votes versus the proportion of Gore votes by county. Are these proportions always close to the 50% proportion for the state?

3.6 In 1996, changes in the United States welfare laws resulted in more monies being spent on noncash assistance (child care, training, etc.) than on cash assistance. A table of the percentages of cash assistance is given in Table 3.5. Make a segmented barplot illustrating the percentages for both. The total spending is approximately $25 billion per year.

Table 3.5 Shift in what welfare provides

	'97	'98	'99	'00	'01	'02
Cash assistance	76%	70%	68%	52%	48%	46%

source: *New York Times* October 13, 2003

3.7 The data set `UScereal (MASS)` contains information about cereals on a shelf of a United States grocery store. Make a table showing the relationship between manufacturer, `mfr`, and shelf placement, `shelf`. Are there any obvious differences between manufacturers?

3.2 Comparing independent samples

In many situations we have two samples that may or may not come from the same population. For example, a medical trial may have a treatment group and a control group. Are any measured effects the same for each? A consumer may be comparing two car companies. From samples, can he tell if the ownership costs will be about the same? When two samples are drawn from populations in such a manner that knowing the outcomes of one sample doesn't affect the knowledge of the distribution of the other sample, we say that they are independent samples. For independent samples, we may be interested in comparing their populations. Are the centers the same? The spreads? Do they have the same shape distribution? In Chapter 7 we use statistical models to help answer such questions. In this section, we learn to explore the relationships graphically to gain insight into the answers.

3.2.1 Side-by-side boxplots

The stem-and-leaf plot and boxplot were very effective at summarizing a distribution. The stem-and-leaf plot was used when the data set was small; the boxplot can be used on larger data sets. By putting them side by side or back to back, we can make comparisons of the two samples.

Table 3.6 Weight loss during ephedra trial (in pounds)

placebo treatment		ephedra treatment
42000	0	0
5	0	679
4443	1	13
775	1	66678
	2	01

Table 3.6 contains hypothetical data on weight loss during a clinical trial of the ephedra supplement. As mentioned in Example 1.3, ephedra is a popular supplement that was forced off the market due to its side effects.

The back-to-back stem-and-leaf plot shows that the ephedra group has a larger center. The question of whether this is "significant" is answered using a t-test, which is covered in Chapter 8.

The `stem()` function doesn't make back-to-back stem-and-leaf plots. If the data set is too large to make a stem-and-leaf plot by hand, side-by-side boxplots are useful for highlighting differences. (These are better named parallel boxplots, as they may be displayed horizontally or vertically.)

The command `boxplot(x,y)` will create side-by-side boxplots from two variables. The `names=` argument is used to label the boxplots in Figure 3.2. The figure shows slightly different distributions, with, perhaps, similar medians.

```
> pl = c(0, 0, 0, 2, 4, 5, 14, 14, 14, 13, 17, 17, 15)
> ep = c(0, 6, 7, 9, 11, 13, 16, 16, 16, 17, 18, 20, 21)
> boxplot(pl,ep, names=c("placebo","ephedra"))
```

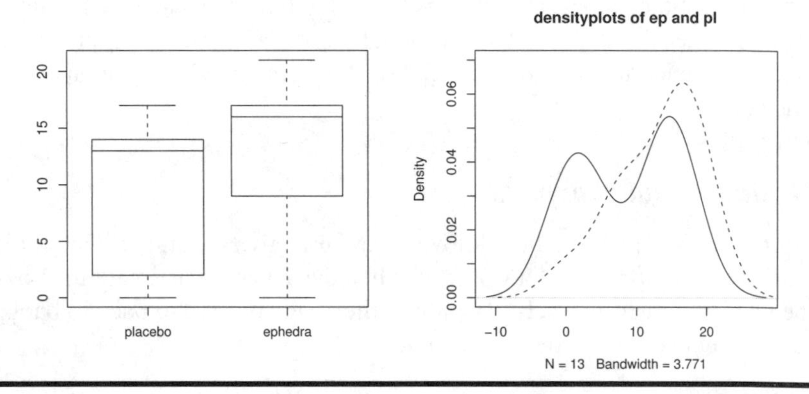

Figure 3.2 Left graphic shows side-by-side boxplots of placebo and ephedra group. Right graphic contains densityplots of the two variables.

3.2.2 Densityplots

We can compare distributions with two histograms, but it is difficult to put both on the same graphic. Densityplots, however, lend themselves readily to this. *

We draw the first densityplot with the `plot()` function and add subsequent ones with the `lines()` function. The argument `lty=` can be set to a value between 1 and 6, to change the type of line drawn for identification purposes. For example, the densityplots in Figure 3.2 are made as follows:

* You *can* compare histograms by recording the graphs. In the Windows GUI, you can turn on recording from the menu bar of the graph window. In general, you can store the current plot in a variable with `recordPlot()`, and view this stored plot with `replayPlot()`.

```
> plot(density(pl),ylim=c(0,0.07), main="densityplots of ep and pl")
> lines(density(ep), lty=2)
```

The argument `ylim=` adjusts the *y*-axis to accommodate both densities. The value was arrived at after we plotted both densities and found the maximum values.

3.2.3 Strip charts

Strip charts can compare distributions effectively when the values are similar and there aren't too many. To create a strip chart with multiple data vectors, we first combine the data vectors into a list with the `list()` function. By using a named list the stripchart will be drawn with labels for the data sets.

```
> stripchart(list(ephedra=ep,placebo=pl), # named list
+ method = "stack",                       # stack multiples
+ pch=16,offset = 1/2, cex=3)             #  big circles -- not squares
```

Figure 3.3 shows the graphic (slightly modified).

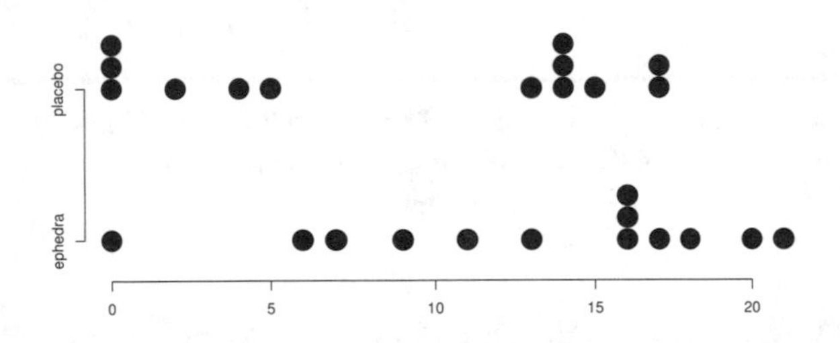

Figure 3.3 Strip chart of placebo and ephedra group

3.2.4 Quantile-quantile plots

The boxplot uses the quartiles (essentially) of a data set to graphically represent a data set succinctly. If we use more of the quantiles, a very clear picture of the data can be had at the expense of a more complicated graph to read. A **quantile-quantile plot** (q-q plot) plots the quantiles of one distribution against the quantiles of another as points. If the distributions have similar shapes, the points will fall roughly along a straight line. If they are different, the points will not lie near a line, in a manner that can indicate why not.

A **normal quantile plot** plots the quantiles of a data set against the quantiles of a benchmark distribution (the normal distribution introduced in Chapter 5). Again, the basic idea is that if the data set is similar to the benchmark one, then the graph will essentially be a straight line. If not, then the line will be "curved" in a manner that can be interpreted from the graph.

Figure 3.4 shows the q-q plot for two theoretical distributions that are clearly not the same shape. Each shaded region is 5% of the total area. The difference in the shapes produces differences in the quantiles that curve the q-q plot.

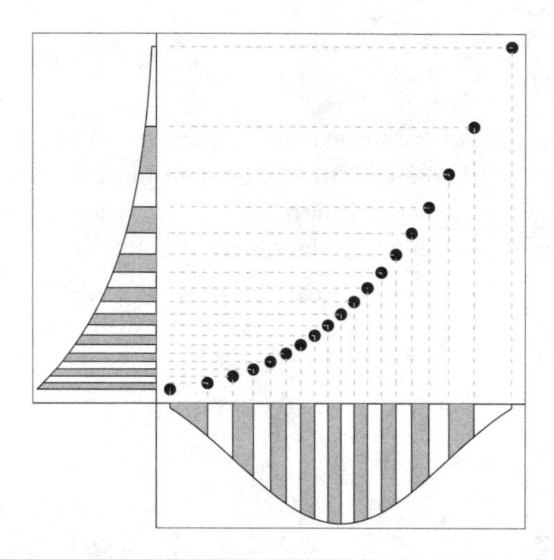

Figure 3.4 Quantile-quantile plot of two distributions. The shaded areas represent similar areas. As the distributions have different shapes, the q-q plot has a curve.

Creating q-q plots The R function to make a q-q plot is `qqplot()`, as in `qqplot(x,y)`. The `qqnorm()` function, as in `qqnorm(x)`, will produce a normal quantile plot. In this case, a reference line may be added with the `qqline()` function, as in `qqline(x)`.

Figure 3.5 shows six normal quantile graphs for data that is a combination of symmetric, or skewed right, and short, normal or long tailed. The combination (normal/symmetric) looks like a straight line. Were we to plot a histogram of this data, we would see the familiar bell-shaped curve. The figure (short/symmetric) shows what happens with short tails. In particular, if the right tail is short, it forces the quantile graph to curve down. In contrast, the graph (long/skewed) curves up, as this data has a long right tail.

3.2.5 Problems

3.8 The use of a cell phone while driving is often thought to increase the chance of an accident. The data set `reaction.time (UsingR)` is simulated data on the time it takes to react to an external event while driving. Subjects with

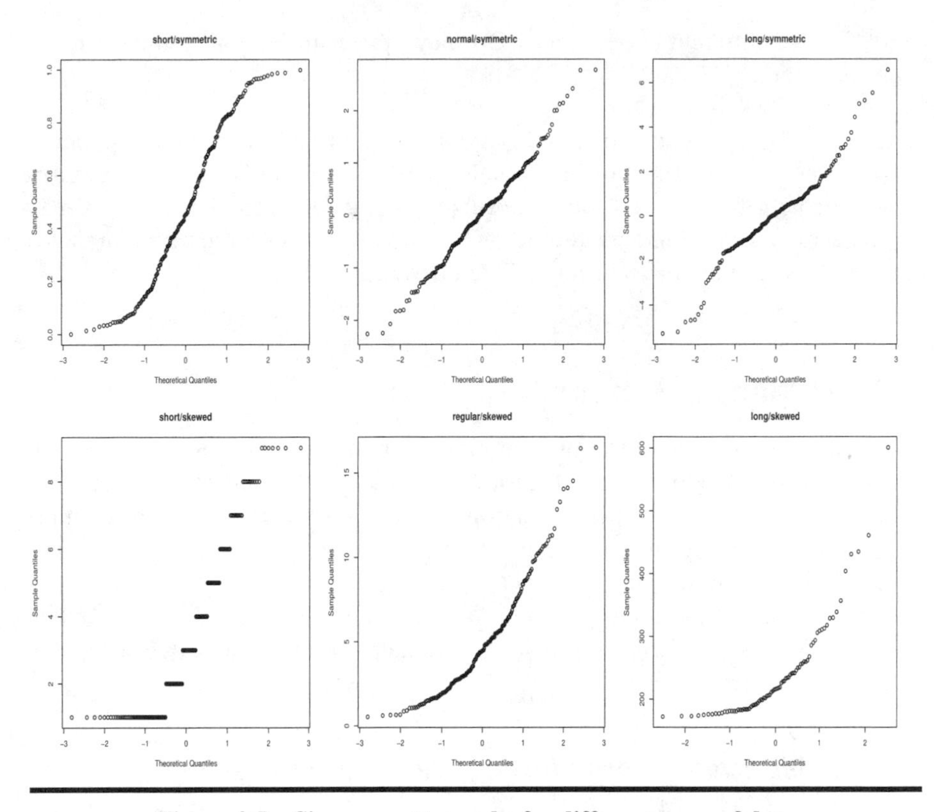

Figure 3.5 Six `qqnorm()` graphs for different types of data

`control == "C"` are not using a cell phone, and those with `control == "T"` are. Their time to respond to some external event is recorded in seconds.

Create side-by-side boxplots of the variable `reaction.time` for the two values of `control`. Compare the centers and spreads.

3.9 For the data set `twins (UsingR)` make a boxplot of the `Foster` and `Biological` variables. Do they appear to have the same spread? The same center?

3.10 The data set `stud.recs (UsingR)` contains 160 SAT scores for incoming college students stored in the variables `sat.v` and `sat.m`. Produce side-by-side densityplots of the data. Do the two data sets appear to have the same center? Then make a quantile-quantile plot. Do the data sets appear to have the same shape?

3.11 For the data set `morley`, make a boxplot of the `Speed` variable for `Expt == 1` and `Expt == 2`. These data sets are the measurements of the speed of light for two different experiments. Do they appear to have the same spread? The same center?

3.12 The data set `normtemp (UsingR)` contains normal body temperature measurements for 130 healthy individuals recorded in the variable `temperature`. The variable `gender` is 1 for a male subject and 2 for a female subject. Break the data up by gender and create side-by-side boxplots. Does it appear that males and females have similar normal body temperatures?

3.3 Relationships in numeric data

There are many scientific relationships between numeric variables. Among them: distance equals rate times time, pressure is proportional to temperature; and demand is inverse to supply. Many relationships are not precisely known, prompting an examination of the data. For instance, is there a relationship between a person's height and weight?

If a bivariate data set has a natural pairing, such as $(x_1,y_1),\ldots,(x_n,y_n)$, then it likely makes sense for us to investigate the data set jointly, as a two-way table does for categorical data.

3.3.1 Using scatterplots to investigate relationships

A **scatterplot** is a good place to start when investigating a relationship between two numeric variables. A scatterplot plots the values of one data vector against another as points (x_i,y_i) in a Cartesian plane.

The `plot()` function will make a scatterplot. A basic template for its usage is

```
plot(x, y)
```

where `x` and `y` are data vectors containing the paired data. The `plot()` function is used to make many types of plots, including densityplots, as seen. For scatterplots, there are several options to `plot()` that can adjust how the points are drawn, whether the points are connected with lines, etc. We show a few examples and then collect them in Table 3.7.

■ **Example 3.2: Home values** Buying a home has historically been a good investment. Still, there are expenses. Typically, a homeowner needs to pay a property tax in proportion to the assessed value of the home. To ensure some semblance of fairness, the assessed values should be updated periodically. In Maplewood, New Jersey, properties were reassessed in the year 2000 for the first time in 30 years. The data set `homedata (UsingR)` contains values for 150 randomly chosen homes. A scatterplot of assessed values should show a relationship, as homes that were expensive in 1970 should still have been expensive in 2000. We can use this data set to get an insight into the change in property values for these 30 years.

The scatterplot is made after loading the data set.

```
> attach(homedata)
> plot(y1970, y2000)                    # make the scatterplot
```

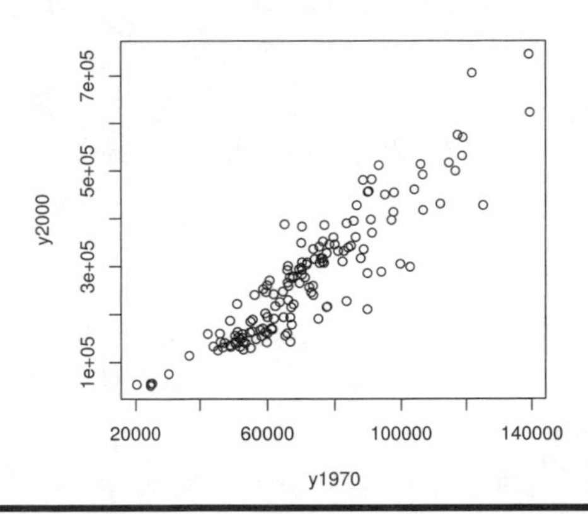

Figure 3.6 Assessed values of homes in Maplewood, N.J. in 1970 and 2000

Figure 3.6 shows the scatterplot. The data falls more or less along a straight line, although with some variation. A few questions immediately come to mind. For instance, what are the distributions of each variable like? What is the change in price?

```
> summary(y1970)
   Min. 1st Qu.  Median    Mean 3rd Qu.    Max.
  20300   57000   68500   71300   83500  139000
> summary(y2000)
   Min. 1st Qu.  Median    Mean 3rd Qu.    Max.
  51600  163000  260000  274000  342000  745000
> summary(y2000/y1970)
   Min. 1st Qu.  Median    Mean 3rd Qu.    Max.
   2.10    2.89    3.80    3.68    4.30    5.97
> detach(homedata)                      # tidy up
```

For the 1970 data, the mean and the median are nearly the same. Not so for the 2000 data. Property values are often skewed right. In this sampling, some houses went up in value just over two times and others nearly six times. On average they went up 3.68 times.

When one is buying a home, it is obviously desirable to figure out which homes are likely to appreciate more than others. ■

■ **Example 3.3: Does the weather predict the stock market?** As a large amount of stock traders work in New York City, it may be true that unseasonably good or bad weather there affects the performance of the stock market. The data set `maydow (UsingR)` contains data for the Dow Jones Industrial Average (DJIA) and maximum temperatures in Central Park for May 2003. This month was unseasonably cool and wet. The data set contains only maximum temperature, so we can ask for that month, whether there was a relationship between maximum temperature and the stock market?

```
> attach(maydow)
> names(maydow)
[1] "Day"       "DJA"       "max.temp"
> plot(max.temp[-1], diff(DJA), main="Max. temp versus daily change")
> detach(maydow)
```

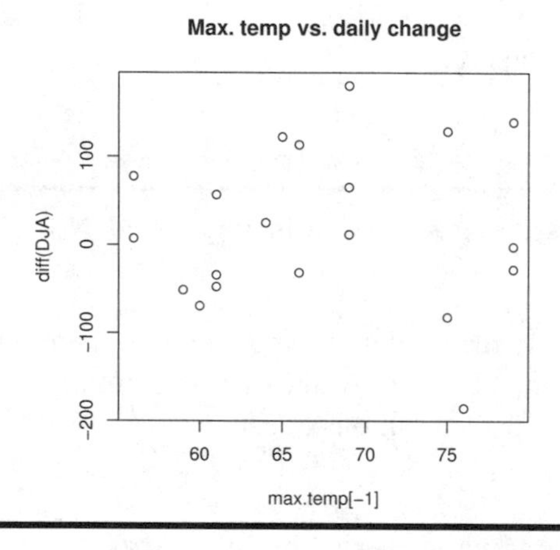

Figure 3.7 Maximum temperature versus daily change in DJIA

Figure 3.7 contains the plot of maximum daily temperature versus daily change in the variable `DJA` calculated using the `diff()` function. We needed to drop the first day's temperature, as we have no difference data for that day. This was done using negative indexing, `max.temp[-1]`. The scatterplot shows no trend. If the temperature does influence the stock market, more data would be needed to see exactly how. ■

■ **Example 3.4: Kids' weights: the relationship between height and weight**
The proportions of the human body have long been of interest to humankind. Even Jonathan Swift wrote in *Gulliver's Travels* (1726),

> Then they measured my right Thumb, and desired no more; for by a mathematical Computation, that twice round the Thumb is once round

the Wrist, and so on to the Neck and the Waist, and by the help of my old Shirt, which I displayed on the Ground before them for a Pattern, they fitted me exactly.

Just as it seems intuitive that the bigger you are the bigger your thumb, it seems clear that the taller you are the heavier you are. What is the relationship between height and weight? Is it linear? Nonlinear? The body mass index (BMI) is a ratio of weight to height squared in the units of kilograms/meters^2. This well-used statistic suggests height and weight are in a squared relationship.

The data set `kid.weights (UsingR)` contains height and weight data for children ages 0 to 12 years. A plot of height versus weight is found with the following. (The `pch=` argument forces the plot character to be "M" for the boys and "F" for the girls.)

```
> attach(kid.weights)
> plot(height, weight, pch=as.character(gender))
> detach(kid.weights)
```

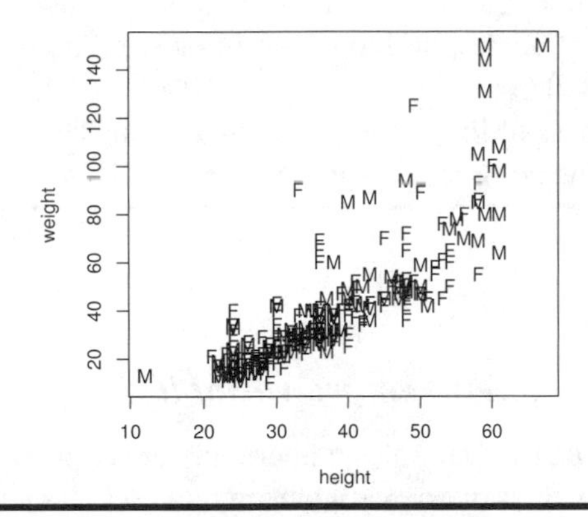

Figure 3.8 Height versus weight for `kid.weights`

Figure 3.8 indicates that weight may be related to the height squared. It certainly does not appear to be a straight-line relationship. ■

Arguments for the `plot()` function The `plot()` function, and other graphing functions like `hist()`, `boxplot()`, and `lines()`, take extra arguments that can control portions of the graphic. Table 3.7 list some of the possible arguments. More details are found on the help pages for the individual functions. Most of these arguments are graphics parameters that may also be queried and set using `par()`. The documentation for these arguments is found in the help page for `par()`. (This function is discussed more fully in Appendix D.) Some of the arguments, such as `main=`, `xlim=`, and `xlab=`, can be used only with plotting functions that set up a plot window (called high-level plotting functions) .

Table 3.7 Useful arguments for `plot()` and other graphic functions

`main=`	Title to put on the graphic.
`xlab=`	Label for the x-axis. Similarly for `ylab=`.
`xlim=`	Specify the x-limits, as in `xlim=c(0,10)`, for the interval $[0,10]$. Similar argument for the y-axis is `ylim=`.
`type=`	Type of plot to make. Use `"p"` for points (the default), `"l"` (ell not one) for lines, and `"h"` for vertical lines.
`bty=`	Type of box to draw. Use `"l"` for "L"-shaped, default is `"o"`, which is "O"-shaped. Details in `?par`.
`pch=`	The style of point that is plotted. This can be a number or a single character. Numbers between 0 and 25 give different symbols. The command `plot(0:25,pch=0:25)` will show those possible.
`cex=`	Magnification factor. Default is 1.
`lty=`	When lines are plotted, specifies the type of line to be drawn. Different numbers correspond to different dash combinations. (See `?par` for full details.)
`lwd=`	The thickness of lines. Numbers bigger than 1 increase the default.
`col=`	Specifies the color to use for the points or lines.

3.3.2 The correlation between two variables

The correlation between two variables numerically describes whether larger- and smaller-than-average values of one variable are related to larger- or smaller-than-average values of the other variable.

Figure 3.9 shows two data sets: the scattered one on the left is weakly correlated; the one on the right with a trend is strongly correlated. We drew horizontal and vertical lines through $(\bar{x},\bar{y})$, breaking the figure into four quadrants. The correlated data shows that larger than average values of the `x` variable are paired with larger-than-average values of the `y` variable, as these points are concentrated in upper-right quadrant and not scattered throughout both right quadrants. Similarly for smaller-than-average values.

For the correlated data, the products $(x_i - \bar{x})(y_i - \bar{y})$ will tend to be positive, as this happens in both the upper-right and lower-left quadrants. This is not the case with the scattered data set. Because of this, the quantity $\sum_i (x_i - \bar{x})(y_i - \bar{y})$ will be useful in describing the correlation between two variables. When the data is uncorrelated, the terms will tend to cancel each other out; for correlated data they will not.

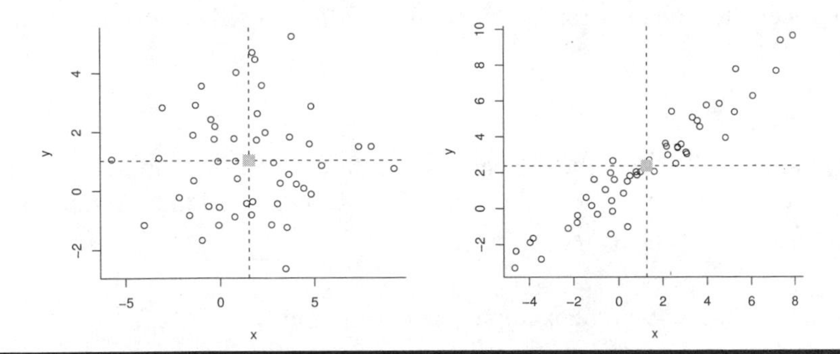

Figure 3.9 Two data sets with horizontal and vertical lines drawn through $(\bar{x}, \bar{y})$. The data set on the left shows weak correlation and data spread throughout the four quadrants of the plot. The data set on the right is strongly correlated, and the data is concentrated into opposite quadrants.

To produce a numeric summary that can be used to compare data sets, this sum is scaled by a term related to the product of the sample standard deviations. With this scaling, the correlation only involves the respective z-scores, and the quantity is always between -1 and 1.

When there is a linear relationship between x and y then values of r^2 close to 1 indicate a strong linear relationship, and values close to 0 a weak linear relationship. (Sometimes r may be close to 0, but a different type of relationship holds.)

The Pearson correlation coefficient

The Pearson correlation coefficient, r, of two data vectors x and y is defined by

$$r = \mathrm{cor}(x,y) = \frac{\sum(x_i - \bar{x})(y_i - \bar{y})}{\sqrt{\sum(x_i - \bar{x})^2 \sum(y_i - \bar{y})^2}} = \frac{1}{n-1}\sum\left(\frac{x_i - \bar{x}}{s_x}\right)\left(\frac{y_i - \bar{y}}{s_y}\right). \quad (3.1)$$

The value of r is between -1 and 1.

In R this is found with the `cor()` function, as in `cor(x,y)`.

We look at the correlations for the three data sets just discussed. First we attach the variable names, as they have been previously detached.

```
> attach(homedata); attach(maydow); attach(kid.weights)
```

In Example 3.2, on Maplewood home values, we saw a nearly linear relationship between the 1970 assessed values and the 2000 ones. The correlation in this case is

```
> cor(y1970,y2000)
[1] 0.9111
```

In Example 3.3, where the temperature's influence on the Dow Jones average was considered, no trend was discernible. The correlation in this example is

```
> cor(max.temp[-1],diff(DJA))
[1] 0.01029
```

In the height-and-weight example, the correlation is

```
> cor(height,weight)
[1] 0.8238
```

The number is close to 1, but we have our doubts that a linear relationship a correct description.

The Spearman rank correlation

If the relationship between the variables is not linear but is increasing, such as the apparent curve for the height-and-weight data set, we can still use the correlation coefficient to understand the strength of the relationship. Rather than use the raw data for the calculation, we use the **ranked data**. That is, the data is ordered from smallest to largest, and a data point's rank is its position after sorting, with 1 being the smallest and n the largest. Ties are averaged. The **Spearman rank correlation** is the Pearson correlation coefficient computed with the ranked data.

The `rank()` function will rank the data.

```
> x = c(30,20,7,42,50,20)
> rank(x)                          # ties are averaged
[1] 4.0 2.5 1.0 5.0 6.0 2.5
```

The first three numbers are interpreted as: 30 is the fourth smallest value, 20 is tied for second and third, and 7 is the smallest.

Computing the Spearman correlation is done with `cor()` using the argument `method="spearman"` (which can be abbreviated). It can also be done directly combining `cor()` with `rank()`.

For our examples, the correlations are as follows:

```
## homedata example. r = 0.9111
> cor(rank(y1970), rank(y2000))
[1] 0.907
## Dow Jones example. r = 0.01029
> cor(max.temp[-1], diff(DJA), method="spearman") # slight?
[1] 0.1316
## height and weight example. r = 0.8238
> cor(height,weight, m="s")     # abbreviated
[1] 0.8822
> detach(homedata); detach(maydow); detach(kid.weights)
```

The data on home values is basically linear, and there the Spearman correlation actually went down. For the `height`-versus-`weight` data, the Spearman correlation coefficient increases as expected, as the trend there appears to be more quadratic than linear.

3.3.3 Problems

3.13 For the `homedata (UsingR)` data set, make a histogram and density estimate of the multiplicative change in values (the variable `y2000/y1970`). Describe the shape, and explain why it is shaped thus. (Hint: There are two sides to the tracks.)

3.14 The `galton (UsingR)` data set contains measurements of a child's height and an average of his or her parents' heights (analyzed by Francis Galton in 1885). Find the Pearson and Spearman correlation coefficients.

3.15 The data set `normtemp (UsingR)` contains body measurements for 130 healthy, randomly selected individuals. The variable `temperature` measures normal body temperature, and the variable `hr` measures resting heart rate. Make a scatterplot of the two variables and find the Pearson correlation coefficient.

3.16 The data set `fat (UsingR)` contains several measurements of 252 men. The variable `body.fat` contains body-fat percentage, and the variable `BMI` records the body mass index (weight divided by height squared). Make a scatterplot of the two variables and then find the correlation coefficient.

3.17 The data set `twins (UsingR)` contains IQ scores for pairs of identical twins who were separated at birth. Make a scatterplot of the variables `Foster` and `Biological`. Based on the scatterplot, predict what the Pearson correlation coefficient will be and whether the Pearson and Spearman coefficients will be similar. Check your guesses.

3.18 The `state.x77` data set contains various information for each of the fifty United States. We wish to explore possible relationships among the variables. First, we make the data set easier to work with by turning it into a data frame.

```
> x77 = data.frame(state.x77)
> attach(x77)
```

Now, make scatterplots of `Population` and `Frost`; `Population` and `Murder`; `Population` and `Area`; and `Income` and `HS.Grad`. Do any relationships appear linear? Are there any surprising correlations?

3.19 The data set `nym.2002 (UsingR)` contains information about the 2002 New York City Marathon. What do you expect the correlation between age and finishing time to be? Find it and see whether you were close.

3.20 For the data set `state.center` do this plot:

```
> with(state.center,plot(x,y))
```

Can you tell from the shape of the points what the data set is?

3.21 The `batting (UsingR)` data set contains baseball statistics for the 2002 major league baseball season. Make a scatterplot to see whether there is any trend. What is the correlation between the number of strikeouts (`SO`) and the number of home runs (`HR`)? Does the data suggest that in order to hit a lot of home runs one should strike out a lot?

3.22 The `galton (UsingR)` data set contains data recorded by Galton in 1885 on the heights of children and their parents. The data is discrete, so a simple scatterplot does not show all the data points. In this case, it is useful to "jitter" the points a little when plotting by adding a bit of noise to each point. The `jitter()` function will do this. An optional argument, `factor=`, allows us to adjust the amount of jitter. Plot the data as below and find a value for `factor=` that shows the data better.

```
> attach(galton)
> plot(jitter(parent,factor=1),jitter(child,factor=1))
```

3.4 Simple linear regression

In this section, we introduce the simple linear regression model for describing paired data sets that are related in a linear manner. When we say that variables x and y have a linear relationship *in a mathematical sense* we mean that $y = mx + b$, where m is the slope of the line and b the intercept. We call x the independent variable and y the dependent one.

In statistics, we don't assume these variables have an exact linear relationship: rather, the possibility for noise or error is taken into account.

In the **simple linear regression model** for describing the relationship between x_i and y_i, an error term is added to the linear relationship:

$$y_i = \beta_0 + \beta_1 x_i + \varepsilon_i. \tag{3.2}$$

The value ε_i is an **error term**, and the coefficients β_0 and β_1 are the **regression coefficients.**† The data vector x is called the **predictor variable** and y the

† These are Greek letters: ε is *epsilon* and β is *beta*.

response variable. The error terms are unknown, as are the regression coefficients. The goal of linear regression is to estimate the regression coefficients in a reasonable manner from the data.

The term "linear" applies to the way the regression coefficients are used. The model $y_i = \beta_0 + \beta_1 x_i^2 + \varepsilon_i$ would also be considered a linear model. The term "simple" is used to emphasize that only one predictor variable is used, in contrast with the multiple regression model, which is discussed in Chapter 10.

Estimating the intercept β_0 and the slope β_1 gives an estimate for the underlying linear relationship. We use "hats" to denote the estimates. The estimated regression line is then written

$$\widehat{y} = \widehat{\beta_0} + \widehat{\beta_1} x.$$

For each data point x_i we have a corresponding value, $\widehat{y_i} = \widehat{\beta_0} + \widehat{\beta_1} x_i$, with $(x_i, \widehat{y_i})$ being a point on the estimated regression line.

We refer to $\widehat{y_i}$ as the predicted value for y_i, and to the estimated regression line as the prediction line. The difference between the true value y_i and this predicted value is the **residual**, e_i:

$$e_i = y_i - \widehat{y_i} = \text{actual} - \text{predicted}.$$

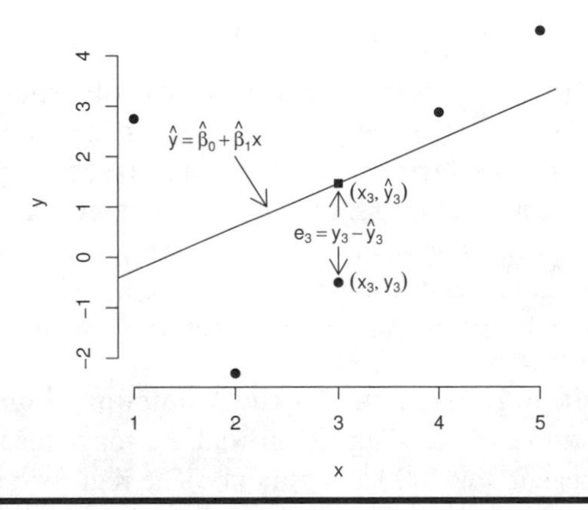

Figure 3.10 Prediction line with residual for (x_3, y_3) indicated

Geometrically, the residual is the signed vertical distance of the point (x_i, y_i) to the prediction line as seem in Figure 3.10. If the estimated line is a good one, these distances should be small. The **method of least squares** chooses the line (equivalently the coefficients) so that the sum of the *squared* residuals is *as small* as possible. This is a tractable problem and its solution gives

$$\widehat{\beta}_1 = \frac{\sum(x_i - \bar{x})(y_i - \bar{y})}{\sum(x_i - \bar{x})^2}, \tag{3.3}$$

$$\widehat{\beta}_0 = \bar{y} - \widehat{\beta}_1 \bar{x}.$$

Interpreting, the regression line goes through the point $(\bar{x}, \bar{y})$ and has slope given by $\widehat{\beta}_1$.

3.4.1 Using the regression model for prediction

One of the primary uses of simple linear regression is to make predictions for the response value for new values of the predictor. For example, high school GPAs may be used by colleges during the admission process to predict college GPAs. Once the coefficients are estimated, the value of $\widehat{y}$ is used for the prediction.

3.4.2 Finding the regression coefficients using `lm()`

The regression coefficients could certainly be found directly from the formulas, but we would like to have some convenient function do it for us directly. R provides the `lm()` function for linear models. The most basic usage is

```
lm( model.formula )
```

The `model.formula` is a formula that represents the simple linear regression model. The notation for this is `y ~ x`. The `~` in this notation is read "is modeled by," so the **model formula** `y ~ x` would be read "y is modeled by x." The model formula implicitly assumes an intercept term and a linear model. The model formula approach is quite flexible, as we will see. We approach the notation step by step, on a need-to-know basis. A comprehensive description is contained in the manual *An Introduction to R* that accompanies R.

■ **Example 3.5: The regression line for the Maplewood home data**

The data set `homedata (UsingR)` showed a strong linear trend between the 1970 assessments and the 2000 assessments. The regression coefficients are found from the data as follows:

```
> attach(homedata)
> lm(y2000 ~ y1970)
Call:
lm(formula = y2000 ~ y1970)
Coefficients:
(Intercept)        y1970
  -1.13e+05     5.43e+00
```

The value of $\widehat{\beta}_0$ is indicated by `(Intercept)`, and the value of $\widehat{\beta}_1$ appears under the variable name `y1970`.

It is recommended that the results of the modeling be stored, as there are several ways of extracting more information than is initially shown. For example, we assign the results of the `homedata` model to the variable `res`.

```
> res = lm(y2000 ~ y1970)          # type res to see default output
```

The intercept is *negative* \$113,000 and the slope is 5.43. Such a big negative intercept might seem odd. Did we make a mistake? We doublecheck using the formulas:

```
> sxy = sum((y1970 - mean(y1970)) * (y2000 - mean(y2000)))
> sx2 = sum( (y1970 - mean(y1970))^2 )
> sxy/sx2
[1] 5.429
> mean(y2000) - sxy/sx2 * mean(y1970)
[1] -113153
```

The negative intercept should be a warning not to use this model for prediction with a really low 1970 home value. In general, predictions should be restricted to the range of the predictor variable.

Adding the regression line to a scatterplot: `abline()`

Adding a regression line to the scatterplot is facilitated with the convenient, but oddly named, `abline()` function. (Read it "a-b-line.")

```
> plot(y1970,y2000, main="-113,000 + 5.43 x")
> abline(res)
```

The output of `lm()`, stored in `res`, is plotted by `abline()`. We see in Figure 3.11 that the data tends to cluster around the regression line, although there is much variability.

In addition to adding regression lines, the `abline()` function can add other lines to a graphic. The line $y = a + bx$ is added with `abline(a,b)`; the horizontal line $y = c$ is added with `abline(h=c)`; and the vertical line $x = c$ with `abline(v=c)`.

Using the regression line for predictions

One of the uses of the regression line is to predict the y value for a given x value. For example, the year-2000 predicted value of a house worth \$50,000 dollars in 1970 is found from the regression line with

$$\widehat{y} = \widehat{\beta}_0 + \widehat{\beta}_1 \cdot 50,000.$$

That is, the y-value on the prediction line for the given value of x. This is

```
> -113000 + 5.43 * 50000
[1] 158500
```

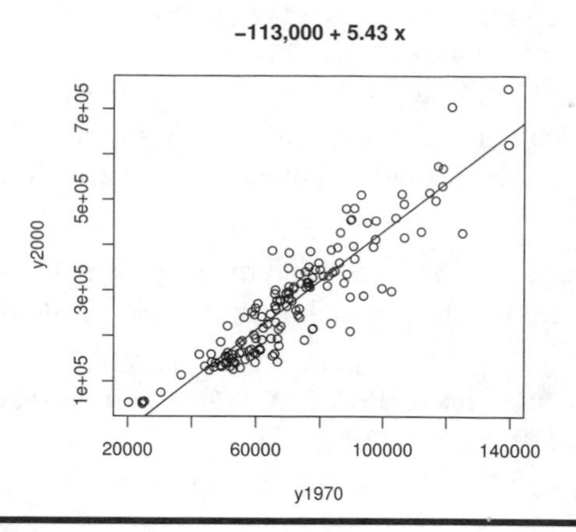

Figure 3.11 Home-data scatterplot with least-squares regression line

The previous calculation can be done without having to type in the coefficients, possibly adding round-off error to the answers. The coefficients are returned by the function `coef()`. Multiplying by a vector of the type $(1,x)$ and adding will produce $\widehat{y}$:

```
> betas = coef(res)
> sum(betas * c(1,50000))          # beta0 * 1 + beta1 * 50000
[1] 158308                         # no rounding in betas this way
```

There are other useful extractor functions, such as `coef()`. For example, the function `residuals()` returns the residuals, and `predict()` will perform predictions as above. To illustrate for the data point $(55100, 130200)$ we find the predicted and residual value.

To specify the x value desired for prediction to `predict()` requires a data frame with properly named variables. Data frames will be discussed more fully in the next chapter. For this usage, we note that the function `data.frame()` will create a data frame, and the names are set with the format `name=values`.

To use `predict()` with $x_i = 55,100$ is done with

```
> predict(res, data.frame(y1970=55100))
[1] 185997
```

The residual can then be computed by subtraction:

```
> 130200 - predict(res, data.frame(y1970=55100))
[1] -55797
```

The residual is also returned by `residuals()` after finding out which index corresponds to the data point:

```
> residuals(res)[which(y1970 == 55100 & y2000 == 130200)]
  6688
-55797
```

We needed both conditions, as there are two homes with an assessed value of \$55,100 in 1970. ■

More on model formulas Model formulas can be used with many R functions—for instance, the `plot()` function. The `plot()` function is an example of a generic function in R. For these functions, different implementations are used based on the first argument. When the first argument of the `plot()` function is a model formula containing numeric predictor and response variables, a scatterplot is created. Previously, we've seen that when the argument is the output of the `density()` function a densityplot is produced. Other usages will be introduced in the sequel. The scatterplot and regression line could then be made as follows:

```
> plot(y2000 ~ y1970)
> res = lm(y2000 ~ y1970)
> abline(res)
```

A small advantage to this usage is that the typing can be reused with the history mechanism. This could also be achieved by saving the model formula to a variable.

More importantly, the model formula offers some additional flexibility. With model formula, the argument `data=` can usually be used to attach a data frame temporarily. This convenience is similar to that offered more generally by the function `with()`. Both styles provide an environment where R can reference the variables within a data frame by their name, avoiding the trouble of attaching and detaching the data frame. Equally useful is the argument `subset=`, which can be used to restrict the rows that are used in the data. This argument can be specified by a logical condition or a specification of indices.

We will use both of these arguments in the upcoming examples.

3.4.3 Transformations of the data

As the old adage goes, "If all you have is a hammer, everything looks like a nail." The linear model is a hammer of sorts; we often try to make the problem at hand fit the model. As such, it sometimes makes sense to transform the data to make the linear model appropriate.

■ **Example 3.6: Kids' weights: Is weight related to height squared?**

In Figure 3.8, the relationship between height and weight is given for the `kid.weights (UsingR)` data set. In Example 3.4, we mentioned that the BMI suggests a relationship between height squared and weight. We model this as follows:

```
> height.sq = kid.weights$height^2
> plot(weight ~ height.sq, data=kid.weights)
> res = lm(weight ~ height.sq, data=kid.weights)
> abline(res)
> res
```

```
Call:
lm(formula = weight ~ height.sq, data = kid.weights)

Coefficients:
(Intercept)    height.sq
     3.1089       0.0244
```

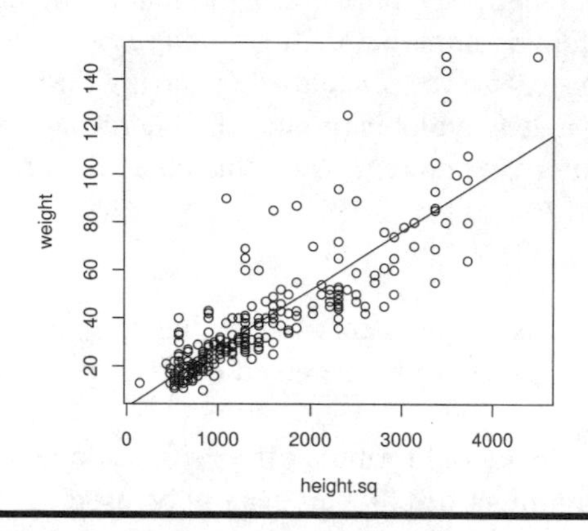

Figure 3.12 Height squared versus weight

Figure 3.12 shows a better fit with a linear model than before. However, the BMI is not constant during a person's growth years, so this is not exactly the expected relationship. ■

Using a model formula with transformations If we had tried the above example using this model formula, we'd be in for a surprise:

```
> plot(weight ~ height^2, data = kid.weights)   # not as expected
> res = lm(weight ~ height^2, data=kid.weights)
> abline(res)
```

The resulting graph would look identical to the graph of height versus weight in Figure 3.8 and not the graph of height squared versus weight in Figure 3.12.

The reason for this is that the model formula syntax uses the familiar math notations `*,/,^` differently. To use them in their ordinary sense, we need to *insulate* them in the formulas with the `I()` function, as in:

```
> plot(weight ~ I(height^2), data = kid.weights)
> res = lm(weight ~ I(height^2), data = kid.weights)
> abline(res)
```

3.4.4 Interacting with a scatterplot

When looking at a scatterplot we see the trend of the data as well as individual data points. If one of these data points stands out, how can it be identified? Which index in the data set corresponds to it? What are its x-and y-coordinates? If the data set is small, the answers can be identified by visual inspection of the data. For larger data sets, better methods are available.

The `R` function to identify points on a scatterplot by their corresponding index is `identify()`. A template for its usage is

```
identify(x, y, labels=..., n=...)
```

In order to work, `identify()` must know about the points we want to identify. These are specified as variables and not as a model formula. The value `n=` specifies the number of points to identify. By default, `identify()` identifies points with each mouse click until instructed to stop. (This varies from system to system. Typically it's a right-click in Windows, a middle-click in Linux, and the escape key in Mac OS X.) As points are identified, `R` will put the index of the point next to it. The argument `labels=` allows for the placement of other text. The `identify()` function returns the indices of the selected points.

For example, if our plot is made with `plot(x,y)`, `identify(x,y,n=1)` will identify the closest point to our first mouse click on the scatterplot by its index, whereas `identify(x,y,labels=names(x))` will let us identify as many points as we want, labeling them by the names of `x`.

The function `locator()` will locate the (x,y) coordinates of the points we select with our mouse. It is called with the number of points desired, as with `locator(2)`. The return value is a list containing two data vectors, `x` and `y`, holding the x and y positions of the selected points.

■ **Example 3.7: Florida 2000** The `florida (UsingR)` data set contains county-by-county vote counts for the 2000 United States presidential election in the state of Florida. This election was *extremely* close and was marred by several technical issues, such as poorly designed ballots and outdated voting equipment. As an academic exercise only, we might try to correct for one of these issues statistically in an attempt to divine the true intent of the voters.

As both Pat Buchanan and George Bush were conservative candidates (Bush was the Republican and Buchanan was an Independent), there should be some relationship between the number of votes for Buchanan and those for Bush. A scatterplot (Figure 3.13) is illuminating. There are two outliers. We identify the outliers as follows:

```
> plot(BUCHANAN ~ BUSH, data=florida)
> res = lm(BUCHANAN ~ BUSH, data=florida)     # store it
> abline(res)
> with(florida, identify(BUSH,BUCHANAN,n=2,labels=County))
[1] 13 50
> florida$County[c(13,50)]
[1] DADE        PALM BEACH
67 Levels: ALACHUA BAKER BAY BRADFORD BREVARD ... WASHINGTON
```

(We use both `with()` and the dollar-sign notation instead of attaching the data frame.)

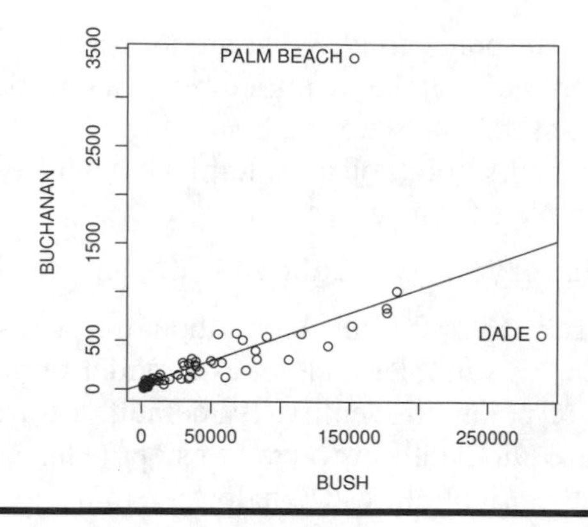

Figure 3.13 Scatterplot of Bush and Buchanan votes by county in Florida

There is a strong linear relationship with two outliers. If the relationship were exactly linear without intercept, this would say that Buchanan always had the same percentage of conservative votes.

Palm Beach County's infamous "butterfly ballot" design was believed to have caused many people to cast votes incorrectly. Suppose this were true. How many votes might this have cost Al Gore, the Democrat? Say that the extra Buchanan votes were to go to Gore. How many extra Buchanan votes were there? One way to estimate the amount is to use the regression line to make a prediction based on the number of Bush votes for that county.

The predicted amount and residual for Palm Beach are found as follows:

```
> with(florida, predict(res, data.frame(BUSH = BUSH[50])) )
[1] 796.8
> residuals(res)[50]
  50
2610
```

This simple analysis indicates that Buchanan received 2,610 of Gore's votes—many more than the 567 that decided the state and the presidency. (The *Palm Beach Post*, using different data, concluded that Gore lost 6,607 votes when voters marked more than one name on the butterfly ballot.) ■

3.4.5 Outliers in the regression model

For the simple linear regression model, there are two types of outliers. For the individual variables, there can be outliers in the univariate sense—a data point that doesn't fit the pattern set by the bulk of the data. In addition, there can be outliers in the regression model. These would be points that are far from the trend or pattern of the data. In the Florida 2000 example, both Dade County and Palm

Beach County are outliers in the regression.

■ **Example 3.8: Emissions versus GDP** The `emissions (UsingR)` data set contains data for several countries on CO_2 emissions and per-capita gross domestic product (GDP). A scatterplot with a regression line indicates one isolated point that seems to "pull" the regression line upward. The regression line found without this point has a much different slope.

```
> f = CO2 ~ perCapita                # save formula
> plot(f, data = emissions)
> abline(lm(CO2 ~ perCapita, data = emissions))
> abline(lm(f, data = emissions, subset = -1), lty=2)
```

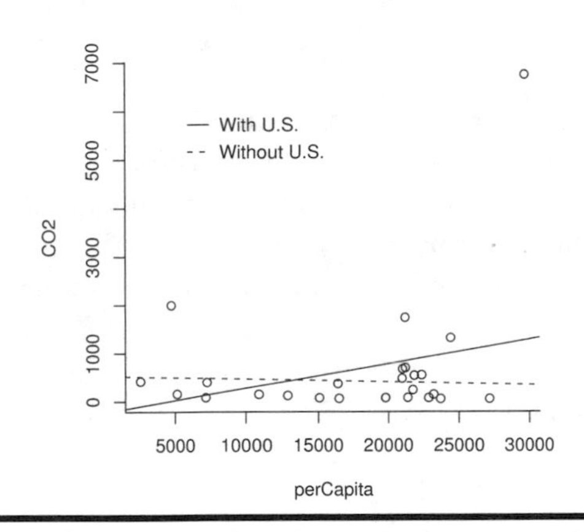

Figure 3.14 Emissions data with and without the United States data point

(In this example, we save the model formula for reuse and take advantage of the `subset=` argument with model formulas.)

The isolated point is the United States. This point is an outlier for the CO_2 variable separately, but not for the per-capita GDP. It is an outlier in the bivariate sense, as it stands far off from the trend set by the rest of the data. In addition, it is an influential observation, as its presence dramatically affects the regression line. ■

3.4.6 Resistant regression lines: `lqs()` *and* `rlm()`

In the previous example, the regression line was computed two ways. In Figure 3.14 the two lines are definitely different. Our eye should be able to see that the outlier pulls the initial regression line up quite a bit; hence the first one has a bigger slope.

Just like the mean and the standard deviation, the regression coefficients are subject to strong influences from outliers. For the mean and standard deviation we discussed resistant alternatives. such as the median or the IQR. As well, several resistant alternatives exist for the regression line.

Least-trimmed squares

The trimmed mean finds the mean after first trimming off values from the left and right of the distribution. This makes a measure of center that is less sensitive to outliers. The method of **least-trimmed squares** is similar (but computationally much more difficult).

The least-squares regression line is found conceptually by calculating for each line the sum of the squared residuals and then minimizing this value over all possible lines. Least-trimmed squares does the same, only the sum of the squared residuals is replaced by the sum of the q smallest squared residuals, where q is roughly $n/2$.

The least-trimmed squares algorithm is implemented in the `lqs()` function from the `MASS` package. This package is not loaded automatically. The default algorithm for the `lqs()` function is least-trimmed squares. As with `lm`, the data is specified using a model formula.

To illustrate on the emissions data, we add a least-trimmed squares line with line type 3:

```
> library(MASS)                  # load library if not already done
> abline(lqs(f,data=emissions), lty=3)
```

Resistant regression using `rlm()`

Alternatively, we can use the `rlm()` function, also from the `MASS` package, for resistant regression. It is not as resistant to outliers as `lqs()` by default but can be made so with the `method="MM"`. As with `lm()` and `lqs()`, the function is called using a model formula.

```
> abline(rlm(f, data=emissions, method="MM"), lty=4)
```

R's programmers strive to use a consistent interface to this type of function. Thus, it is no more difficult to find any of these regression lines, though the mathematics behind the calculations can be much harder.

Adding legends to plots

This example currently has a scatterplot and four different regression lines. We've been careful to draw each line with a different line type, but it is hard to tell which line is which without some sort of legend. The `legend()` function will do this for us.

To use `legend()`, we need to specify where to draw the legend, what labels to place, and how things are marked. The placement can be specified in (x,y) coordinates or done with the mouse using `locator(n=1)`. The labels are specified with the `legend=` argument. Markings may be done with different line types (`lty=`), as above where we used line types 1-4; with different colors (`col=`); or even with different plot characters (`pch=`).

To add a legend to the plot shown in Figure 3.15, we issue the following commands:

```
> the.labels = c("lm","lm w/o 1","least trimmed squares",
+ "rlm with MM")
> the.ltys = 1:4
> legend(5000,6000,legend=the.labels,lty=the.ltys)
```

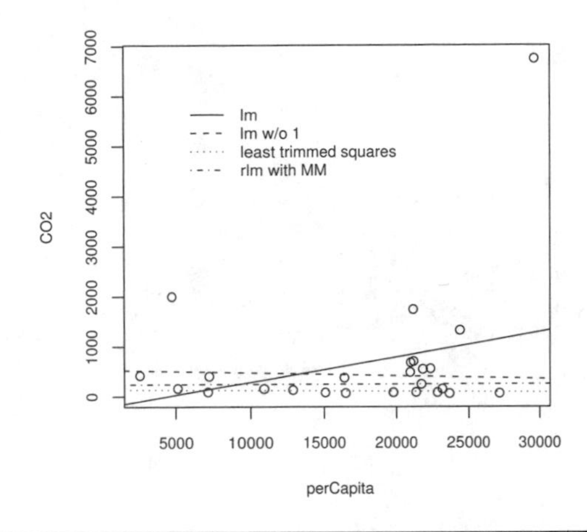

Figure 3.15 Emissions data with four different regression lines

3.4.7 Trend lines

If a scatterplot suggests some relationship, but not an obvious transformation to make a linear relationship, we can still superimpose a "trend line" on top of the data using one of the many scatterplot smoothing techniques available in R. These produce a smooth curve summarizing the relationship between the two variables.

The `stats` package provides several functions for creating trend lines. The `scatter.smooth()` function uses the `loess()` function from the same package to plot both the scatterplot and a trend line. Additionally, `smooth.spline()` will fit the data using cubic splines, and the `supsmu()` function will perform Friedman's "super smoother" algorithm.

■ **Example 3.9: Five years of temperature data** Weather data should show seasonal trends. The data set `five.yr.temperature (UsingR)` has five years of New York City temperature data. A scatterplot shows a periodic, sinusoidal pattern. In Figure 3.16, three trend lines are shown, although two are nearly identical.

```
> attach(five.yr.temperature)
> scatter.smooth(temps ~ days,col=gray(.75),bty="n")
> lines(smooth.spline(temps ~ days), lty=2, lwd=2)
> lines(supsmu(days, temps), lty=3, lwd=2)
> legend(locator(1),lty=c(1,2,3),lwd=c(1,2,2),
+   legend=c("scatter.smooth","smooth.spline","supsmu"))
> detach(five.yr.temperature)
```

■

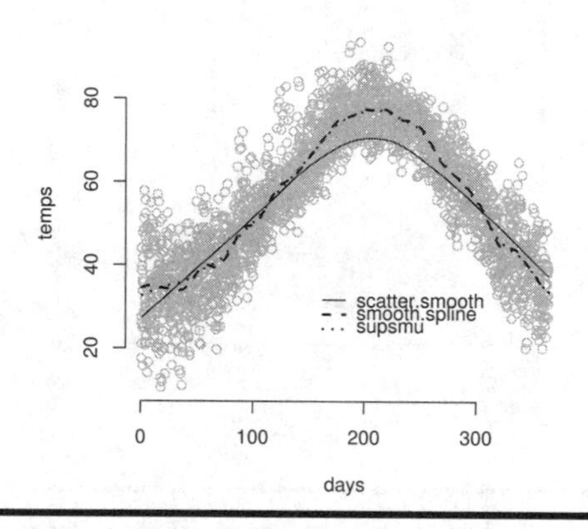

Figure 3.16 Temperature data with three trend lines

3.4.8 Problems

3.23 Try to establish the relationship that twice around the thumb is once around the wrist. Measure some volunteers' thumbs and wrists and fit a regression line. What should the slope be? While you are at it, try to find relationships between the thumb and neck size, or thumb and waist. What do you think: Did Gulliver's shirt fit well?

3.24 The data set `fat (UsingR)` contains ten body circumference measurements. Fit a linear model modeling the circumference of the `abdomen` by the circumference of the `wrist`. A 17-cm wrist size has what predicted abdomen size?

3.25 The data set `wtloss` (`MASS`) contains measurements of a patient's weight in kilograms during a weight-rehabilitation program. Make a scatterplot showing how the variable `Weight` decays as a function of `Days`.

1. What is the Pearson correlation coefficient of the two variables?
2. Does the data appear appropriate for a linear model? (A linear model says that for two comparable time periods the same amount of weight is expected to be lost.)
3. Fit a linear model. Store the results in `res`. Add the regression line to your scatterplot. Does the regression line fit the data well?
4. Make a plot of the residuals, `residuals(res)`, against the `Days` variable. Comment on the shape of the points.

3.26 The data frame `x77` contains data from each of the fifty United States. First coerce the `state.x77` variable into a data frame with

```
> x77 = data.frame(state.x77)
```

For each of the following models, make a scatterplot and add the regression line.

1. The model of illiteracy rate (`Illiteracy`) modeled by high school graduation rate `HS.Grad`.
2. The model of life expectancy (`Life.Exp`) modeled by (`Murder()`) the murder rate.
3. The model of income (`Income`) modeled by the illiteracy rate (`Illiteracy`).

Write a sentence or two describing any relationship. In particular, do you find it as expected or is it surprising?

3.27 The data set `batting (UsingR)` contains baseball statistics for the year 2002. Fit a linear model to runs batted in (`RBI`) modeled by number of home runs (`HR`). Make a scatterplot and add a regression line. In 2002, Mike Piazza had 33 home runs and 98 runs batted in. What is his predicted number of RBIs based on his number of home runs? What is his residual?

3.28 In the American culture, it is not considered unusual or inappropriate for a man to date a younger woman. But it is viewed as inappropriate for a man to date a *much* younger woman. Just what is too young? Some say anything less than half the man's age plus seven. This is tested with a survey of ten people, each indicating what the cutoff is for various ages. The results are in the data set `too.young (UsingR)`. Fit the regression model and compare it with the rule of thumb by also plotting the line $y = 7 + (1/2)x$. How do they compare?

3.29 The data set `diamond (UsingR)` contains data about the price of 48 diamond rings. The variable `price` records the price in Singapore dollars and the variable `carat` records the size of the diamond. Make a scatterplot of `carat` versus `price`. Use `pch=5` to plot with diamonds. Add the regression line and predict the amount a one-third carat diamond ring would cost.

3.30 The data set `Animals` (`MASS`) contains the body weight and brain weight of several different animals. A simple scatterplot will not suggest the true relationship, but a log-transform of both variables will. Do this transform and then find the slope of the regression line.

Compare this slope to that found from a robust regression model using `lqs()`. Comment on any differences.

3.31 To gain an understanding of the variability present in a measurement, a researcher may repeat or replicate a measurement several times. The data set `breakdown (UsingR)` includes measurements in minutes of the time it takes an insulating fluid to break down as a function of an applied voltage. The relationship calls for a log-transform.

Plot the voltage against the logarithm of time. Find the coefficients for simple linear regression and discuss the amount of variance for each level of the voltage.

3.32 The `motors` (`MASS`) data set contains measurements on how long, in hours, it takes a motor to fail. For a range of temperatures, in degrees Celsius, a number of motors were run in an accelerated manner until they failed, or until time was cut off. (When time is cut off the data is said to have been *censored*.) The data shows a relationship between increased temperature and shortened life span.

The commands

```
> data(motors, package="MASS")
> plot(time ~ temp, pch=cens, data=motors)
```

produce a scatterplot of the variable `time` modeled by `temp`. The `pch=cens` argument marks points that were censored with a square; otherwise a circle is used. Make the scatterplot and answer the following:

1. How many different temperatures were used in the experiment?
2. Does the data look to be a candidate for a linear model? (You might want to consider why the data point (150,8000) is marked with a square.)
3. Fit a linear model. What are the coefficients?
4. Use the linear model to make a prediction for the accelerated lifetime of a motor run at a temperature of 210°C.

3.33 The data set `mw.ages (UsingR)` contains census 2000 data on the age distribution of residents of Maplewood, New Jersey. The data is broken down by male and female.

Attach the data set and make a plot of the `Male` and `Female` variables added together. Connect the dots using the argument `type="l"`. For example, with the command `plot(1:103,Male + Female,type="l")`.

Next, layer on top two trend lines, one for male and one for female, using the `supsmu()` function. What age group is missing from this town?

Chapter 4

Multivariate Data

Multivariate data can be summarized and viewed in ways that are similar to those discussed for bivariate and univariate data, although differences exist, as there are many more possible relationships. These differences can be handled by looking at all the variables simultaneously, or by holding some variables constant while we look at others.

The tools used are similar to those for bivariate data, though if we enhance our data-manipulation skills the work will be easier. This chapter includes more details on R's data frames, lists, and model formula notation. Also included is an introduction to R's `lattice` graphics package, which greatly enhances certain explorations of multivariate data.

4.1 Viewing multivariate data

In this chapter we look at three typical examples of multivariate data. The first summarizes survey results (categorical data), the second compares independent samples; and the third searches for relationships among many different variables.

4.1.1 Summarizing categorical data

Just as we used tables to summarize bivariate data, we use them for multivariate data as well. However, as a table shows only two relationships at once, we will need to use several when looking at three or more relationships.

■ **Example 4.1: Student expenses** The `student.expenses (UsingR)` data set contains the results of a simple survey. Students were asked which of five different expenses they incur. The data is in Table 4.1.

Even a small table like this contains too much information for us to identify

Table 4.1 Student expenses survey

Student	cell.phone	cable.tv	dial.up	cable.modem	car
1	Y	Y	Y	N	Y
2	Y	N	N	N	N
3	N	N	N	N	Y
4	Y	Y	N	Y	Y
5	N	N	N	N	N
6	Y	N	Y	N	Y
7	Y	N	N	Y	N
8	N	N	N	N	Y
9	Y	Y	N	N	Y
10	Y	N	Y	N	N

any trends quickly. We would like to be able to summarize such data flexibly. We've used the `table()` function to make two-way tables. We can use it here as well. Let's look at the relationship between having a cell phone and a car:

```
> library(UsingR)                    # once per session
> attach(student.expenses)
> names(student.expenses)
[1] "cell.phone" "cable.tv"    "dial.up"     "cable.modem"
[5] "car"
> table(cell.phone,car)
          car
cell.phone N Y
         N 1 2
         Y 3 4
```

In this small sample, almost all the students have at least one of the two, with both being most common.

Three-way contingency tables (or, more generally, n-**way contingency tables**) show relationships among three (or n) variables using tables. We fix the values of the extra variable(s) while presenting a two-way table for the main relationship. To investigate whether paying for a cable modem affects the relationship between having a cell phone and a car, we can use a three-way table.

```
> table(cell.phone,car,cable.modem)
, , cable.modem = N


          car
cell.phone N Y
         N 1 2
         Y 2 3


, , cable.modem = Y

```

```
          car
cell.phone N Y
         N 0 0
         Y 1 1
```

It appears that paying for a cable modem requires cutbacks elsewhere.

The `table()` function uses the first two variables for the main tables and the remaining variables to construct different tables for each combination of the variables. In this case, two other tables for the levels of `cable.modem`.

Flattened contingency tables

This type of information is usually formatted differently to save space. In Table 4.2 the tables are set side-by-side, with the column headings layered to indicate the levels of the conditioning factor.

Table 4.2 Student expenses

	Modem			
	N		Y	
	Car		Car	
Cell phone	N	Y	N	Y
N	1	2	0	0
Y	2	3	1	1

This layout can be achieved with the `ftable()` (flatten table) function. Its simplest usage is to call it on the result of a `table()` command. For example:

```
> ftable(table(cell.phone,car,cable.modem))
...                                    # not side-by-side
```

This isn't quite what we want, as Table 4.2 has only one variable for the row. This is done by specifying the desired row variables or column variables. The column variables are set with the argument `col.vars=`. This argument expects a vector of variable names or indices.

```
> ftable(table(cell.phone,car,cable.modem),
+ col.vars=c("cable.modem","car")) # specify column variables
           cable.modem N   Y
           car         N Y N Y
cell.phone
N                      1 2 0 0
Y                      2 3 1 1
> detach(student.expenses)
```

■

4.1.2 Comparing independent samples

When we have data for several variables of the same type, we often want to compare their centers, spreads, or distributions. This can be done quite effectively using boxplots.

■ **Example 4.2: Taxi-in-and-out times at Newark Liberty International Airport** The data set ewr (UsingR) contains taxi-in-and-out data for airplanes landing at Newark Liberty airport. The data set contains monthly averages for eight major carriers. An examination of this data allows us to see which airlines take off and land faster than the others. We treat the data as a collection of independent samples and use boxplots to investigate the differences. Figure 4.1 shows the boxplots.

```
> attach(ewr)
> names(ewr)
 [1] "Year"     "Month"   "AA"      "CO"      "DL"      "HP"
 [7] "NW"       "TW"      "UA"      "US"      "inorout"
> boxplot(AA,CO,DL,HP,NW,TW,US,US)
> detach(ewr)
```

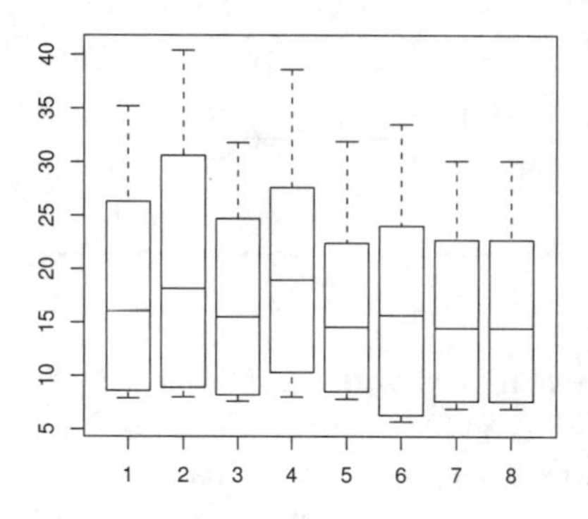

Figure 4.1 Taxi-in-and-out times at Newark Liberty airport

From the boxplots we see that the second airline (Continental) appears to be the worst, as the minimum, maximum, and median amount of time are all relatively large. However, the fourth (America West) has the largest median.

The boxplot() function will make a boxplot for each data vector it is called with. This is straightforward to use but has many limitations. It is tedious and prone to errors, as much needs to be typed. As well, adding names to the boxplots must be done separately. More importantly, it is a chore to do other things with the data. For example, the variable inorout indicates whether the time is for taxi in or taxi out. Taxi-in times should all be about the same, as airplanes usually land and go to their assigned gate with little delay. Taxi-out times are more likely to vary, as the queue to take off varies in length depending on the time of day. In the next section we see how to manipulate data to view this difference. ■

4.1.3 Comparing relationships

Scatterplots are used to investigate relationships between two variables. They can also be used when there are more than two variables. We can make multiple scatterplots, or plot multiple relationships on the same scatterplot using different plot characters or colors to distinguish the variables.

■ **Example 4.3: Birth variables** The data set `babies` (`UsingR`) contains several variables collected from new mothers as part of a study on child health and development. The variables include gestation period, maternal age, whether and how much the mother smokes, and other factors, such as mother's level of education. In all, there are 23 variables.

R has a built-in function for creating scatterplots of all possible pairs of variables. This graphic is called a **scatterplot matrix** and is made with the `pairs()` function, as in `pairs(babies)`. For the `babies` data set this command will create over 500 graphs, as there are so many variables. We hold off on using `pairs()` until we see how to extract subsets of the variables in a data frame.

We can still explore the data with scatterplots, using different colors or plotting characters to mark the points based on information from other factors. In this way, we can see more than two variables at once. For example, the plot of `gestation` versus `weight` in Figure 4.2 shows a definite expected trend: the longer the gestation period the more time a baby has to increase its birth weight. Do other factors, such as maternal smoking or maternal weight, affect this relationship?

To plot with different plot characters, we set the `pch=` argument using another variable to decide the plot character. First we recode the data with `NA`, as the data set uses 999 for missing data (cf. `?babies`).

```
> attach(babies)
> gestation[gestation == 999] = NA # 999 is code for NA
> plot(gestation,wt)              # scatterplot
> plot(gestation,wt,pch=smoke)    # different plot characters
> table(smoke)                    # values of plot characters
smoke
  0   1   2   3   9
544 484  95 103  10
> legend(locator(1),
+ legend=c("never","yes","until pregnant","long ago","unknown"),
+ pch=c(0:3,9))
```

The `table()` function was used to find out the range of values of `smoke`. We consulted the help page (`?babies`) to find out what the values mean.

Figure 4.2 is a little too crowded to tell if any further relationship exists.

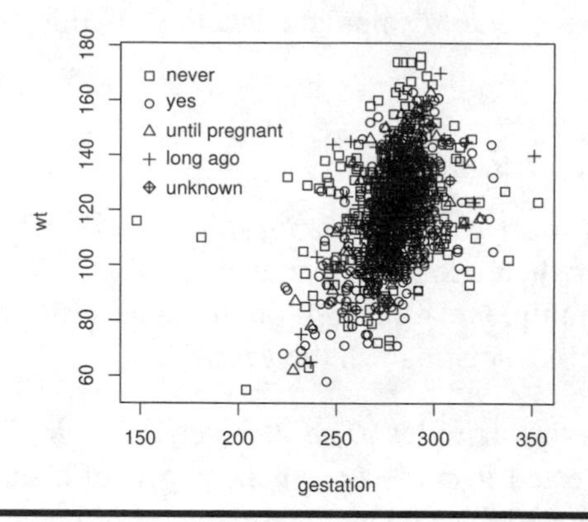

Figure 4.2 Scatterplot of gestation versus weight by smoke factor

Sometimes different colors will help where different plot characters don't. To change colors we set the `col=` argument in `plot()`. We use `rainbow()` to create five colors and then extract these using the values of `smoke` as indices. We want our indices to be 1, 2, 3, 4, 5, so we change the "9" to a "4" and then add 1. This is done as follows:

```
> smoke[smoke == 9] = 4
> plot(gestation,wt, col = rainbow(5)[smoke+1])
```

If we make the scatterplot, it shows that changing colors tells more of a story than changing the plot characters. Still, no additional trends show up. ■

What might be useful are different scatterplots for each level of the `smoke` factor. This can be done by subsetting, as is described next, or by using the `lattice` package described later on.

Plotting additional points and functions

Figure 4.2 is made all at once. There are times when we would like to add new points or lines to an existing plot, as we did in the previous chapter when we added a regression line to a graph. To do this, it helps to understand that R's plotting functions come in two types: "high-level" plot functions, like `plot()`, and "low-level" functions, like `abline()`. The difference between the two is that the high-level ones set up a graphic window and produce a graphic, while the low-level ones add to the current graphic window. Table 4.3 collects many useful plotting functions used in the examples.

We redo Example 4.3 plotting just a few variables separately. For fun, we will add regression lines.

First we need to make a plot. The following makes one for the occurrences where smoke has a value of `0`.

Table 4.3 Various plotting functions for creating or adding to figures

`plot()`	When used for scatterplots, will plot points by default. Use argument `type="l"` to produce lines. High-level function, used to make many types of figures.
`points()`	A low-level plot function with arguments similar to `plot()`.
`lines()`	Similar to `points()` but connects points with lines segments.
`abline()`	Function for adding lines to a figure. The arguments a= and b= will plot the line $y = a + bx$, the arguments h= and v= will plot horizontal or vertical lines.
`curve()`	A high- or low-level plot function for adding the graph of a function of x. When argument add=TRUE is given, will draw graph on the current figure using the current range of x values. If add=TRUE is not given, it will produce a new graph over the range specified with `from=` and `to=`. The defaults are 0 and 1. The function to be graphed may be specified by name or written as a function of x.
`rug()`	Adds lines along the x- or y-axis to show data values in a univariate data set. By default, the lines are drawn on the x-axis; use `side=2` to draw on the y-axis.
`arrows()`	Adds arrows to a figure.
`text()`	Adds text to a figure at specified points.
`title()`	Adds labels to a figure. Argument `main=` will set main title, `sub=` the subtitle, `xlab=` and `ylab=` will set x and y labels.
`legend()`	Adds a legend to a figure.

```
> gestation[gestation == 999] = NA
> f = wt[smoke == 0] ~ gestation[smoke == 0] # save typing
> plot(f, xlab="gestatation", ylab="wt")
> abline(lm(f))
```

We stored the model formula to save on typing.

To add to the graphic, we use `points()` with a similar syntax:

```
> f1 = wt[smoke == 1] ~ gestation[smoke == 1]
> points(f1, pch=16)
> abline(lm(f1), cex=2, lty=2)
> legend(150,175, legend=c("0 = never smoked","1 = smokes now"),
+ pch=c(1,16), lty=1:2)
> detach(babies)
```

The choice of plot character allows us to see the different data sets easily, as illustrated in Figure 4.3.

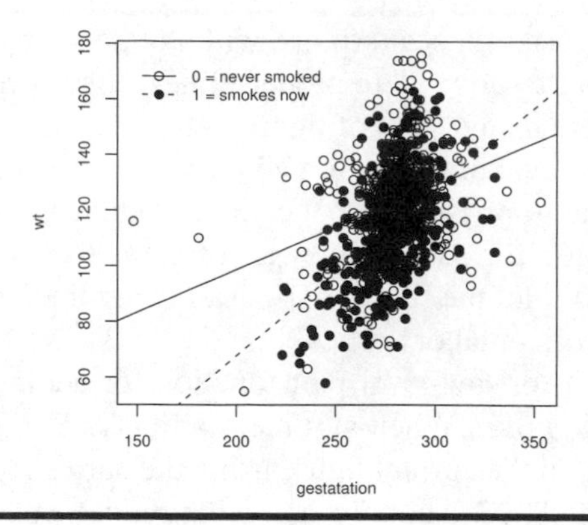

Figure 4.3 Scatterplot of gestation versus weight with smoking status added

4.1.4 Problems

4.1 The `samhda (UsingR)` data set contains variables from a Substance Abuse and Mental Health survey for American teens. Make a three-way contingency table of the variables `gender`, `amt.smoke`, and `marijuana`. Use the levels of `gender` to break up the two-way tables. Are there any suspected differences?

Repeat using the variable `live.with.father` in place of `gender`.

4.2 The data set `Cars93 (MASS)` contains information on numerous cars. We want to investigate, using tables, any relationships between miles per gallon, price, and type. Before doing so, we turn numeric variables into categorical ones using `cut()`.

```
> library(MASS)                    # loads Cars93 data set
> mpg = with(Cars93,cut(MPG.city,c(0,17,25,55)))
> names(mpg) = c("bad","decent","excellent")
> price = with(Cars93,cut(Price,c(0,10,20,62)))
> names(price) = c("cheap","mid-priced","expensive")
```

Make the above conversions, then make a flattened contingency table of `mpg`, `price`, and `Type`. Do you see any patterns?

4.3 In the previous exercise, variables in the data set `Cars93 (MASS)` were investigated with tables. Now, make a scatterplot of the variables `MPG.city` and `Price`, marking the points according to their `Type`. Do you see any trend?

4.4 For the `carsafety (UsingR)` data set, make a scatterplot of the variable `Driver.deaths` versus `Other.deaths`. Use `pch=as.numeric(type)` to change the plot character based on the value of `type`. Label any outliers with their make or model using `identify()`. Do you notice any trends?

4.5 The `cancer (UsingR)` data set contains survival times for cancer patients organized by the type of cancer. Make side-by-side boxplots of the variables `stomach`, `bronchus`, `colon`, `ovary`, and `breast`. Which type has the longest tail? Which has the smallest spread? Are the centers all similar?

4.6 The data set `UScereal (MASS)` lists facts about specific cereals sold in a United States supermarket. For this data set investigate the following:

1. Is there a relationship between manufacturer (`mfr`), and vitamin type (`vitamins`) by shelf location (`shelf`)? Do you expect one? Why?
2. Look at the relationship between `calories` and `sugars` with a scatterplot. Identify the outliers. Are these also fat-laden cereals?
3. Now look at the relationship between `calories` and `sugars` with a scatter-plot using different size points given by `cex=2*sqrt(fat)`. (This is called a bubble plot. The area of each bubble is proportional to the value of `fat`.) Describe any additional trend that is seen by adding the bubbles.

Can you think of any other expected relationships between the variables?

4.2 R basics: data frames and lists

Multivariate data consists of multiple data vectors considered as a whole. We are free to work with our data as separate variables, but there are many advantages to combining them into a single data object. This makes it easier to save our work, is convenient for many functions, and is much more organized. In R these "objects" are usually data frames.

A data frame is used to store rectangular grids of data. Usually each row corresponds to measurements on the same subject, statistical unit, or experimental unit. Each column is a data vector containing data for one of the variables. The collection of entries need not all be of the same type (e.g., numeric, character, or logical), but each column must contain the same type of entry as they are data vectors. A rectangular collection of values, all of the same type, may also be stored in a matrix. Although data frames are not necessarily matrices, as their values need not be numbers, they share the same methods of access.

A list is a more general type of storage than a data frame. Think of a list as a collection of components. Each component can be any R object, such as a vector, a data frame, a function, or even another list. In particular, a data frame is a list with top-level components given by equal-length data vectors. A list is a very flexible type of data object. Many of the functions in R, such as `lm()`, have return values that are lists, although only selected portions may be displayed when the return value is printed.

Lists can be used to store variables of different lengths, such as the `cancer`

(UsingR) data set, but it is usually more convenient to store such data in a data frame with two columns—one column recording the measurements and the other a factor recording which variable the value belongs to.

As a data frame is a special type of list, it can be accessed as either a matrix or a list.

4.2.1 Creating a data frame or list

Data frames are created with the `data.frame()` function, and lists are made with the `list()` function. Data frames are also returned by `read.table()` and `read.csv()`.

For example:

```
> x = 1:2                                 # define x
> y = letters[1:2]                        # y = c("a","b")
> z = 1:3                                 # z has 3 elements, x,y only 2
> data.frame(x,y)                         # rectangular, cols are variables
  x y
1 1 a
2 2 b
> data.frame(x,y,z)                       # not all the same size.
Error in data.frame(x, y, z) : arguments imply differing number
of rows: 2, 3
```

Data frames must have variables of the same length.

Lists are created using the function `list()`.

```
> list(x,y,z)
[[1]]
[1] 1 2
[[2]]
[1] "a" "b"
[[3]]
[1] 1 2 3
```

The odd-looking numbers that appear with the command `list(x,y,z)` specify where the values are stored. The first one, `[[1]]`, says this is for the first top-level component of the list, which is a data vector. The following `[1]` refers to the first entry of this data vector.

One difference that isn't apparent in the output, when using `data.frame` to create a data frame, is that character variables, such as `y`, are coerced to be factors, unless insulated with `I`. This coercion isn't done by `list()`. This can cause confusion when trying to add new values to the variable.

Adding names to a data frame or list

Just like data vectors, both data frames and lists can have a names attribute. These are found and set by the `names()` function or when we define the object. The names of a list refer to the top-level components. For data frames, these top-level components are the variables. In the above examples, the command `data.frame(x,y)` assigns names of `x` and `y` automatically, but the `list()` function does not. If we want to define names when using `data.frame()` or `list()` we can use a `name=value` format, as in

```
> list(x.name=x,"y name"=y)      # quotes may be needed
$x.name
[1] 1 2
$"y name"
[1] "a" "b"
```

The `names()` function can be used to retrieve or assign the names. When assigning names it is used on the left side of the assignment (when using the equals sign). For example:

```
> eg = data.frame(x,y)           # store the data frame
> names(eg)                      # the current names
[1] "x" "y"
> names(eg) = c("x.name","y name") # change the names
> names(eg)                      # names are changed
[1] "x.name" "y name"
```

Data frames can also have their column names accessed with the function `colnames()` and their rows named with `rownames()`. Both can be set at the same time with `dimnames()`. The row names must be unique, and it is recommended that column names be also. These functions are applicable to matrix-like objects.

The size of a data frame or list

Data frames represent a number of variables, each with the same number of entries. The `ewr (UsingR)` is an example. As this data is matrix-like, its size is determined by the number of rows and columns. The `dim()` function returns the size of matrix-like objects:

```
> dim(ewr)                       # number or rows and columns
[1] 46 11
> dim(ewr)[2]                    # number of cols is 2nd
[1] 11
```

Row and column sizes may also be found directly with `nrow()` and `ncol()`.

A list need not be rectangular. Its size is defined by the number of top-level components in it. This is found with the function `length()`. As data frames are lists whose top-level components are data vectors, the length of a data frame is the number of variables it contains.

```
> length(ewr)                    # number of top-level components
[1] 11
```

4.2.2 Accessing values in a data frame

The values of a data frame can be accessed in several ways. We've seen that we can reference a variable in a data frame by name. Additionally, we see how to access elements of each variable, or multiple elements at once.

Accessing variables in a data frame by name

Up to this point, most of the times that we have used the data in a data frame we have "attached" the data frame so that the variables are accessible in our work environment by their names. This is fine when the values will not be modified, but can be confusing otherwise. When R attaches a data frame it makes a copy of the variables. If we make changes to the variables, the data frame is not actually changed. This results in two variables with the same names but different values.

The following example makes a data frame using `data.frame()` and then attaches it. When a change is made, it alters the copy but not the data frame.

```
> x = data.frame(a=1:2,b=3:4)     # make a data frame
> a                               # a is not there
Error: Object "a" not found
> attach(x)                       # now a and b are there
> a                               # a is a vector
[1] 1 2
> a[1] = 5                        # assignment
> a                               # a has changed
[1] 5 2
> x                               # not x though
  a b
1 1 3
2 2 4
> detach(x)                       # remove x
> a                               # a is there and changed
[1] 5 2
> x                               # x is not changed
  a b
1 1 3
2 2 4
```

The `with()` function and the `data=` argument were mentioned in Chapter 1 as alternatives to attaching a data frame. We will use these when it is convenient.

Accessing a data frame using `[,]` notation

When we use a spreadsheet, we refer to the cell entries by their column names and rows number. Data-frame entries can be referred to by their column names (or numbers) and/or their row names (or numbers).

Entries of a data vector are accessed with the `[]` notation. This allows us to specify the entries we want by their indices or names. If `df` is the data frame, the basic notation is

```
df[row,column]
```

There are two positions to put values, though we may leave them blank on purpose. In particular, the value of `row` can be a single number (to access that row), a vector of numbers (to access those rows), a single name or vector of names (to match the row names), a logical vector of the appropriate length, or left blank to match all the rows. Similarly for the value of `column`. As with data vectors, rows and columns begin numbering at 1.

In this example, we create a simple data frame with two variables, each with three entries, and then we add row names. Afterward, we illustrate several styles of access.

```
> df = data.frame(x=1:3,y=4:6)  # add in column names
> rownames(df) = c("row 1","row 2","row 3") # add row names
> df
      x y
row 1 1 4
row 2 2 5
row 3 3 6
> df[3,2]                       # row=3,col=2
[1] 6
> df["row 3","y"]               # by name
[1] 6
> df[1:3,1]                     # rows 1, 2 and 3; column 1
[1] 1 2 3
> df[1:2,1:2]                   # rows 1 and 2, columns 1 and 2
      x y
row 1 1 4
row 2 2 5
> df[,1]                        # all rows, column 1, returns vector
[1] 1 2 3
> df[1,]                        # row 1, all columns
      x y
row 1 1 4
> df[c(T,F,T),]                 # rows 1 and 3 (T = TRUE)
      x y
row 1 1 4
row 3 3 6
```

■ **Example 4.4: Using data-frame notation to simplify tasks**

The data-frame notation allows us to take subsets of the data frames in a natural and efficient manner. To illustrate, let's consider the data set `babies` (`UsingR`) again. We wish to see if any relationships appear between the gestation time (`gestation`), birth weight (`wt`), mother's age (`age`), and family income (`inc`).

Again, we need to massage the data to work with R. Several of these variables have a special numeric code for data that is not available (`NA`). Looking at the documentation of `babies (UsingR)` (with `?babies`), we see that `gestation` uses 999, `age` uses 99, and `income` is really a categorical variable that uses 98 for "not available."

We can set these values to `NA` as follows:

```
## bad idea, doesn't change babies, only copies
> attach(babies)
> gestation[gestation == 999] = NA
```

```
> age[age == 99] = NA
> inc[inc == 98] = NA
> pairs(babies[,c("gestation","wt","age","inc")])
```

But the graphic produced by `pairs()` won't be correct, as we didn't actually change the values in the data frame `babies`; rather we modified the local copies produced by `attach()`.

A better way to make these changes is to find the indices that are not good for the variables `gestation`, `age`, and `inc` and use these for extraction as follows:

```
> rm(gestation); rm(age); rm(inc) # clear out copies
> detach(babies); attach(babies)  # really clear out
> not.these = (gestation == 999) | (age == 99) | (inc == 98)
## A logical not and named extraction
> tmp = babies[!not.these, c("gestation","age","wt","inc")]
> pairs(tmp)
> detach(babies)
```

The `pairs()` function produces the scatterplot matrix (Figure 4.4) of the new data frame `tmp`. We had to remove the copies of the variables `gestation`, `age`, and `inc` that were created in the previous try at this. To be sure that we used the correct variables, we detached and reattached the data set. Trends we might want to investigate are the relationship between gestation period and birth weight and the relationship of income and age. ■

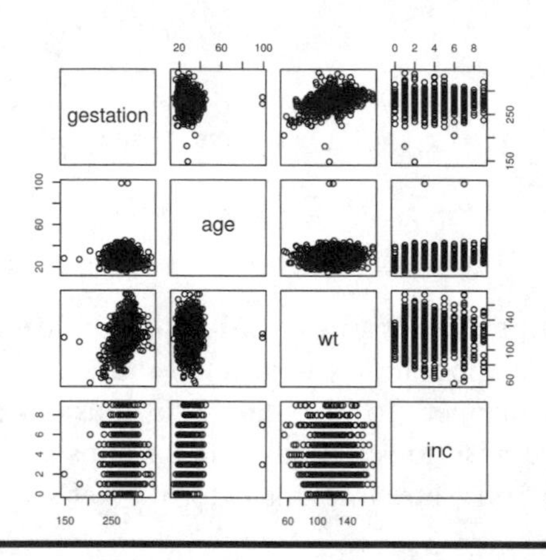

Figure 4.4 Scatterplot matrix of four variables from `babies`

■ **Example 4.5: Accessing a data frame with logical vectors** The data set `ewr (UsingR)` contains the taxi-in and taxi-out times at Newark Liberty airport. In Example 4.2 we noted that it would be nice to break up the data based on the variable `inorout`. We do this by extracting just those rows that have an `inorout` value of `in` (or `out`). Noting that columns 3 through 10 are for the airlines, we can construct the side-by-side boxplots with these commands:

```
> attach(ewr)
> boxplot(ewr[inorout == "in",  3:10], main="Taxi in")
> boxplot(ewr[inorout == "out", 3:10], main="Taxi out")
> detach(ewr)
```

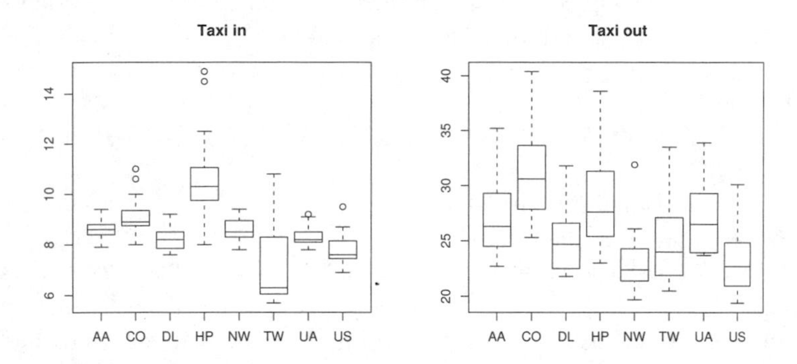

Figure 4.5 Taxi in and out times by airline

The "rows" argument to the `ewr` data frame is a logical vector of the correct length.

From Figure 4.5, as expected, we see that the airlines have less spread when taxiing in and that the taxi-in times are much shorter than the out times, as there is no takeoff queue to go through.

The argument to `boxplot()` is a data frame and not a data vector. If the first argument to `boxplot()` is a list, then each component is plotted with separate boxplots. As a data frame is a list, this behavior makes using `boxplot()` much easier than typing in all the variable names. When a boxplot is called with a list, any names for the list are used to label the boxplot. ■

The `subset()` function An alternate to using using the `[,]` notation to extract values from a data frame is to use the `subset()` function:

```
new.df = subset(old.df, subset=..., select=...)
```

The `subset=` argument works as it does in the formula interface by using a logical condition to restrict the rows used; `select=`, on the other hand, is used to restrict the columns when given a vector of the desired variable names or indices. (See `?subset` for more clever uses.) When we use `subset()`, the variable names are searched within the data frame. There is no need to attach the data frame or use `with()`.

The previous examples could have been completed as follows:

```
> ewr.in  = subset(ewr,subset= inorout == "in",  select=3:10)
> ewr.out = subset(ewr,subset= inorout == "out", select=3:10)
```

```
> boxplot(ewr.in,  main="Taxi in")
> boxplot(ewr.out, main="Taxi out")
```

■ **Example 4.6: Sorting a data frame by one of its columns**

The `sort()` function can be used to arrange the values of a data vector in increasing or decreasing order. For example, sorting the miles per gallon variable (`mpg`) of the `mtcars` data frame is done as follows:

```
> attach(mtcars)
> sort(mpg)
 [1] 10.4 10.4 13.3 14.3 14.7 15.0 15.2 15.2 15.5 15.8 16.4 17.3
...
```

Or from largest to smallest with

```
> sort(mpg, decreasing=TRUE)
 [1] 33.9 32.4 30.4 30.4 27.3 26.0 24.4 22.8 22.8 21.5 21.4 21.4
...
```

Often, we would like to sort a data frame by one of its columns and have the other columns reflect this new order. For this, `sort()` will not work; rather we need to work a bit harder. The basic idea is to rearrange, or *permute*, the row indices so that the new data frame is sorted as desired. The function that does the rearranging is `order()`. If `x` and `y` are data vectors, then the command `order(x)` will return the indices of the sorted values of `x`, so `x[order(x)]` will be `x` sorted from smallest to largest. The command `order(x,y)` will find the order of `x` and break ties using the order of `y`. More variables are possible. The command `order(x,decreasing=TRUE)` will sort in decreasing order.

To illustrate, without the output, we show how to sort the data set `mtcars` in various ways:

```
> mtcars[order(mpg),]              # sort by miles per gallon
> mtcars[order(mpg, decreasing=TRUE), ] # best mpg first
> rownames(mtcars[order(mpg, decreasing=TRUE), ]) # only names
> mtcars[order(cyl,hp),]          # by cylinders then horsepower
> detach(mtcars)
```

■

Accessing a list

The `[,]` notation for accessing a data frame is inherited from the notation for a matrix, which in `R` is a rectangular collection of values of the same type. Data frames, which are also lists, can be accessed using the notation for lists described next.

List access using `[[]]` The double-square-bracket notation, `[[]]`, is used to access list components, as in `lst[[position]]`. It can match using a number for `position` or a name (if the list has names). If position is a vector, then recursive indexing occurs (cf. `?Extract`).

For example:

```
> lst = list(a=1:2,b=letters[1:3],c=FALSE)
> lst[[1]]                          # first component
[1] 1 2
> lst[['a']]                        # by name
[1] 1 2
```

The $ notation The convenient syntax `lst$varname` is a shortcut for the notation `lst[["varname"]]`. The variable name may need to be quoted.

```
> lst = list(one.to.two = 1:2, "a-e"=letters[1:5])
> lst$one.to.two                    # access with $
[1] 1 2
> lst$o                             # unique shortening
[1] 1 2
> lst$a-e                           # needs quotes to work
Error: Object "e" not found
> lst$"a-e"
[1] "a" "b" "c" "d" "e"
```

This example also illustrates that R will match names even if they are not exactly the same. It will match, as long as we use enough letters to identify that name uniquely among the possible names. When the name includes blanks or other special characters, it needs to be quoted.

Lists as vectors The vector notation, `[]`, can be used with lists just as it is with vectors. It returns a list of those components specified by the indices. When `lst` is a list, `lst[1]` returns a list with just the first component of `lst`, whereas `lst[[1]]` returns the first component.

For example, we define a list and then access the first element as a vector:

```
> lst = list(a=1:2,b=letters[1:2],c=FALSE)
> lst[1]
$a
[1] 1 2
```

A list with just the first component is returned as indicated by the `$a` that is printed in the output. This is the name of that component. When a list prints `[[1]]`, it indicates that the first component has no name.

The vector notation for data frames is different from that for lists, as the return value remains a data frame. Suppose we do the above with a data frame using just the first two variables.

```
> df = data.frame(a=1:2,b=letters[1:2])
> df[1]
  a
1 1
2 2
```

If `df` were treated as a list, and not a data frame, then the return value would be a list with the specified top-level components. The different formatting of the

Table 4.4 Different ways to access a data frame

mtcars	mpg	cyl	disp	hp	drat	wt	qsec	vs
Mazda RX4	21.0	6	160.0	110	3.90	2.620	16.46	0
Mazda RX4 Wag	21.0	6	160.0	110	3.90	2.875	17.02	0
...								
Fiat 128	32.4	4	78.7	66	4.08	2.200	19.47	1
Honda Civic	**30.4**	4	75.7	52	4.93	1.615	18.52	1
Toyota Corolla	33.9	4	71.1	65	4.22	1.835	19.90	1
...								
Maserati Bora	15.0	8	301.0	335	3.54	3.570	14.60	0
Volvo 142E	21.4	4	121.0	109	4.11	2.780	18.60	1

To access the row "Honda Civic"	
`mtcars['Honda Civic',]`	By row name
`mtcars['Honda',]`	Can shorten the name if unique match
`mtcars[19,]`	It is also the 19th row in the data set
To access the column "mpg"	
`mtcars[,'mpg']`	By column name
`mtcars[,1]`	It is column 1
`mtcars$mpg`	list access by name
`mtcars[['mpg']]`	Alternate list access. Note, `mtcars['mpg']` is not a vector but a data frame.
To access the value "30.4"	
`mtcars['Honda','mpg']`	By name (with match)
`mtcars[19,1]`	By row and column number
`mtcars$mpg[19]`	`mtcars$mpg` is a vector, this is the 19th entry.

output indicates that this isn't the case. In the data-frame case, this list is then coerced into a data frame. The vector notation specifies the desired variables.

Table 4.4 summarizes the various ways to access elements of a data frame.

4.2.3 Setting values in a data frame or list

We've seen that we can't change a data frame's values by attaching it and then assigning to the variables, as this modifies the copy. Rather, we must assign values to the data frame directly. Setting values in a data frame or list is similar to setting them in a data vector. The basic expressions have forms like

```
df[rows,cols] = values,
lst$name = value,   or
lst$name[i] = value.
```

In the `[,]` notation, if `values` does not have the same size and type as the values that we are replacing, recycling will be done; if the length of `values` is too big, however, an error message is thrown. New rows and columns can be created by assigning to the desired indices. Keep in mind that the resulting data frame cannot have any empty columns (holes).

```
> df = data.frame(a=1:2,b=3:4)      # with names
> df[1,1] = 5                       # first row, first column
> df[,2] = 9:10                     # all rows, second column
> df[1:2,3:4] = cbind(11:12,13:14) # rows and columns at once
> df                                # new columns added
  a  b  c  d
1 5  9 11 13
2 2 10 12 14
> df[1:2, 10:11] = cbind(11:12,13:14) # would create a hole
Error in "[<-.data.frame"('*tmp*', ...
        new columns would leave holes after existing columns
> df[,2:3] = 0                      # recycling occurs
> df
  a b c  d
1 5 0 0 13
2 2 0 0 14
```

Using $ with a list refers to a data vector that can be set accordingly, either all at once or position by position, as with:

```
> lst = list(a=1:2,b=1:4,c=c("A","B","C"))
> lst$a =  1:5                       # replace the data vector
> lst$b[3] = 16                      # replace single element
> lst$c[4] = "D"                     # appends to the vector
> lst
$a
[1] 1 2 3 4 5
$b
[1]  1  2 16  4
$c
[1] "A" "B" "C" "D"
```

The `c()` function can be used to combine lists using the top-level components. This can be used with data frames with the same number of rows, but the result is a list, not a data frame. It can be turned into a data frame again by using `data.frame()`.

4.2.4 Applying functions to a data frame or list

In Chapter 3 we noted that `apply()` could be used to apply a function to the rows or columns of a matrix. The same can be done for a data frame, as it is matrix-like. Although many functions in R adapt themselves to do what we would want, there are times when ambiguities force us to work a little harder by using this technique. For example, if a data frame contains just numbers, the function `mean()` will find the mean of each variable, whereas `median()` will find the median of the entire data set. We illustrate on the `ewr (UsingR)` data set:

```
> df = ewr[ , 3:10]                 # make a data frame of the times
> mean(df)                          # mean is as desired
```

```
   AA    CO    DL    HP    NW    TW    UA    US
17.83 20.02 16.63 19.60 15.80 16.28 17.69 15.49
> median(df)                      # median is not as desired
Error in median(df) : need numeric data
> apply(df,2,median)              # median of columns
   AA    CO    DL    HP    NW    TW    UA    US
16.05 18.15 15.50 18.95 14.55 15.65 16.45 14.45
```

We can apply functions to lists as well as matrices with the `lapply()` function or its user-friendly version, `sapply()`. Either will apply a function to each top-level component of a list or the entries of a vector. The `lapply()` function will return a list, whereas `sapply()` will simplify the results into a vector or matrix when appropriate.

For example, since a data frame is also a list, the median of each variable above could have been found with

```
> sapply(df,median)
   AA    CO    DL    HP    NW    TW    UA    US
16.05 18.15 15.50 18.95 14.55 15.65 16.45 14.45
```

(Compare this to the output of `lapply(df,median)`.)

4.2.5 Problems

4.7 Use the data set `mtcars`.

1. Sort the data set by weight, heaviest first.
2. Which car gets the best mileage (largest `mpg`)? Which gets the worst?
3. The cars in rows `c(1:3,8:14,18:21,26:28,30:32)` were imported into the United States. Compare the variable `mpg` for imported and domestic cars using a boxplot. Is there a difference?
4. Make a scatterplot of weight, `wt`, versus miles per gallon, `mpg`. Label the points according to the number of cylinders, `cyl`. Describe any trends.

4.8 The data set `cfb (UsingR)` contains consumer finance data for 1,000 consumers. Create a data frame consisting of just those consumers with positive `INCOME` and negative `NETWORTH`. What is its size?

4.9 The data set `hall.fame (UsingR)` contains numerous baseball statistics, including Hall of Fame status, for 1,034 players.

1. Make a histogram of the number of home runs hit (`HR`).
2. Extract a data frame containing at bats (`AB`), hits (`hits`), home runs (`HR`), and runs batted in (`RBI`) for all players who are in the Hall of Fame. (The latter can be found with `Hall.Fame.Membership != "not a member"`.) Save the data into the data frame `hf`.
3. For the new data frame, `hf`, make four boxplots using the command:

```
boxplot(lapply(hf,scale))
```

(The `scale()` function allows all four variables to be compared easily.) Which of the four variables has the most skew?

Use matrix notation, list notation, or the `subset()` function to do the above.

4.10 The data set `dvdsales (UsingR)` can be viewed graphically with the command

```
> barplot(t(dvdsales), beside=TRUE)
```

1. Remake the barplots so that the years increase from left to right.
2. Which R commands will find the year with the largest sales?
3. Which R commands will find the month with the largest sales?

4.11 Use the data set `ewr (UsingR)`. We extract just the values for the times with `df=ewr[,3:10]`. The mean of each column is found by using `mean(df)`. How would you find the mean of each row? Why might this be interesting?

4.12 The data set `u2 (UsingR)` contains the time in seconds for albums released by the band U2 from 1980 to 1997. The data is stored in a list.

1. Make a boxplot of the song lengths by album. Which album has the most spread? Are the means all similar?
2. Use `sapply()` to find the mean song time for each album. Which album has the shortest mean? Repeat with the median. Are the results similar?
3. What are the three longest songs? The `unlist()` function will turn a list into a vector. First unlist the song lengths, then sort.

Could you use a data frame to store this data?

4.13 The data set `normtemp (UsingR)` contains measurements for 130 healthy, randomly selected individuals. The variable `temperature` contains body temperature, and `gender` contains the gender, coded `1` for male and `2` for female. Make layered densityplots of `temperature`, splitting the data by `gender`. Do the two distributions look to be the same?

4.14 What do you think this notation for data frames returns: `df[,]`?

4.3 Using model formula with multivariate data

In Example 4.5 we broke up a variable into two pieces based on the value of a second variable. This is a common task and works well. When the value of the second variable has many levels, it is more efficient to use the model-formula notation. We've already seen other advantages to this approach, such as being able to specify a data frame to find the variables using `data=` and to put conditions on the rows considered using `subset=`.

4.3.1 Boxplots from a model formula

In Example 4.4 the `inc` variable is discrete and not continuous, as it has been turned into a categorical factor by binning, using a bin size of $2,500. Rather than plot `gestation` versus `inc` with a scatterplot, as is done in a panel of Figure 4.4, a boxplot would be more appropriate. The boxplot allows us to compare centers and spreads much more easily. As there are nine levels to the income variable, we wouldn't want to specify the data with commands like `gestation[inc == 1]`. Instead, we can use the model formula `gestation ~ inc` with `boxplot()`. We read the formula as `gestation` is modeled by `inc`, which is interpreted by `boxplot()` by splitting the variable `gestation` into pieces corresponding to the values of `inc` and creating boxplots for each.

We use this approach three times. The last appears in Figure 4.6, where the argument `varwidth=TRUE` is specified to show boxes with width depending on the relative size of the sample.

```
> boxplot(gestation ~ inc, data=babies) # not yet
> boxplot(gestation ~ inc, subset = gestation != 999 & inc != 98,
+ data=babies)                   # better
> boxplot(gestation ~ inc, subset = gestation != 999 & inc != 98,
+ data=babies,varwidth=TRUE,     # variable width to see sizes
+ xlab="income level", ylab="gestation (days)")
```

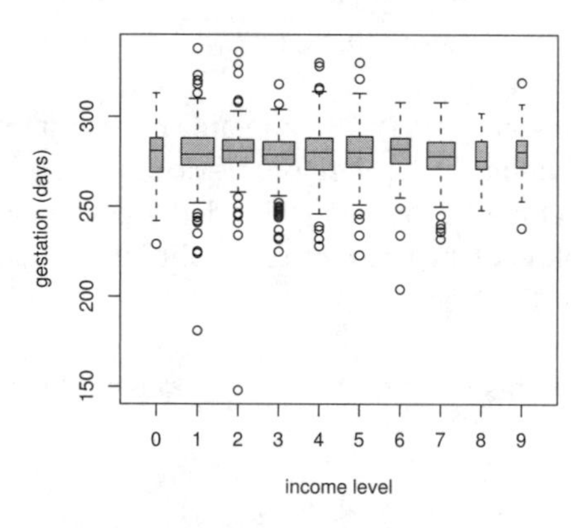

Figure 4.6 Boxplot of gestation times for income levels

4.3.2 The `plot()` function with model formula

Both `boxplot()` and `plot()` are generic functions allowing the programmers of R to write model-formula interfaces to them. The `boxplot()` function always draws boxplots, but we've seen already that the `plot()` function can draw many types of plots (depending on its first argument). In Example 3.6 the `plot()` command was called with a model formula of the type `numeric ~ numeric`, resulting in a scatterplot. If `x` and `y` are paired, numeric data vectors, then the model formula `y ~ x` represents the model $y_i = \beta_0 + \beta_1 x_i + \varepsilon_i$. The typical plot for viewing this type of model is the scatterplot.

A model formula of the type `numeric ~ factor` represents the statistical model $x_{ij} = \mu_i + \varepsilon_{ij}$. That is, for each level i of the factor, there is a sample, $x_{i1}, x_{i2}, \ldots, x_{in_i}$, with mean described by μ_i.

As this model says something about the means of the different samples, multiple boxplots are useful for viewing the data. A boxplot allows us to compare the medians, which are basically the mean if the data is not skewed. Consequently, it is the plot made when the `plot()` function encounters such a model formula. That is, if `x` is a numeric data vector and `f` a factor indicating which group the corresponding element of `x` belongs to, then the command `plot(x ~ f)` will create side-by-side boxplots of values of `x` split up by the levels of `f`.

For example, Figure 4.6 could also have been made with the commands

```
> plot(gestation ~ factor(inc), data=babies, varwidth=TRUE,
+ subset = gestation != 999 & inc !=98,
+ xlab="income level", ylab="gestation (days)")
```

The function `factor()` explicitly makes `inc` a factor and not a numeric data vector. Otherwise, the arguments are identical to those to the `boxplot()` function that created Figure 4.6.

4.3.3 Creating contingency tables with `xtabs()`

We saw in Example 4.1 how to make a three-way contingency table using the `table()` function starting with raw data. What if we had only the count data? How could we enter it in to make a three-way contingency table? We can enter the data in as a data frame and then use `xtabs()` to create contingency tables. The `xtabs()` function offers a formula interface as an alternative to `table()`.

The function `as.data.frame()` inverts what `xtabs()` and `table()` do. It will create a data frame with all possible combinations of the variable levels and a count of frequencies when called on a contingency table.

■ **Example 4.7: Seat-belt usage factors** The three-way table in Table 4.5 shows percentages of seat-belt usage for two years, broken down by type of law enforcement and type of car. Law enforcement is primary if a driver can be pulled over and ticketed for not wearing a seat belt, and secondary if a driver can be ticketed for this offense only if pulled over for another infraction. This data comes from a summary of the 2002 Moving Traffic Study as part of NOPUS (`http://www.nhtsa.gov`), which identified these two factors as the primary factors in determining seat-belt usage.

We show how to enter the data into `R` using a data frame, and from there

Table 4.5 Seat-belt data by type of law and vehicle

Enforcement	primary		secondary	
Year	2001	2002	2001	2002
Car type				
passenger	71	82	71	71
pickup	70	71	50	55
van/SUV	79	83	70	73

[a] source: 2002 NOPUS.

recreate the table.

Though this is not count data, the data entry is similar. First we turn the contingency table into a data frame by creating the proper variables. We enter the percentages in column by column. After doing this, we create variables for `car`, `year`, and `enforcement` that have values matching the percentages. Using `rep()` greatly reduces the work.

```
> percents = c(71,70,79,82,71,83,71,50,70,71,55,73)
> car = rep(c("passenger","pickup","van/suv"), 4)
> year = rep( rep(2001:2002,c(3,3)), 2)
> enforcement = rep(c("primary","secondary"), c(6,6))
> seatbelts = data.frame(percents, car, year, enforcement)
> seatbelts
  percents       car year enforcement
1       71 passenger 2001     primary
2       70    pickup 2001     primary
...
12      73   van/suv 2002   secondary
```

The `xtabs()` function creates contingency tables using a model-formula interface. The formula may or may not have a response. If no response is present, `xtabs()` will tabulate the data as though it has not been summarized in the way `table()` does. We have the equivalent of summarized data stored in `percents`, so we will use this variable as the response. The cross-classifying variables on the right side of the model formula are "added" in with +. The first two variables form the main tables; any additional ones are used to break up the contingency tables. For example:

```
> tab = xtabs(percents ~ car + year + enforcement, data = seatbelt)
> tab
, , enforcement = primary

           year
car          2001 2002
  passenger 71   82
  pickup    70   71
  van/suv   79   83
```

```
, , enforcement = secondary

            year
car          2001 2002
  passenger 71   71
  pickup    50   55
  van/suv   70   73
```

The `ftable()` command can again be used to flatten the tables. We specify the column variables to make the output look like Table 4.5.

```
> ftable(tab, col.vars=c("enforcement","year"))
          enforcement primary        secondary
          year           2001 2002      2001 2002
car
passenger                  71   82        71   71
pickup                     70   71        50   55
van/suv                    79   83        70   73
```

■

4.3.4 Manipulating data frames: `split()` and `stack()`

When a formula interface isn't available for a function, the `split()` function can be used to split up a variable by the levels of some factor. If `x` stores the data and `f` is a factor indicating which sample the respective data values belong to, then the command `split(x,f)` will return a list with top-level components containing the values of `x` corresponding to the levels of `f`. For example, the command `boxplot(split(x,f))` produces the same result as the command `boxplot(x ~ f)`.

Applying a single function to each component of the list returned by `split()` may also be done with the `tapply()` function. The basic format is

```
tapply(x, f, function)
```

The function can be a named function, such as `mean`, or one we define ourselves, as will be discussed in Chapter 6.

Inverse to `split()` is `stack()`, which takes a collection of variables stored in a data frame or list and stacks them into two variables. One contains the values, and the other indicates which variable the data originally came from. The function `unstack()` reverses this process and is similar to `split()`, except that it returns a data frame (and not a list), if possible.

For example, the data set `cancer (UsingR)` contains survival times for different types of cancer. The data is stored in a list, not a data frame, as the samples do not have the same length. We can create a data object, for which the model formula will work using `stack()`.

```
> cancer
$stomach
 [1]  124   42   25   45  412   51 1112   46  103  876  146  340
[13]  396
...
$breast
 [1] 1235   24 1581 1166   40  727 3808  791 1804 3460  719
> stack(cancer)
   values       ind
1     124   stomach
2      42   stomach
...
63   3460    breast
64    719    breast
```

The variable names in the output of `stack()` are always `values` to store the data and `ind` to indicate which sample the data is from. When we use `stack()`, it is important that the variables in the data frame or list have names, so that `ind` can indicate which variable the data is from.

4.3.5 Problems

4.15 The data set `MLBattend (UsingR)` contains attendance data for major league baseball between the years 1969 and 2000. For each `year`, make boxplots of `attendance`. Can you pick out two seasons that were shortened by strikes? (There were three, but the third is hard to see.)

4.16 The data set `MLBattend (UsingR)` contains several variables concerning attendance at major league baseball games from 1969 to 2000. Compare the mean number of runs scored per team for each league before and after 1972 (when the designated hitter was introduced). Is there a difference? Hint: the function `tapply()` can be used, as in

```
> tapply(runs.scored,league,mean)
   AL    NL
713.3 675.4
```

However, do this for the data before and after 1972.

4.17 The data set `npdb (UsingR)` contains malpractice-award information for the years 2000 to 2003 in the United States. The variable `ID` contains an identification number unique to a doctor. The command `table(table(ID))` shows that only 5% of doctors are involved in multiple awards. Perhaps these few are the cause of the large insurance payouts? How can we check graphically?

We'll make boxplots of the total award amount per doctor broken down by the number of awards that doctor has against him and investigate. First though, we need to manipulate the data.

1. The command `tmp = split(award,ID)` will form the list `tmp` with each

element corresponding to the awards for a given doctor. Explain what these commands do: `sapply(tmp,sum)` and `sapply(tmp,length)`.

2. Make a data frame with the command

```
> df = data.frame(sum = sapply(x,sum), number = sapply(x,length))
```

With this, create side-by-side boxplots of the total amount by a doctor broken down by the number of awards.

What do you conclude about these 5% being the main cause of the damages?

4.18 The data set `morley` contains measurements of the speed of light. Make side-by-side boxplots of the `Speed` variable for each experiment recorded by `Expt`. Are the centers similar or different? Are the spreads similar or different?

4.19 For the data set `PlantGrowth`, make boxplots of the `weight` variable for each level of `group`. Are the centers similar or different? Are the spreads similar or different?

4.20 For the data set `InsectSprays`, make boxplots of the `count` variable for levels `C`, `D`, and `E`. Hint: These can be found with a command such as

```
> spray %in% c("C","D","E")
```

Use this with the model notation and the argument `subset=` when making the boxplots.

4.21 The `pairs()` function also has a model-formula interface. We can redo Example (4.4) with the command

```
> pairs(~ gestation + age + wt + inc, data = babies,
+ subset = gestation < 999 & age < 99 & inc < 98)
```

For the `UScereal` (`MASS`) data set, use the formula interface to make a scatterplot matrix of the variables `calories`, `carbo`, `protein`, `fat`, `fibre`, and `sugars`. Which relationships show a linear trend?

4.4 Lattice graphics

In Figure 4.2 various colors and plotting characters were used to show whether the third variable, `smoke`, affected the relationship between gestation time and birth weight. As we noted, the figure was a bit crowded for this approach. A better solution would be to create a separate scatterplot for each level of the third variable. These graphs can be made using the `lattice` graphics package.

The add-on package `lattice` is modeled after Cleveland's Trellis graphics concepts and uses the newer, low-level plotting commands available in the `grid` package. These two are recommended packages and should be part of a standard R installation.

The graphics shown below are useful and easy to create. Many other usages are possible. If the package is loaded (with `library(lattice)`), a description of lattice graphics is available in the help system under `?Lattice`. The help page `?xyplot` also contains extensive documentation.*

The basic idea is that the graphic consists of a number of panels. Each panel corresponds to some value(s) of a conditioning variable. The lattice graphing functions are called using the model-formula interface. The formulas have the format

```
response ~ predictor | condition
```

The `response` variable is not always present. For univariate graphs, such as histograms, it is not given; for bivariate graphs, such as scatterplots, it is. The optional `condition` variable is either a factor or a numeric value. If it is a factor, there is a separate panel for each level. If it is numeric, "shingles" are created that split up the range of the variable to make several panels. Each panel uses the same scale for the axes, allowing for easy comparison between the graphics.

Before beginning, we load the `lattice` package and override the default background color, as it is not the best for reproduction.

```
> library(lattice)                # load in the package
> trellis.device(bg="white")      # set background to white.
```

This can also be achieved by setting the lattice "theme," using `options()`:

```
> options(lattice.theme = "col.whitebg")
```

If we desire, this command can be placed into a startup file† to change the default automatically.

What follows are directions for making several types of graphs using the `lattice` package.

Histograms Histograms are univariate. The following command shows histograms of birth weight (`wt`) for different levels of the factor `smoke`. Note that the response variable is left blank.

```
> histogram( ~ wt | factor(smoke), data=babies,
+ subset = wt != 999, type="density")
```

The last argument, `type="density"`, makes the total area of the histogram add to 1. (The argument `prob=TRUE` was used with `hist()`.)

Densityplots The density estimate is an alternative to a histogram. Density estimates are graphed with the `densityplot()` function, as in

* Some online documentation exists in Volume 2/2 of the R News newsletter (http://cran.r-project.org/doc/Rnews) and the grid package author's home page (http://www.stat.auckland.ac.nz/~paul/grid/grid.html).

† See Appendix A for information about startup files.

```
densityplot( ~ wt | factor(smoke), data=babies)
```

Boxplots Boxplots can be univariate or multivariate.

The relationship between gestation time and income was investigated in Figure 4.6. A similar graph can be made with the function `bwplot()`. (A boxplot is also known as a box-and-whisker plot, hence the name.)

```
> bwplot(gestation ~ factor(inc), data=babies,
+ subset = gestation != 999 )
```

Figure 4.7 shows this broken down further by the smoking variable.

```
> bwplot(gestation ~ factor(inc) |factor(smoke), data=babies,
+ subset = gestation != 999)
```

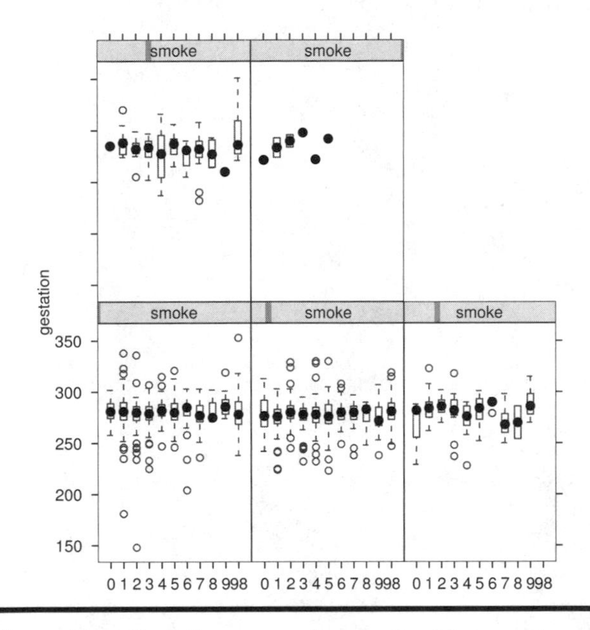

Figure 4.7 Gestation versus income level by smoking status

Scatterplots The lattice function `xyplot()` produces scatterplots. As these are graphs of bivariate relationships, a response variable is needed. This example will plot the relationship between gestation time and weight for each level of the factor `smoke`.

```
> xyplot(wt ~ gestation | factor(smoke), data = babies,
+ subset = (wt != 999 & gestation != 999))
```

Scatterplots with regression line The last example can be improved if we add a regression line, as was done using `abline()`. However, the panels of the scatterplot do not allow this type of command directly. Rather, lines are added as the panels are drawn. To override the default graphic, we specify a panel-drawing function. The following command creates a panel function called `plot.regression`. (User-defined functions will be discussed in Chapter 6.)

```
> plot.regression = function(x,y) {
+ panel.xyplot(x,y)                  # make x-y plot
+ panel.abline(lm(y ~ x))            # add a regression line
+ }
```

This function is used as the value of the `panel=` argument:

```
> xyplot(wt ~ gestation | factor(smoke), data = babies,
+ subset = (wt != 999 & gestation != 999),
+ panel = panel.regression)       # a new panel function
```

Figure 4.8 contains the graphic. We might ask, "Are the slopes similar?" and, "What would it mean if they aren't?"

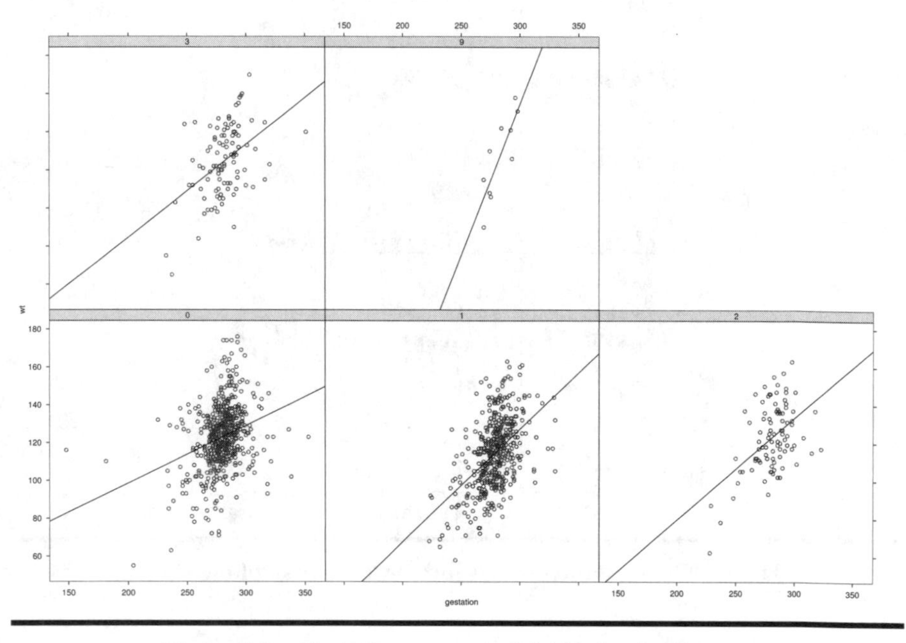

Figure 4.8 Gestation versus weight for levels of `smoke`

4.4.1 Problems

4.22 The `kid.weights (UsingR)` data set contains measurements for several children. There is a clear relationship between height and weight. Break the data down by the `age` of the child. As `age` is numeric, it helps to turn it into a factor: `cut(age/12,3*(0:4))`. Do you see the same trend for all the age groups?

4.23 For the `kid.weights (UsingR)` data set, explore the relationship of weight and gender for the age ranges 0-3, 3-6, 6-9, and 9-12. Is the relationship always the same?

4.24 The `female.inc (UsingR)` data set contains income and race data for females in the United States for the year 2001. Make a boxplot of the `income` variable broken up by the `race` variable. Do there appear to be any major differences among the races? In addition, compute the summary statistics for each racial group in the data.

4.25 The data set `ToothGrowth` contains measurements of tooth growth for different dosages of a supplement. Use the function `bwplot()` to make boxplots of `len` for each level of `dose` broken up by the levels of `supp`. You need to make `dose` a factor first. Also, repeat the graphic after reversing the role of the two factors.

4.26 The `carsafety (UsingR)` data set contains data on accident fatalities per million vehicles for several types of cars. Make boxplots of `Driver.deaths` broken down by `type`. Which type of car has the largest spread? Repeat using the variable `Driver.deaths + Other.deaths`. Which has the largest spread? Are there any trends?

4.27 The data set `Orange` contains data on the growth of five orange trees. Use `xyplot()` to make a scatterplot of the variable `circumference` modeled by `age` for each level of `Tree`. Are the growth patterns similar?

4.28 The data set `survey` (`MASS`) contains survey information on students.

1. Make scatterplots using `xyplot()` of the writing-hand size (`Wr.Hnd`) versus non-writing-hand size (`NW.Hnd`), broken down by gender (`Sex`). Is there a trend?
2. Make boxplots using `bwplot()` of `Pulse()` for the four levels of `Smoke` broken down by gender `Sex`. Are there any trends? Differences?

Do you expect any other linear relationships among the variables?

4.5 Types of data in R

(This section may be skipped initially. It is somewhat technical and isn't used directly in the remainder of the text.)

The basic structures for storing data in R are data vectors for univariate data, matrices and data frames for rectangular data, and lists for more general needs. Each data vector must contain the same type of data. The basic types are numeric, logical, and character.

Many objects in R also have a class attribute given by the class() function. It is the class of an object that is used by R to give different meanings to generic functions, such as plot() and summary().

4.5.1 Factors

Factors should also be considered to be another storage type, as they are handled differently than a data vector. Recall, factors keep track of categorical data, as can a data vector, yet unlike other data types, their values can come only from the specified levels of the factor. Manipulating the levels requires knowing how to do a few things: creating a factor, listing the levels, adding a level, dropping levels, and ordering levels.

Creating factors

Factors are made with factor() or as.factor(). These functions coerce the data into a factor and define the levels of the new factor. For example:

```
> x = 1:3; fac=letters[1:3]
> factor(x)                    # default uses sorted order
[1] 1 2 3
Levels: 1 2 3
> factor(fac)                  # same with characters
[1] a b c
Levels: a b c
```

When a factor is printed, the levels also appear.

It is important to realize that these are factors and not numbers. For example, we can't add factors.

```
> x + factor(x)
[1] NA NA NA
Warning message:
"+" not meaningful for factors in: Ops.factor(x, factor(x))
```

Adding levels: levels= and levels()

When defining a factor, the levels= argument can be used to specify the levels.

```
> factor(x,levels=1:10)        # add more levels than 3
[1] 1 2 3
Levels: 1 2 3 4 5 6 7 8 9 10
> factor(x,levels=fac)         # get NA if not in levels
[1] <NA> <NA> <NA>
Levels: a b c
```

The values of data vector being coerced into a factor should match those specified to levels=.

The levels() function can list the levels of a factor directly and allows us to change or add to the levels.

```
> x = factor(1:5)
> x
[1] 1 2 3 4 5
Levels: 1 2 3 4 5
> levels(x)                        # character vector
[1] "1" "2" "3" "4" "5"
```

The `levels()` function can also change the levels of a factor. As `levels()` changes an attribute of the data vector, like `names()`, it appears on the other side of the assignment. For example:

```
> x = factor(1:3)
> x
[1] 1 2 3
Levels: 1 2 3
> levels(x) = letters[1:3]
> x
[1] a b c
Levels: a b c
> levels(x) = 1:10                 # add more levels
> x
[1] 1 2 3
Levels: 1 2 3 4 5 6 7 8 9 10
```

The number of new levels must be at least as great as the current number.

Dropping levels of factors

If we take a subset from a factor, we may want the levels to be shortened to match only those values in the subset. By default, our subset will contain all the old levels. Releveling can be done in a few ways. The `levels()` function won't work, as that expects the same or more levels. However, `factor()` will work:

```
> x = factor(letters[1:5])
> x
[1] a b c d e
Levels: a b c d e
> x[1:3]
[1] a b c
Levels: a b c d e
> factor(x[1:3])
[1] a b c
Levels: a b c
```

A more direct way to relevel is to use the `drop=TRUE` argument when doing the extraction:

```
> x[1:3,drop=TRUE]
[1] a b c
Levels: a b c
```

Ordering levels in a factor

The ordering of the levels is usually done alphabetically (according to the result of `sort(unique(x))`. If we want a specific ordering of the levels, we can set the levels directly in the `levels=` argument of `factor()` or using `levels()`. For example:

```
> l = letters[1:5]                 # the first 5 letters
> factor(l)                        # order of levels is by ASCII
[1] a b c d e
Levels: a b c d e
> factor(l,levels=rev(l))          # specific order of levels
[1] a b c d e
Levels: e d c b a
```

If we want a certain level to be considered first, the `relevel()` function can be used, as in `relevel(factor,ref=...)`.

4.5.2 Coercion of objects

Coercion is the act of forcing one data type into another. This is typically done with an "`as.`" function, such as `as.data.frame()`, although we'll see that some coercion happens implicitly.

Coercing different data types

We use the language data vector to refer to an indexed set of data of length n all of the same type. The types we have encountered are "numeric" for storing numbers, "character" for storing strings, "factor" for storing factors, and "logical" for storing logical expressions. We can convert from one type to another with the "`as.`" functions, such as `as.numeric()` or `as.character()`. We can check the type with the corresponding "`is.`" functions. For example:

```
> x = 1:5
> c(is.numeric(x),is.character(x),is.factor(x),is.logical(x))
[1]  TRUE FALSE FALSE FALSE
> x = as.character(x)
> x
[1] "1" "2" "3" "4" "5"
> c(is.numeric(x),is.character(x),is.factor(x),is.logical(x))
[1] FALSE  TRUE FALSE FALSE
> x = as.factor(x)
> x
[1] 1 2 3 4 5
Levels: 1 2 3 4 5
> c(is.numeric(x),is.character(x),is.factor(x),is.logical(x))
[1] FALSE FALSE  TRUE FALSE
> as.logical(x)
[1] NA NA NA NA NA
```

Each type prints differently. When coercion fails, a value of `NA` is returned. The coercion to logical is picky. Values like 0 or "F" or "FALSE" will coerce to `FALSE`, but not values like "0" (a character) or "f."

A caveat: although we use the term "data vector" to describe these data sets, the `is.vector()` function does not consider factors to be vectors.

Coercing factors to other types

Coercing factors can be tricky, as they print differently from how they are stored. An artificial example will illustrate:

```
> f = factor(letters[1:5])
> f
[1] a b c d e
Levels: a b c d e
> unclass(f)                        # shows how f is stored
[1] 1 2 3 4 5
attr(,"levels")
[1] "a" "b" "c" "d" "e"
> as.vector(f)                      # coerce to vector type
[1] "a" "b" "c" "d" "e"
> as.integer(f)                     # coerce to integers
[1] 1 2 3 4 5
```

The `unclass()` function shows that a factor is stored using an internal coding involving numbers. The attribute `"levels"` gives the levels. The coercion to a vector gives a character vector in this case; the coercion to an integer returns the internal codes.

The final one can cause confusion. Consider this example:

```
> g = factor(2:4)
> g
[1] 2 3 4
Levels: 2 3 4
> as.numeric(g)
[1] 1 2 3
> as.numeric(as.character(g))
[1] 2 3 4
```

The `as.numeric()` command by itself returns the codes when applied to a factor. To get the levels as numbers, we convert to character, then to numeric:

```
> as.numeric(as.character(x))
[1] 2 3 4
```

As a convenience, when factors are used to label graphs, say with the `labels=` argument of the `text()` function, this conversion is done automatically.

Coercing vectors, data frames, and lists

There are "`as.`" functions to coerce data storage from one type to another. But they can't do all the work.

Coercing a vector to a data frame If we want to coerce a vector into a data frame we can do it with the function `as.data.frame()`. If `x` is a vector, then `as.data.frame(x)` will produce a data frame with one column vector. By default, strings will be coerced to factors.

We may want to create a matrix from our numbers before coercing to a data frame. This can be achieved by setting the `dim()` attribute of the data vector. The `dim()` function takes a vector where each component specifies the size of that dimension. As usual, rows first, then columns.

```
> x = 1:8
> dim(x) = c(2,4)                    # 2 rows 4 columns
> x                                  # column by column
     [,1] [,2] [,3] [,4]
[1,]    1    3    5    7
[2,]    2    4    6    8
> as.data.frame(x)                    # turn matrix to data frame
  V1 V2 V3 V4
1  1  3  5  7
2  2  4  6  8
```

Coercing a data frame or list into a vector To coerce a list or data frame into a vector, we should start out with all the same type of data, otherwise the data will be coerced implicitly. The `unlist()` function, when applied to a list, will form a vector of all the "atomic" components of the list, recursively traversing through a list to do so. When applied to a data frame it goes column by column creating a vector. For example:

```
> x = 1:8;dim(x) = c(2,4);df = data.frame(x)
> df
  X1 X2 X3 X4
1  1  3  5  7
2  2  4  6  8
> unlist(df)
X11 X12 X21 X22 X31 X32 X41 X42
  1   2   3   4   5   6   7   8
```

Chapter 5

Describing populations

Statistical inference is the process of forming judgments about a population based on a sample from the population. In this chapter we describe populations and samples using the language of probability.

5.1 Populations

In order to make statistical inferences based on data we need a probability model for the data. Consider a univariate data set. A single data point is just one of a possible range of values. This range of values will be called the population. We use the term **random variable** to be a random number drawn from a population. A data point will be a realization of some random variable. We make a distinction between whether or not we have observed or realized a random variable. Once observed, the value of the random variable is known. Prior to being observed, it is full of potential—it can be any value in the population it comes from. For most cases, not all values or ranges of values of a population are equally likely, so to fully describe a random variable prior to observing it, we need to indicate the probability that the random variable is some value or in a range of values. We refer to a description of the range and the probabilities as the **distribution of a random variable**.

By probability we mean some number between 0 and 1 that describes the likelihood of our random variable having some value. Our intuition for probabilities may come from a physical understanding of how the numbers are generated. For example, when tossing a fair coin we would think that the probability of heads would be one-half. Similarly, when a die is rolled the probability of rolling a ⚃ would be one-sixth. These are both examples in which all outcomes are equally likely and finite in number. In this case, the probability of some event, a collection of outcomes, is the number of outcomes in the event divided by the total number of outcomes. In particular, this says the probability of any event is between 0 and 1.

For situations where our intuition comes about by performing the same action over and over again, our idea of the probability of some event comes from a proportion of times that event occurs. For example, the batting average of a baseball player is a running proportion of a player's success at bat. Over the course of a season, we expect this number to get closer to the probability that an official at bat will be a success. This is an example in which a *long-term frequency* is used to give a probability.

For other populations, the probabilities are simply assigned or postulated, and our model is accurate as far as it matches the reality of the data collected.

We indicate probabilities using a $\mathsf{P}()$ and random variables with letters such as X. For example, $\mathsf{P}(X \leq 5)$ would mean the probability the random variable X is less than or equal to 5.

5.1.1 Discrete random variables

Numeric data can be discrete or continuous. As such, our model for data comes in the same two flavors.

Let X be a discrete random variable. The range of X is the set of all k where $\mathsf{P}(X = k) > 0$. The distribution of X is a specification of these probabilities. Distributions are not arbitrary, as for each k in the range, $\mathsf{P}(X = k) > 0$ and $\mathsf{P}(X = k) \leq 1$. Furthermore, as X has some value, we have $\sum_k \mathsf{P}(X = k) = 1$.

Here are a few examples for which the distribution can be calculated.

■ **Example 5.1: Number of heads in two coin tosses** If a coin is tossed two times we can keep track of the outcome as a pair. (H,T), for example, denotes "heads" then "tails." The set $\{(H,H),(H,T),(T,H),(T,T)\}$ contains all possible outcomes. If X is the number of heads, then X is either 0, 1, or 2. Intuitively, we know that for a fair coin all the outcomes have the same probability, so $\mathsf{P}(X = 0) = 1/4$, $\mathsf{P}(X = 1) = 1/2$, and $\mathsf{P}(X = 2) = 1/4$. ■

■ **Example 5.2: Picking balls from a bag** Imagine a bag with N balls, of which R are red and $N - R$ are green. We pick a ball, note its color, replace the ball, and repeat. Let X be the number of red balls. As in the previous example, X is 0, 1, or 2. The probability that $X = 2$ is intuitively $(R/N) \cdot (R/N)$ as R/N is the probability of picking a red ball on any one pick. The probability that $X = 0$ is $((N-R)/N)^2$ by the same reasoning, and as all probabilities add to 1, $\mathsf{P}(X = 1) = 2(R/N)((N-R)/N)$. This specifies the distribution of X.

The binomial distribution describes the result of selecting n balls, not two. ■

The intuition that leads us to multiply two probabilities together is due to the two events being independent. Two events are independent if knowledge that one occurs doesn't change the probability of the other occurring. Two events are disjoint if they can't both occur for a given outcome. Probabilities add with disjoint events.

■ **Example 5.3: Specifying a distribution** We can specify the distribution of a discrete random variable by first specifying the range of values and then assigning to each k a number $p_k = \mathsf{P}(X = k)$ such that $\sum p_k = 1$ and $p_k \geq 0$. To visualize a physical model where this can be realized, imagine making a pie chart with areas proportional to p_k, placing a spinner in the middle, and spinning. The ending position determines the value of the random variable. ■

Figure 5.1 shows a **spike plot** of a distribution and a spinner model to realize values of X. A spike plot shows the probabilities for each value in the range of X as spikes, emphasizing the discreteness of the distribution. The spike plot is made with the following commands:

```
> k = 0:4
> p = c(1,2,3,2,1)/9
> plot(k,p,type="h",xlab="k", ylab="probability",ylim=c(0,max(p)))
> points(k,p,pch=16,cex=2)         # add the balls to top of spike
```

The argument `type="h"` plots the vertical lines of the spike plot.

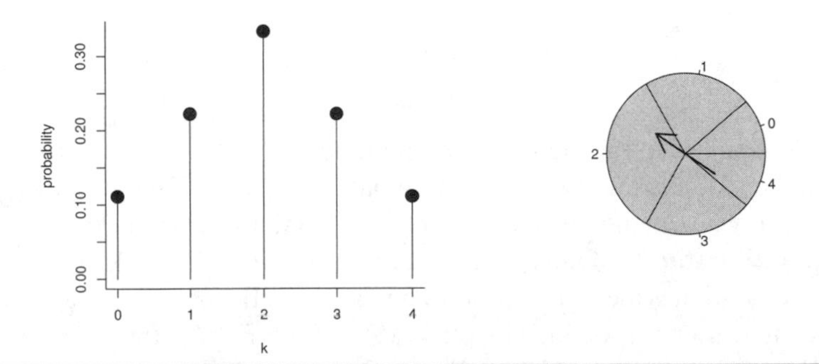

Figure 5.1 Spike plot of distribution of X and a spinner model to realize values of X with the specified probabilities

Using `sample()` *to generate random values*

R will generate observations of a discrete random variable with the `sample()` function. If the data vector `k` contains the values we are sampling from, and `p` contains the probabilities of each value being selected, then the command `sample(k,size=1,prob=p)` will select one of the values of `k` with probabilities specified by p.

For example, the number of heads for two coin tosses can be simulated as follows:

```
> k = 0:2
> p = c(1,2,1)/4
> sample(k,size=1,prob=p)
[1] 0
> sample(k,size=1,prob=p)
[1] 2
```

The default probabilities for `prob=` make each value of `k` equally likely. We can use this to simulate rolling a pair of dice and adding their values:

```
>  sample(1:6,size=1) + sample(1:6,size=1)
[1] 12
>  sample(1:6,size=1) + sample(1:6,size=1)
[1] 5
```

The mean and standard deviation of a discrete random variable

For a data set, the mean and standard deviation are summaries of center and spread. For random variables these concepts carry over, though the definitions are different.

The **population mean** is denoted by μ (the Greek letter *mu*). If X is a random variable with this population, then the mean is also called the "the expected value of X" and is written $\mathsf{E}(X)$. A formula for the expected value of a discrete random variable is

$$\mu = \mathsf{E}(X) = \sum_k k\mathsf{P}(X = k). \tag{5.1}$$

This is a weighted average of the values in the range of X.

On the spike plot, the mean is viewed as a balancing point if there is a weight assigned to each value in the range of X proportional to the probability.

The **population standard deviation** is denoted by σ (the Greek letter *sigma*). The standard deviation is the square root of the variance. If X is a discrete random variable, then its variance is defined by $\sigma^2 = \mathsf{VAR}(X) = \mathsf{E}((X-\mu)^2)$. This is the expected value of the random variable $(X-\mu)^2$. That is, the population variance measures spread in terms of the expected squared distance from the mean.

5.1.2 Continuous random variables

Continuous data is modeled by continuous random variables. For a continuous random variable X, it is not fruitful to specify probabilities like $\mathsf{P}(X = k)$ for each value in the range, as the range has too many values. Rather, we specify probabilities based on the chance that X is in some interval. For example, $\mathsf{P}(a < X \leq b)$, which would be the chance that the random variable is more than a but less than or equal to b.

Rather than try to enumerate values for all a and b, these probabilities are given in terms of an area for a specific picture.

A function $f(x)$ is the **density** of X if, for all b, $\mathsf{P}(X \leq b)$ is equal to the area under the graph of f and above the x-axis to the left of b. Figure 5.2 shows this by shading the area under f to the left of b. Although in most cases computing these areas is more advanced than what is covered in this text, we can find their values for many different densities using the appropriate group of functions in R.

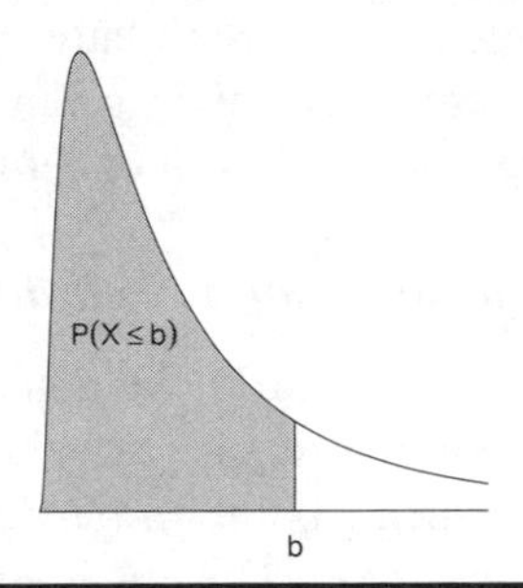

Figure 5.2 $\mathsf{P}(X \leq b)$ **is defined by the area to left of** b **under the density of** X

Using our intuitive notions of probability, for $f(x)$ to be a density of X the total area under $f(x)$ should be 1 and $f(x) \geq 0$ for all x. Otherwise, some intervals could get negative probabilities. Areas can also be broken up into pieces, as Figure 5.3 illustrates, showing $\mathsf{P}(a < X \leq b) = \mathsf{P}(X \leq b) - \mathsf{P}(X \leq a)$. This reasoning also gives the useful complement rule: for any b, $\mathsf{P}(X \leq b) = 1 - \mathsf{P}(X > b)$.

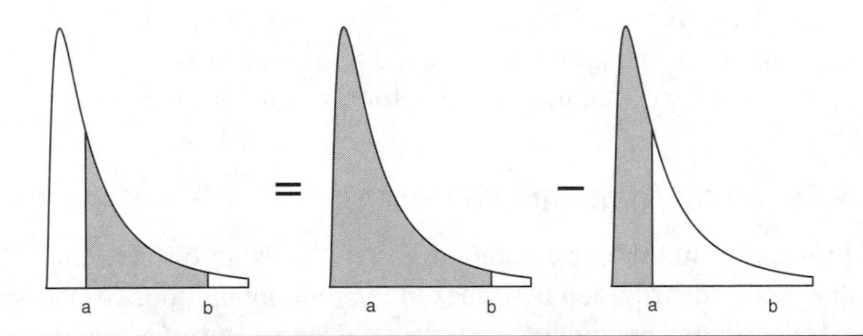

Figure 5.3 **Shaded areas can be broken into pieces and manipulated. This illustrates** $\mathsf{P}(a < X \leq b) = \mathsf{P}(X \leq b) - \mathsf{P}(X \leq a)$.

For example, the uniform distribution on $[0,1]$ has density $f(x) = 1$ on the interval $[0,1]$ and is 0 otherwise. Let X be a random variable with this density.

Then $\mathsf{P}(X \leq b) = b$ if $0 \leq b \leq 1$, as the specified area is a rectangle with length b and height 1. As well, $\mathsf{P}(X > b) = 1 - b$ for the same reason. Clearly, we have $\mathsf{P}(X \leq b) = 1 - \mathsf{P}(X > b)$.

The p.d.f. and c.d.f.

For a discrete random variable it is common to define a function $f(k)$ by $f(k) = \mathsf{P}(X = k)$. Similarly, for a continuous random variable X, it is common to denote the density of X by $f(x)$. Both usages are called **p.d.f.**'s. For the discrete case, p.d.f. stands for probability *distribution* function, and for the continuous case, probability *density* function. The cumulative distribution function, **c.d.f.**, is $F(b) = \mathsf{P}(X \leq b)$. In the discrete case this is given by $\sum_{k \leq b} \mathsf{P}(X = k)$, and in the continuous case it is the area to the left of b under the density $f(x)$.

The mean and standard deviation of a continuous random variable

The concepts of the mean and standard deviation apply for continuous random variables, although their definitions require calculus. The intuitive notion for the mean of X is that it is the balancing point for the density of X. The notation μ or $\mathsf{E}(X)$ is used for the mean, and σ or $\mathsf{SD}(X)$ is used for the standard deviation.

If X has a uniform distribution on $[0,1]$, then the mean is 1/2. This is clearly the balancing point of the graph of the density, which is constant on the interval. The variance can be calculated to be 1/12, so σ is about .289.

Quantiles of a continuous random variable

The quantiles of a data set roughly split the data by proportions. Let X be a continuous random variable with positive density. Referring to Figure 5.2, we see that for any given area between 0 and 1 there is a b for which the area to the right of b under f is the desired amount. That is, for each p in $[0,1]$ there is a b such that $\mathsf{P}(X \leq b) = p$. This defines the p-quantile or $100 \cdot p$ percentile of X. The quantile function is inverse to the c.d.f., as it returns the x value for a given area, whereas the c.d.f. returns the area for a given x value.

5.1.3 Sampling from a population

Our probability model for a data point is that it is an observation of a random variable whose distribution describes the parent population. To perform statistical inference about a parent population, we desire a **sample** from the population. That is, a sequence of random variables $X_1, X_2, \ldots, X_n$. A sequence is **identically distributed** if each random variable has the same distribution. A sequence is **independent** if knowing the value of some of the random variables does not give additional information about the distribution of the others. A sequence that is both independent and identically distributed is called an **i.i.d.** sequence, or a **random sample**.

Toss a coin n times. If we let X_i be 1 for a heads on the ith coin toss and 0 otherwise, then clearly $X_1, X_2, \ldots, X_n$ is an *i.i.d.* sequence. For the spinner analogy of generating discrete random variables, the different numbers will be *i.i.d.* if the spinner is spun so hard each time that it forgets where it started and is equally likely to stop at any angle.

If we get our random numbers by randomly selecting from a finite population, then the values will be independent if the sampling is done with replacement. This might seem counterintuitive, as there is a chance a member is selected more than once, so the values seem dependent. However, the distribution of a future observation is not changed by knowing a previous observation.

Random samples generated by `sample()`

The `sample()` function will take samples of size `n` from a discrete distribution by specifying `size=n`. The sample will be done with replacement if we specify `replace=TRUE`. This is important if we want to produce an *i.i.d.* sample. The default is to sample without replacement.

```
## toss a coin 10 times. Heads=1, tails=0
> sample(0:1,size=10,replace=TRUE)
 [1] 0 0 1 1 1 1 1 0 1 0
> sample(1:6,size=10,replace=TRUE) ## roll a die 10 times
 [1] 1 4 2 2 2 1 4 6 4 4
## sum of dice roll 10 times
> sample(1:6,size=10,replace=TRUE) + sample(1:6,size=10,replace=TRUE)
 [1]  7  7  7  9 12  4  7  9  5  4
```

■ Example 5.4: Public-opinion polls as random samples

The goal of a public-opinion poll is to find the proportion of a target population that shares a given attitude. This is achieved by selecting a sample from the target population and finding the sample proportion who have the given attitude. A public-opinion poll can be thought of as a random sample from a target population if each person polled is randomly chosen from the entire population with replacement. Assume we know that the target population of 10,000 people has 6,200 that would answer "yes" to our survey question. Then a sample of size 10 could be generated by

```
> sample(rep(0:1,c(3200,6800)),size=10,replace=T)
 [1] 1 0 1 0 1 1 1 1 1 0
```

The `rep()` function produces 10,000 values: 3,200 0's and 6,800 1's.

The target population is different from the "population," or distribution, of the random variables. For the responses, the possible values are coded with a 0 or 1 with respective probabilities $1 - p$ and p. Using this distribution, a random sample can also be produced by specifying the probabilities using `prob=`:

```
> sample(0:1,size=10,replace=T,prob=c(1-.62,.62))
 [1] 0 1 0 1 0 0 0 1 1 0
```

■

5.1.4 Sampling distributions

A statistic is a value derived from a random sample. Examples are the sample mean, $\bar{X} = (X_1 + X_2 + \cdots + X_n)/n$, and the sample median. Since a statistic depends on a random sample, it, too, is a random variable. To emphasize this, we use a capital $\bar{X}$. The distribution of a statistic is known as its **sampling distribution**.

The sampling distribution of a statistic can be quite complicated. However, for many common statistics, properties of the sampling distribution are known and are related to the population parameters. For example, the sample mean of a random sample has

$$\mathsf{E}(\bar{X}) = \mu_{\bar{X}} = \mu \quad \text{and} \quad \mathsf{SD}(\bar{X}) = \sigma_{\bar{X}} = \frac{\sigma}{\sqrt{n}}.$$

That is, the mean of $\bar{X}$ is the same as the mean of the parent population, and the standard deviation of $\bar{X}$ is related to the standard deviation of the parent population, but it differs as it is smaller by a factor of $1/\sqrt{n}$. These facts allow us to use $\bar{X}$ to make inferences about the population mean.

5.1.5 Problems

5.1 Toss two coins. Let X be the resulting number of heads and Y be the number of tails. Find the distribution of each.

5.2 Roll a pair of dice. Let X be the largest value shown on the two dice. Use `sample()` to simulate five values of X.

5.3 The National Basketball Association lottery to award the first pick in the draft is held by putting 1,000 balls into a hopper and selecting one. The teams with the worst records the previous year have a greater proportion of the balls. The data set `nba.draft (UsingR)` contains the ball allocation for the year 2002. Use `sample()` with `Team` as the data vector and `prob=Balls` to simulate the draft. What team do you select? Repeat until Golden State is chosen. How long did it take?

5.4 Let $f(x) = x$ for $0 \leq x \leq \sqrt{2}$ be the p.d.f. of a triangular random variable X. Find $\mathsf{P}(X \leq b)$ for $0 \leq b \leq \sqrt{2}$.

5.5 Let X have the uniform distribution on $[0, 1]$. That is, it has density $f(x) = 1$ for $0 \leq x \leq 1$. For $0 \leq p \leq 1$ find the quantile function that returns b, where $\mathsf{P}(X \leq b) = p$.

5.6 Repeat the previous problem for the triangular distribution with density

$f(x) = x$ for $0 \leq x \leq \sqrt{2}$.

5.7 Toss two coins. Let X be the number of heads and Y the number of tails. Are X and Y independent?

5.2 Families of distributions

In statistics there are a number of distributions that come in families. Each family is described by a function that has a number of parameters characterizing the distribution. For example, the uniform distribution is a continuous distribution on the interval $[a,b]$ that assigns equal probability to equal-sized areas in the interval. The parameters are a and b, the endpoints of the intervals.

5.2.1 The `d`, `p`, `q`, and `r` functions

R has four types of functions for getting information about a family of distributions. The "d" functions return the p.d.f. of the distribution, whereas the "p" functions return the c.d.f. of the distribution. The "q" functions return the quantiles, and the "r" functions return random samples from a distribution.

These functions are all used similarly. Each family has a name and some parameters. The function name is found by combining either d, p, q, or r with the name for the family. The parameter names vary from family to family but are consistent within a family.

For example, the uniform distribution on $[a,b]$ has two parameters. The family name is `unif`. In R the parameters are named `min=` and `max=`.

```
> dunif(x=1, min=0, max=3)
[1] 0.3333
> punif(q=2, min=0, max=3)
[1] 0.6667
> qunif(p=1/2, min=0, max=3)
[1] 1.5
> runif(n=1, min=0, max=3)
[1] 1.260
```

The above commands are for the uniform distribution on $[0,3]$. They show that the density is $1/3$ at $x = 1$ (as it is for all $0 \leq x \leq 3$); the area to the left of 2 is $2/3$; the median or .5-quantile is 1.5; and a realization of a random variable is 1.260. This last command will vary each time it is run.

It is useful to know that the arguments to these functions can be vectors, in which case other arguments are recycled. For example, multiple quantiles can be found at once. These commands will find the quintiles:

```
> ps = seq(0,1,by=.2)                  # vector
> names(ps) = as.character(seq(0,100,by=20)) # give names
```

```
> qunif(ps, min=0, max=1)
  0  20  40  60  80 100
0.0 0.2 0.4 0.6 0.8 1.0
```

This command, on the other hand, will find five uniform samples from five different distributions.

```
> runif(5, min=0, max=1:5)       # recycle min,
[1] 0.6331 0.6244 1.9252 2.8582 3.0076
```

5.2.2 Binomial, normal, and some other named distributions

There are a few basic distributions that are used in many different probability models: among them are the Bernoulli, binomial, and normal distributions.

Bernoulli random variables

A **Bernoulli random variable** X is one that has only two values: 0 or 1. The distribution of X is characterized by $p = \mathsf{P}(X = 1)$. We use $\mathsf{Bernoulli}(p)$ to refer to this distribution. Often the term "success" is given to the event when $X = 1$ and "failure" to the event when $X = 0$. If we toss a coin and let X be 1 if a heads occurs, then X is a Bernoulli random variable where the value of p would be $1/2$ if the coin is fair. A sequence of coin tosses would be an *i.i.d.* sequence of Bernoulli random variables, also known as a sequence of Bernoulli trials.

A Bernoulli random variable has a mean $\mu = p$ and a variance $\sigma^2 = p(1-p)$.

In R, the `sample()` command can be used to generate random samples from this distribution. For example, to generate ten random samples when $p = 1/4$ can be done with

```
> n = 10; p = 1/4
> sample(0:1, size=n, replace=TRUE,prob=c(1-p,p))
[1] 0 0 0 0 0 0 0 1 0 0
```

Binomial random variables

A **binomial random variable** X counts the number of successes in n Bernoulli trials. There are two parameters that describe the distribution of X: the number of trials, n, and the success probability, p. Let $\mathsf{Binomial}(n, p)$ denote this distribution. The possible range of values for X is $0, 1, \ldots, n$. The distribution of X is known to be

$$\mathsf{P}(X = k) = \binom{n}{k} p^k (1-p)^{n-k}.$$

The term $\binom{n}{k}$ is called the binomial coefficient and is defined by

$$\binom{n}{k} = \frac{n!}{(n-k)!k!},$$

where $n!$ is the factorial of n, or $n \cdot (n-1) \cdots 2 \cdot 1$. By convention, $0! = 1$. The binomial coefficient, $\binom{n}{k}$, counts the number of ways k objects can be chosen from n distinct objects and is read "n choose k." The `choose()` function finds the binomial coefficients.

The mean of a $\mathsf{Binomial}(n, p)$ random variable is $\mu = np$, and the standard deviation is $\sigma = \sqrt{np(1-p)}$.

In R the family name for the binomial is `binom`, and the parameters are labeled `size=` for n and `prob=` for p.

■ **Example 5.5: Tossing ten coins** Toss a coin ten times. Let X be the number of heads. If the coin is fair, X has a $\mathsf{Binomial}(10, 1/2)$ distribution.

The probability that $X = 5$ can be found directly from the distribution with the `choose()` function:

```
> choose(10,5) * (1/2)^5 * (1/2)^(10-5)
[1] 0.2461
```

This work is better done using the "d" function, `dbinom()`:

```
> dbinom(5, size=10, prob=1/2)
[1] 0.2461
```

The probability that there are six or fewer heads, $\mathsf{P}(X \leq 6) = \sum_{k \leq 6} \mathsf{P}(X = k)$, can be given either of these two ways:

```
> sum(dbinom(0:6,size=10,prob=1/2))
[1] 0.8281
> pbinom(6,size=10,p=1/2)
[1] 0.8281
```

If we wanted the probability of seven or more heads, we could answer using $\mathsf{P}(X \geq 7) = 1 - \mathsf{P}(X \leq 6)$, or using the extra argument `lower.tail=FALSE`. This returns $\mathsf{P}(X > k)$ rather than $\mathsf{P}(X \leq k)$.

```
> sum(dbinom(7:10,size=10,prob=1/2))
[1] 0.1719
> 1 - pbinom(6,size=10,p=1/2)
[1] 0.1719
> pbinom(6,size=10,p=1/2, lower.tail=FALSE) # k = 6 not 7!
[1] 0.1719
```

A spike plot (Figure 5.4) of the distribution can be produced using `dbinom()`:

```
> heights = dbinom(0:10,size=10,prob=1/2)
> plot(0:10, heights, type="h",
+ main="Spike plot of X", xlab="k", ylab="p.d.f.")
> points(0:10, heights, pch=16,cex=2)
```

■

■ **Example 5.6: Binomial model for a public-opinion poll** In a public-opinion poll, the proportion of "yes" respondents is used to make inferences about the population proportion. If the respondents are chosen by sampling with replacement from the population, and the "yes" responses are coded by a 1 and the "no"

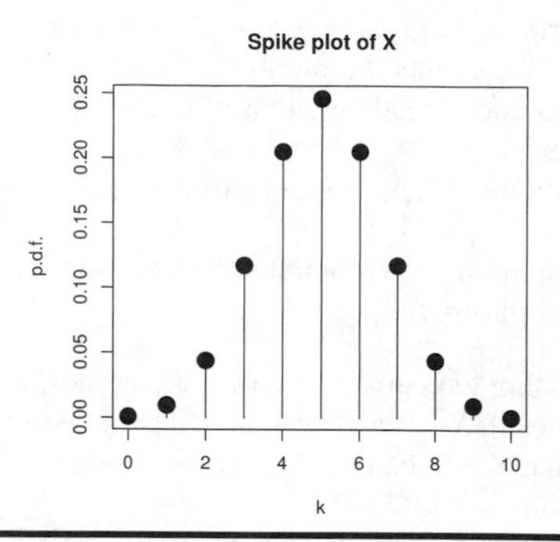

Figure 5.4 Spike plot of Binomial$(10, 1/2)$ **distribution**

responses by a 0, then the sequence of responses is an *i.i.d.* Bernoulli sequence with parameter p, the population proportion. The number of "yes" respondents is then a Binomial(n, p) random variable where n is the size of the sample.

For instance, if it is known that 62% of the population would respond favorably to the question were they asked, and a sample of size 100 is asked, what is the probability that 60% or less of the sample responded favorably?

```
> pbinom(60, size=100, prob=0.62)
[1] 0.3759
```

■

Normal random variables

The normal distribution is a continuous distribution that describes many populations in nature. Additionally, it is used to describe the sampling distribution of certain statistics. The normal distribution is a family of distributions with density given by

$$f(x|\mu,\sigma) = \frac{1}{\sqrt{2\pi\sigma^2}} e^{-\frac{1}{2\sigma^2}(x-\mu)^2}.$$

The two parameters are the mean, μ, and the standard deviation, σ. We use Normal(μ, σ) to denote this distribution, although many books use the variance, σ^2, for the second parameter.

The R family name is `norm` and the parameters are labeled `mean=` and `sd=`.

Figure 5.5 shows graphs of two normal densities, $f(x|\mu = 0, \sigma = 1)$ and $f(x|\mu = 4, \sigma = 1/2)$. The curves are symmetric and bell-shaped. The mean, μ, is a point of symmetry for the density. The standard deviation controls the spread of the curve. The distance between the inflection points, where the curves change from opening down to opening up, is two standard deviations.

The figure also shows two shaded areas. Let Z have $\text{Normal}(0,1)$ distribution and X have $\text{Normal}(4,1/2)$ distribution. Then the left shaded region is $\text{P}(Z \leq 1.5)$ and the right one is $\text{P}(X \leq 4.75)$. The random variable Z is called a **standard normal**, as it has mean 0 and variance 1. A key property of the normal distribution is that for any normal random variable the *z*-score, $(X-\mu)/\sigma$, is a standard normal. This says that areas are determined by *z*-scores. In Figure 5.5 the two shaded areas are the same, as each represents the area to the left of 1.5 standard deviations above the mean.

We can verify this with the "p" function:

```
> pnorm(1.5,  mean=0,sd=1)
[1] 0.9332
> pnorm(4.75, mean=4,sd=1/2)    # same z-score as above
[1] 0.9332
```

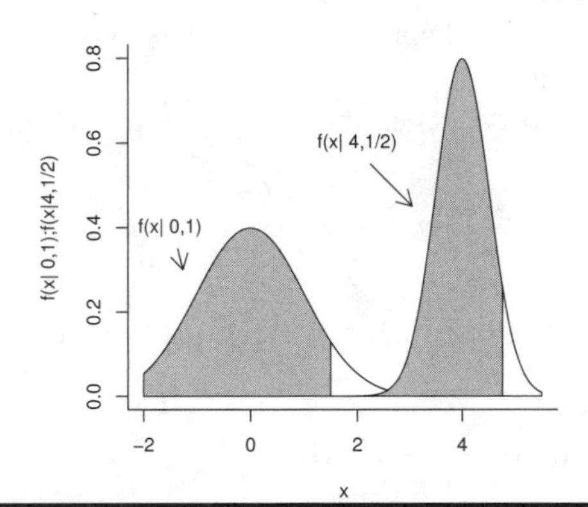

Figure 5.5 Two normal densities: the standard normal, $f(x|0,1)$, and $f(x|4,1/2)$. For each, the shaded area corresponds to a *z*-score of $3/2$ or less.

It is useful to know some areas for the normal distribution based on *z*-scores. For example, the IQR is the range of the middle 50%. We can find this for the standard normal by breaking the total area into quarters.

```
> qnorm(c(.25,.5,.75))
[1] -0.6745  0.0000  0.6745
```

We use `qnorm()` to specify the area we want. The mean and standard deviation are taken from the defaults of 0 and 1. For any normal random variable, this says the IQR is about 1.35σ.

How much area is no more than one standard deviation from the mean? We use `pnorm()` to find this:

```
> pnorm(1)-pnorm(-1)
[1] 0.6827
```

We see that roughly 68% of the area is in this range. For two and three standard deviations the numbers are 95% and 99.7%. We illustrate two ways to find these:

```
> 1 - 2*pnorm(-2)                  # subtract area of two tails
[1] 0.9545
> diff(pnorm(c(-3,3)))             # use diff to subtract
[1] 0.9973
```

This says that 95% of the time a normal random variable is within two standard deviations of the mean, and 99.7% of the time it is within three standard deviations of its mean. These three values, 68%, 95%, and 99.7%, are useful to remember as rules of thumb.

■ **Example 5.7: What percent of men are at least 6 feet tall?** Many distributions in nature are well approximated by the normal distribution. For example, the population of heights for adult males within an ethnic class. Assume for some group the mean is 70.2 inches, and the standard deviation is 2.89 inches. What percentage of adult males are taller than 6 feet? What percentage are taller than 2 meters? Assuming the model applies for all males, what does it predict for the tallest male on the planet?

We convert 6 feet into 72 inches and use `pnorm` to see that 73% are 6 feet or shorter:

```
> mu = 70.2; sigma = 2.89
> pnorm(72,mean=mu,sd=sigma)
[1] 0.7332
```

To answer the question for meters we convert to metric. Each inch is 2.54 centimeters, or 0.0254 meters.

```
> conv = 0.0254
> pnorm(2/conv,mean = mu, sd = sigma)
[1] 0.9984
```

That is, fewer than 1% are 2 meters or taller.

Finally, the tallest man could be found using quantiles. There are roughly 2.5 billion males, so the tallest man would be in the top 1/(2.5 billion) quantile:

```
> p = 1 - 1/2500000000
> qnorm(p,mu,sigma)/12
[1] 7.33
```

This predicts 7 feet 4 inches, not even the tallest in the NBA. Expecting a probability model with just two parameters to describe a distribution like this completely is asking too much. ■

■ **Example 5.8: Testing the rules of thumb** We can test the rules of thumb using random samples from the normal distribution as provided by `rnorm()`.

First we create 1,000 random samples and assign them to `res`:

```
> mu = 100; sigma = 10
> res = rnorm(1000,mean=mu,sd=sigma)
> k = 1;sum(res > mu - k*sigma & res < mu + k*sigma)
```

```
[1] 694
> k = 2;sum(res > mu - k*sigma & res < mu + k*sigma)
[1] 958
> k = 3;sum(res > mu - k*sigma & res < mu + k*sigma)
[1] 998
```

Our simulation has 69.4%, 95.8%, and 99.8% of the data within 1, 2, and 3 standard deviations of the mean. If we repeat this simulation, the answers will likely differ, as the 1,000 random numbers will vary each time. ■

5.2.3 Popular distributions to describe populations

Many populations are well described by the normal distribution; others are not. For example, a population may be multimodal, not symmetric, or have longer tails than the normal distribution. Many other families of distributions have been defined to describe different populations. We highlight a few.

Uniform distribution

The uniform distribution on $[a,b]$ is useful to describe populations that have no preferred values over their range. For a finite range of values, the `sample()` function can choose one with equal probabilities. The uniform distribution would be used when there is a range of values that is continuous.

The density is a constant on $[a,b]$. As the total area is 1, the height is $1/(b-a)$. The mean is in the middle of the interval, $\mu = (a+b)/2$. The variance is $(b-a)^2/12$. The distribution has short tails.

As mentioned, the family name in R is `unif`, and the parameters are `min=` and `max=` with defaults 0 and 1. We use $\mathsf{Uniform}(a,b)$ to denote this distribution. The left graphic in Figure 5.6 shows a histogram and boxplot of 25 random samples from $\mathsf{Uniform}(0,10)$. On the histogram are superimposed the empirical density and the population density. The random sample is shown using the `rug()` function.

```
> res = runif(50, min=0, max=10)
## fig= setting uses bottom 35% of diagram
> par(fig=c(0,1,0,.35))
> boxplot(res,horizontal=TRUE, bty="n", xlab="uniform sample")
## fig= setting uses top 75% of figure
> par(fig=c(0,1,.25,1), new=TRUE)
> hist(res, prob=TRUE, main="", col=gray(.9))
> lines(density(res),lty=2)
> curve(dunif(x, min=0, max=10), lwd=2, add=TRUE)
> rug(res)
```

(We overlaid two graphics by using the `fig=` argument to `par()`. This parameter sets the portion of the graphic device to draw on. You may manually specify the range on the x-axis in the histogram using `xlim=` to get the axes to match. Other layouts are possible, as detailed in the help page `?layout`.)

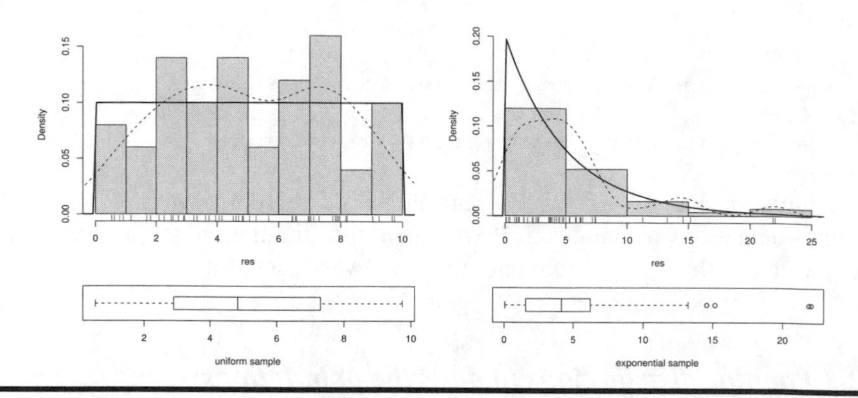

Figure 5.6 Histogram and boxplot of 50 samples from the Uniform(0,10) **distribution and the** Exponential(1/5) **distribution. Both empirical densities and population densities are drawn.**

Exponential distribution

The exponential distribution is an example of a skewed distribution. It is a popular model for populations such as the length of time a light bulb lasts. The density is $f(x|\lambda) = \lambda e^{-\lambda x}$, $x \geq 0$. The parameter λ is related to the mean by $\mu = 1/\lambda$ and to the standard deviation by $\sigma = 1/\lambda$.

In R the family name is `exp` and the parameter is labeled `rate=`. We refer to this distribution as Exponential(λ).

The right graphic of Figure 5.6 shows a random sample of size 50 from the Exponential(1/5) distribution, made as follows:

```
> res = rexp(50, rate=1/5)
## boxplot
> par(fig=c(0,1,0,.35))
> boxplot(res, horizontal=TRUE, bty="n", xlab="exponential sample")
## histogram
> par(fig=c(0,1,.25,1), new=TRUE)
## store values, then find largest y one to set ylim=
> tmp.hist=hist(res, plot=FALSE)
> tmp.edens = density(res)
> tmp.dens = dexp(0, rate=1/5)
> y.max = max(tmp.hist$density, tmp.edens$y, tmp.dens)
## make plots
> hist(res, ylim=c(0,y.max), prob=TRUE, main="", col=gray(.9))
> lines(density(res), lty=2)
> curve(dexp(x, rate=1/5), lwd=2, add=TRUE)
> rug(res)
```

Plotting the histogram and then adding the empirical and population densities as shown may lead to truncated graphs, as the y-limits of the histogram may not be large enough. In the above, we look first at the maximum y-values of the histogram and the two densities. Then we set the `ylim=` argument in the call to `hist()`. Finding the maximum value differs in each case. For the `hist()` function, more is returned than just a graphic. We store the result and access the `density` part with `tmp.hist$density`. For the empirical density, two named parts of the return value are `x` and `y`. We want the maximum of the `y` value. Finally, the population density is maximal at 0, so we simply use the `dexp()` function at 0 to give this. For other densities, we may need to find the maximum by other means.

Lognormal distribution

The lognormal distribution is a heavily skewed continuous distribution on the positive numbers. A lognormal random variable, X, has its name as $\log(X)$ is normally distributed. Lognormal distributions describe populations such as income distribution.

In R the family name is `lnorm`. The two parameters are labeled `meanlog=` and `sdlog=`. These are the mean and standard deviation of $\log(X)$, not of X.

Figure 5.7 shows a sample of size 50 from the lognormal distribution, with parameters `meanlog=0` and `sdlog=1`.

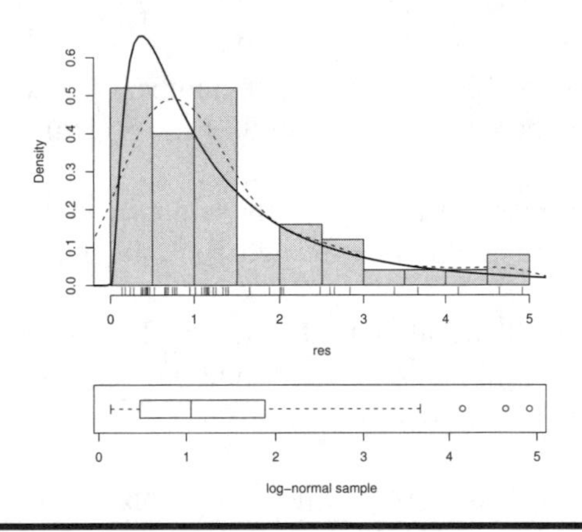

Figure 5.7 Histogram and boxplot of 50 samples from lognormal distribution with `meanlog=0` and `sdlog=1`

5.2.4 Sampling distributions

The following three distributions are used to describe sampling distributions. These are the t-distribution, the F-distribution, and the chi-squared distribution (sometimes written using the Greek symbol χ).

The family names in R are `t`, `f`, and `chisq`. Their parameters are termed "degrees of freedom" and are related to the sample size when used as sampling distributions. For the t and chi-squared distributions, the degrees-of-freedom argument is `df=`. For the F-distribution, as two degrees of freedom are specified, the arguments are `df1=` and `df2=`.

For example, values l and r for each distribution containing 95% of the area can be found as follows:

```
> qt(c(.025,.975), df=10)            # 10 degrees of freedom
[1] -2.228  2.228
> qf(c(.025,.975), df1=10, df2=5) # 10 and 5 degrees of freedom
[1] 0.2361 6.6192
> qchisq(c(.025,.975), df=10)       # 10 degrees of freedom
[1]  3.247 20.483
```

5.2.5 Problems

5.8 A die is rolled five times. What is the probability of three or more rolls of four?

5.9 Suppose a decent bowler can get a strike with probability $p = .3$. What is the chance he gets 12 strikes in a row?

5.10 A fair coin is tossed 100,000 times. The number of heads is recorded. What is the probability that there are between 49,800 and 50,200 heads?

5.11 Suppose that, on average, a baseball player gets a hit once every three times she bats. What is the probability that she gets four hits in four at bats?

5.12 Use the binomial distribution to decide which is more likely: rolling two dice twenty-four times and getting at least one double sixes, or rolling one die four times and getting at least one six?

5.13 A sample of 100 people is drawn from a population of 600,000. If it is known that 40% of the population has a specific attribute, what is the probability that 35 or fewer in the sample have that attribute?

5.14 If Z is $\mathsf{Normal}(0,1)$, find the following:

1. $\mathsf{P}(Z \leq 2.2)$
2. $\mathsf{P}(-1 < Z \leq 2)$
3. $\mathsf{P}(Z > 2.5)$
4. b such that $\mathsf{P}(-b < Z \leq b) = 0.90$.

5.15 Suppose that the population of adult, male black bears has weights that are approximately distributed as $\mathsf{Normal}(350, 75)$. What is the probability that a randomly observed male bear weighs more than 450 pounds?

5.16 The maximum score on the math ACT test is 36. If the average score for all high school seniors who took the exam was 20.6 with a standard deviation of 5.5, what percent received the passing mark of 22 or better? If 1,000,000 students took the test, how many more would be expected to fail if the passing mark were moved to 23 or better? Assume a normal distribution of scores.

5.17 A study found that foot lengths for Japanese women are normally distributed with mean 24.9 centimeters and standard deviation 1.05 centimeters. For this population, find the probability that a randomly chosen foot is less than 26 centimeters long. What is the 95th percentile?

5.18 Assume that the average finger length for females is 3.20 inches, with a standard deviation of 0.35 inches, and that the distribution of lengths is normal. If a glove manufacturer makes a glove that fits fingers with lengths between 3.5 and 4 inches, what percent of the population will the glove fit?

5.19 The term "six sigma" refers to an attempt to reduce errors to the point that the chance of their happening is less than the area more than six standard deviations from the mean. What is this area if the distribution is normal?

5.20 Cereal is sold by weight not volume. This introduces variability in the volume due to settling. As such, the height to which a cereal box is filled is random. If the heights for a certain type of cereal and box have a $\mathsf{Normal}(12, 0.5)$ distribution in units of inches, what is the chance that a randomly chosen cereal box has cereal height of 10.7 inches or less?

5.21 For the `fheight` variable in the `father.son (UsingR)` data set, compute what percent of the data is within 1, 2, and 3 standard deviations from the mean. Compare to the percentages 68%, 95%, and 99.7%.

5.22 Find the quintiles of the standard normal distribution.

5.23 For a $\mathsf{Uniform}(0, 1)$ random variable, the mean and variance are $1/2$ and $1/12$. Find the area within 1, 2, and 3 standard deviations from the mean and compare to 68%, 95%, and 99.7%. Do the same for the $\mathsf{Exponential}(1/5)$ distribution with mean and standard deviation of 5.

5.24 A q-q plot is an excellent way to investigate whether a distribution is approximately normal. For the symmetric distributions $\mathsf{Uniform}(0, 1)$, $\mathsf{Normal}(0, 1)$, and t with 3 degrees of freedom, take a random sample of size 100 and plot a quantile-normal plot using `qqnorm()`. Compare the three and comment on the curve of the plot as it relates to the tail length. (The uniform is short-tailed; the t-distribution with 3 degrees of freedom is long-tailed.)

5.25 For the t-distribution, we can see that as the degrees of freedom get large the density approaches the normal. To investigate, plot the standard normal density with the command

```
> curve(dnorm(x),-4,4)
```

and add densities for the t-distribution with $k = 5, 10, 25, 50$, and 100 degrees of freedom. These can be added as follows:

```
> k = 5; curve(dt(x,df=k), lty=k, add=TRUE)
```

5.26 The mean of a chi-squared random variable with k degrees of freedom is k. Can you guess the variance? Plot the density of the chi-squared distribution for $k = 2, 8, 18, 32, 50$, and 72, and then try to guess. The first plot can be done with `curve()`, as in

```
> curve(dchisq(x,df=2), 0, 100)
```

Subsequent ones can be added with

```
> k=8; curve(dchisq(x,df=k), add=TRUE)
```

5.3 The central limit theorem

It was remarked that for an *i.i.d.* sample from a population the distribution of the sample mean had expected value μ and standard deviation $\sigma/\sqrt{n}$, where μ and σ are the population parameters. For large enough n, we see in this section that the sampling distribution of $\bar{X}$ is normal or approximately normal.

5.3.1 Normal parent population

When the sample $X_1, X_2, \ldots, X_n$ is drawn from a $\mathsf{Normal}(\mu, \sigma)$ population, the distribution of $\bar{X}$ is precisely the normal distribution. Figure 5.8 draws densities for the population, and the sampling distribution of $\bar{X}$ for $n = 5$ and 25 when $\mu = 0$ and $\sigma = 1$.

```
> n=25; curve(dnorm(x,mean=0,sd=1/sqrt(n)), -3,3,
+ xlab="x",ylab="Densities of sample mean",bty="l")
> n=5;  curve(dnorm(x,mean=0,sd=1/sqrt(n)), add=TRUE)
> n=1;  curve(dnorm(x,mean=0,sd=1/sqrt(n)), add=TRUE)
```

The center stays the same, but as n gets bigger, the spread of $\bar{X}$ gets smaller and smaller. If the sample size goes up by a factor of 4, the standard deviation goes down by $1/2$ and the density concentrates on the mean. That is, with greater and greater probability, the random value of $\bar{X}$ is close to the mean, μ, of the parent population. This phenomenon of the sample average concentrating on the mean is known as the **law of large numbers**.

For example, if adult male heights are normally distributed with mean 70.2 inches and standard deviation 2.89 inches, the average height of 25 randomly

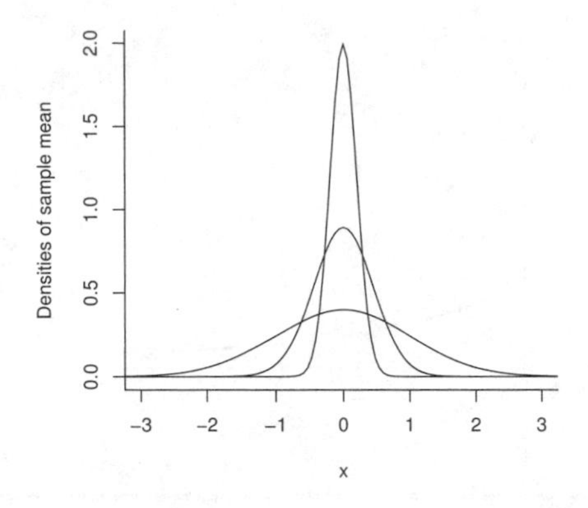

Figure 5.8 Density of $\bar{X}$ for $n = 5$ and $n = 25$ along with parent population $\mathsf{Normal}(0,1)$. **As n increases, the density concentrates on μ.**

chosen males is again normal with mean 70.2 but standard deviation 1/5 as large. The probability that the sample average is between 70 and 71 is found with

```
> mu=70.2; sigma=2.89; n=25
> diff( pnorm(70:71, mu, sigma/sqrt(n)) )
[1] 0.5522
```

Compare this to the probability for a single person

```
> diff( pnorm(70:71, mu, sigma) )
[1] 0.1366
```

5.3.2 Nonnormal parent population

The **central limit theorem** states that for any parent population with mean μ and standard deviation σ, the sampling distribution of $\bar{X}$ for large n satisfies

$$\mathsf{P}\left(\frac{\bar{X}-\mu}{\sigma/\sqrt{n}} \leq b\right) \approx \mathsf{P}(Z \leq b),$$

where Z is a standard normal random variable. That is, for n big enough, the distribution of $\bar{X}$ once standardized is approximately a standard normal distribution. We also refer to this as saying $\bar{X}$ is asymptotically normal.

Figure 5.9 illustrates the central limit theorem for data with an $\mathsf{Exponential}(1)$ distribution. This parent population and simulations of the distribution of $\bar{X}$ for $n = 5$, 25, and 100 are drawn. As n gets bigger, the sampling distribution of $\bar{X}$ becomes more and more bell shaped.

Figure 5.9 was produced by simulating the sampling distribution of $\bar{X}$. Simulations will be discussed in the next chapter.

■ **Example 5.9: Average service time** The time it takes to check out at a

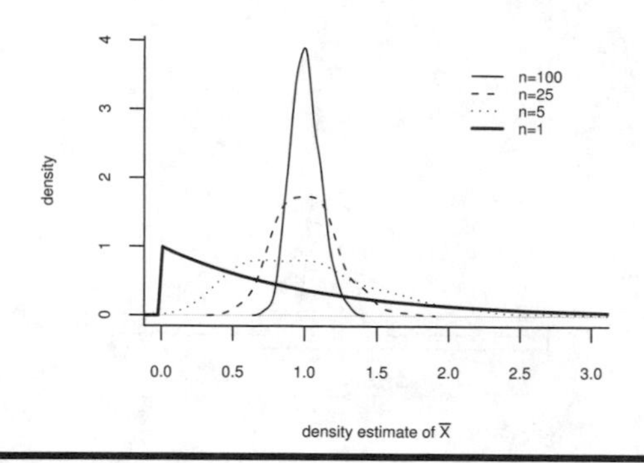

Figure 5.9 Density estimates for $\bar{X}$ when $n = 5, 25, 100$ for an $\mathsf{Exponential}(1)$ **population. As n increases, density becomes bell shaped and concentrates on $\mu = 1$.**

grocery store can vary widely. A certain checker has a historic average of one-minute service time per customer, with a one-minute standard deviation. If she sees 20 customers, what is the probability that her check-out times average 0.9 minutes or less?

We assume that each service time has the unspecified parent population with $\mu = 1$ and $\sigma = 1$ and the sequence of service times is *i.i.d.* As well, we assume that n is large enough that the distribution of $\bar{X}$ is approximately $\mathsf{Normal}(\mu, \sigma/\sqrt{20})$. Then $\mathsf{P}(\bar{X} \leq 0.9)$ is given by

```
> pnorm(.9, mean=1, sd = 1/sqrt(20))
[1] 0.3274
```

■

There are other consequences of the central limit theorem. For example, if we replace σ with the sample standard deviation s when we standardize $\bar{X}$ we still have

$$\mathsf{P}(\frac{\bar{X} - \mu}{s/\sqrt{n}} \leq b) \approx \mathsf{P}(Z \leq b).$$

This fact will be behind many of the statements in the next two chapters. This does not tell us what the sampling distribution is when n is not large; that will be discussed later.

In this next example, we show how the central limit theorem applies to the binomial distribution for large n.

■ **Example 5.10: The normal approximation to the binomial distribution**
For an *i.i.d.* sequence of Bernoulli trials $X_1, X_2, \ldots, X_n$ with success probability p, the sample mean, $\bar{X}$, is simply the number of successes divide by n, or the proportion of successes. We will use the notation $\widehat{p}$ instead of $\bar{X}$ in this case.

The central limit theorem says that $\widehat{p}$ is asymptotically normal with mean p and standard deviation $\sqrt{p(1-p)/n}$.

If X is the number of successes, then X is $\mathsf{Binomial}(n,p)$. Since $X = n\widehat{p}$, we know that X is approximately normal with mean np and variance $\sqrt{np(1-p)}$. That is, a binomial random variable is approximately normal if n is large enough.

Let X have a $\mathsf{Binomial}(30,2/3)$ distribution. Figure 5.10 shows a plot of the distribution over $[10,30]$. The shaded boxes above each integer k have base 1 and height $\mathsf{P}(X=k)$, so their area is equal to $\mathsf{P}(X=k)$. The normal curve that is added to the figure has mean and standard deviation equal to that of X: $\mu = 30\cdot 2/3 = 20$ and $\sigma = \sqrt{30\cdot 2/3\cdot 1/3}$. From the figure, we can see that the area of the shaded boxes, $\mathsf{P}(k \leq 22)$, is well approximated by the area to the left of 22.5 under the normal curve. This says $\mathsf{P}(X \leq 22) \approx \mathsf{P}(Z \leq (22.5-\mu)/\sigma)$ for a standard normal Z. For a general binomial random variable with mean μ and standard deviation σ, the approximation $\mathsf{P}(a \leq X \leq b) \approx \mathsf{P}((a-1/2-\mu)/\sigma \leq Z \leq (b+1/2-\mu)/\sigma)$ is an improvement to the central limit theorem. ■

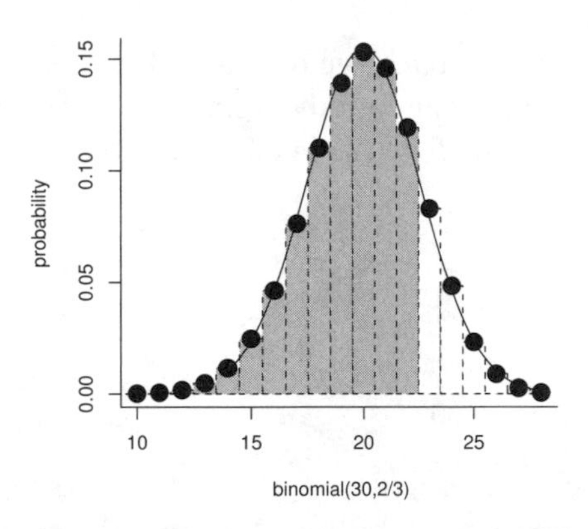

Figure 5.10 Plot of $\mathsf{Binomial}(30,2/3)$ **distribution marked by dots. The area of the rectangle covering** k **is the same as the probability of** k **successes. The drawn density is the normal distribution with the same population mean and standard deviation as the binomial.**

5.3.3 Problems

5.27 Compare the exact probability of getting 42 or fewer heads in 100 coin tosses to the probability given by the normal approximation.

5.28 Historically, a certain baseball player has averaged three hits every ten official at bats (he's a .300 hitter). Assume a binomial model for the number of hits in a 600-at-bat season. What is the probability the player has a batting average higher than .350? Use the normal approximation to answer.

5.29 Assume that a population is evenly divided on an issue ($p = 1/2$). A random sample of size 1,000 is taken. What is the probability the random sample will have 550 or more in favor of the issue? Answer using a normal approximation.

5.30 An elevator can safely hold 3,500 pounds. A sign in the elevator limits the passenger count to 15. If the adult population has a mean weight of 180 pounds with a 25-pound standard deviation, how unusual would it be, if the central limit theorem applied, that an elevator holding 15 people would be carrying more than 3,500 pounds?

5.31 A restaurant sells an average of 25 bottles of wine per night, with a variance of 4. Assuming the central limit theorem applies, what is the probability that the restaurant will sell more than 775 bottles in the next 30 days?

5.32 A traffic officer writes an average of four tickets per day, with a variance of one ticket. Assume the central limit theorem applies. What is the probability that she will write fewer than 75 tickets in a 21-day cycle?

Chapter 6

Simulation

One informal description of insanity is "repeating the same action while expecting a different result." By this notion, the act of simulating a distribution could be considered somewhat insane, as it involves repeatedly sampling from a distribution and investigating the *differences* in the results. But simulating a distribution is far from insane. Simulating a distribution can give us great insight into the distribution's shape, its tails, its mean and variance, etc. We'll use simulation to justify the size of n needed in the central limit theorem for approximate normality of the sample mean. Simulation is useful with such specific questions, as well as with those of a more exploratory nature.

In this chapter, we will develop two new computer skills. First **for loops** will be introduced. These are used to repeat something again and again, such as sampling from a distribution. Then we will see how to define simple functions in R. Defining functions not only makes for less typing; it also organizes your work and train of thought. This is indispensable when you approach larger problems.

6.1 The normal approximation for the binomial

We begin with a simulation to see how big n should be for the binomial distribution to be approximated by the normal distribution. Although we know explicitly the distribution of the binomial, we approach this problem by taking a random sample from this distribution to illustrate the approach of simulation.

To perform the simulation, we will take m samples from the binomial distribution for some n and p. We should take m to be some large number, so that we get a good idea of the underlying population the sample comes from. We will then compare our sample to the normal distribution with $\mu = np$, and $\sigma^2 = np(1-p)$. If the sample appears to come from this distribution; we will say the approximation is valid.

Let $p = 1/2$. We can use the `rbinom()` function to generate the sample of size m. We try $n =$ 5, 15, and 25. In Figure 6.1 we look at the samples with histograms that are overlaid with the corresponding normal distribution.

```
> m = 200; p = 1/2;
> n = 5
> res = rbinom(m,n,p)                        # store results
> hist(res, prob=TRUE, main="n = 5") # don't forget prob=TRUE
> curve(dnorm(x, n*p, sqrt(n*p*(1-p))), add=TRUE) # add density
### repeat last 3 commands with n=15, n=25
```

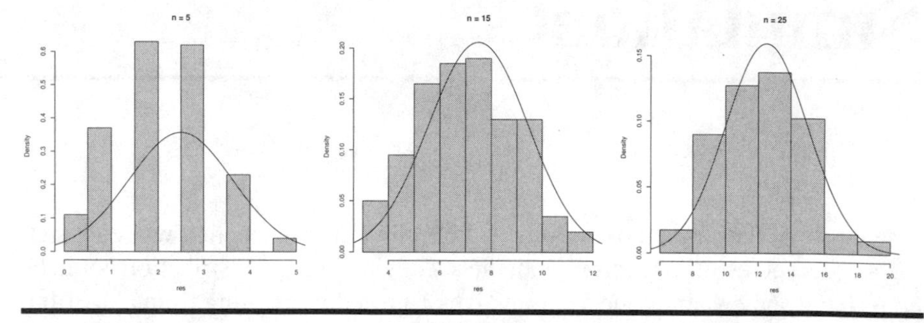

Figure 6.1 Histograms of normal approximation to binomial with $p = 1/2$ and $n =$ 5, 10, and 25

We see from the figure that for $n = 5$ the approximation is not valid at all—the discreteness of the binomial distribution is still apparent. By $n =$ 15 and 25, the approximation looks pretty good. This agrees with the rule of thumb that when np and $n(1-p)$ are both greater than 5 the normal approximation is valid.

A better way to judge normality than with a histogram is with the quantile-normal plot made by `qqnorm()`. If the sampling distribution is normal then this plot will show an approximate straight line.

```
> m = 200; p = 1/5; n = 25
> res = rbinom(m,n,p)
> qqnorm(res)
```

Figure 6.2 shows the graph. The discreteness of the binomial shows through, but we can see that the points are approximately on a straight line, as they should be if the distribution is approximately normal.

6.2 `for` loops

Generating samples from the binomial distribution was straightforward due to the `rbinom()` function. For other statistics, we can generate samples, but perhaps only one at a time. In this case, to get a large enough sample to investigate the sampling distribution, we use a for loop to repeat the sampling.

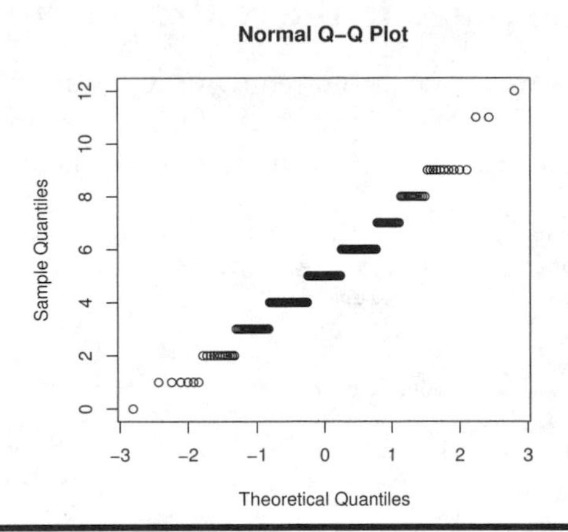

Figure 6.2 Quantile-normal plot of binomial sample for $n = 25$, $p = 1/5$

The central limit theorem tells us that the sampling distribution of $\bar{X}$ is approximately normal if n is large enough. To see how big n needs to be we can repeat the above approach. That is, we find a large sample from the sampling distribution of $\bar{X}$ and compare it to the normal distribution.

Assume our population is Uniform$(0,1)$, and we want to investigate whether $\bar{X} = (1/n)\sum X_i$ is normally distributed when $n = 10$. A single sample from the sampling distribution of $\bar{X}$ can be found with the command `mean(runif(10))`.

To draw repeated samples, we can use a for loop. A for loop will repeat itself in a predictable manner. For example, these commands will repeat the sampling 100 times, storing the answer in the `res` variable.

```
> res = c()
> for(i in 1:100) {
+ res[i] = mean(runif(10))
+ }
```

The variable `res` now holds 100 samples of $\bar{X}$ for $n = 10$ and each X_i being Uniform$(0,1)$.

The basic format of a for loop is

```
for(variable.name in values) {
  block_of_commands
}
```

The keyword `for` is used. The `variable.name` is often something like `i` or `j` but can be any valid name. Inside the block of the `for` loop the variable takes on a different value each time through. The `values` can be a vector or a list. In the example it is `1:100`, or the numbers 1 through 100. It could be something like `letters` to loop over the lowercase letters, or `x` to loop over the values of `x`. When it is a list, the value of `variable.name` loops over the top-level components.

6.3 Simulations related to the central limit theorem

We use a for loop to investigate the normality of $\bar{X}$ for different parent populations and different sample sizes. For example, if the X_i are Uniform$(0, 1)$ we can simulate $\bar{X}$ for $n = 2$, 10, 25, and 100 with these commands:

```
## set up plot window
> plot(0,0,type="n",xlim=c(0,1),ylim=c(0,13.5),
+      xlab="Density estimate",ylab="f(x)")
> m = 500;a=0;b=1
> n = 2
> for (i in 1:m) res[i] = mean(runif(n,a,b))
> lines(density(res),lwd=2)
## repeat last 3 lines with n=10, 25, and 100
```

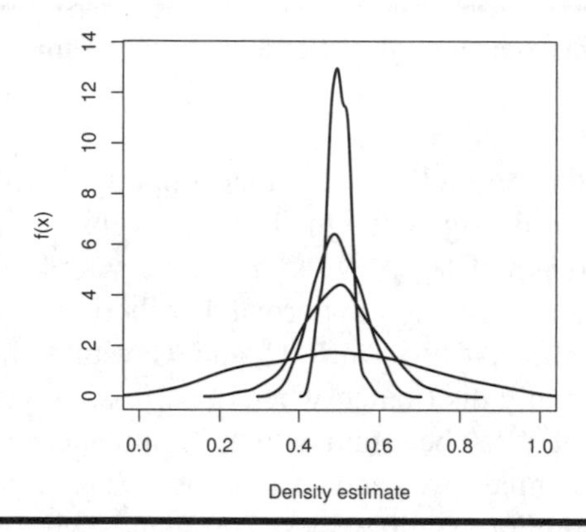

Figure 6.3 Density estimates for $\bar{X}$ for $n = 2, 10, 25$, and 100 with Uniform$(0, 1)$ **data**

In Figure 6.3 a density estimate is plotted for each simulation. Observe how the densities squeeze in and become approximately bell shaped, as expected, even for $n = 10$. As the standard deviation of $\bar{X}$ is $\sigma/\sqrt{n}$, if n goes up four times (from 25 to 100, for example), the standard deviation gets cut in half. Comparing the density estimate for $n = 25$ and $n = 100$, we can see that the $n = 100$ graph has about half the spread.

In this example the `for` loop takes the shortened form

```
for(i in values) a_single_command
```

If there is just a single command, then no braces are necessary. This is convenient when we use the up arrow to edit previous command lines.

In the problems, you are asked to simulate $\bar{X}$ for a variety of parent populations to verify that the more skewed the data is, the larger n must be for the normal distribution to approximate the sampling distribution of $\bar{X}$.

6.4 Defining a function

In the previous examples, we have found a single sample of $\bar{X}$ using a command like

```
> mean(runif(n))
```

This generates *n i.i.d.* samples from the uniform distribution and computes their sample mean. It is often convenient to define functions to perform tasks that require more than one step. Functions can simplify our typing, organize our thoughts, and save our work for reuse. This section covers some of the basics of functions—their basic structure and the passing of arguments. More details are available in Appendix E.

A basic function in R is conceptually similar to a mathematical function. In R, a function has a name (usually), a rule (the body of the function), a way of defining the inputs (the arguments to a function), and an output (the last command evaluated).

Functions in R are created with the `function()` keyword. For example, we define a function to find the mean of a sample of size 10 from the Exponential(1) distribution as follows:

```
> f = function() {
+ mean(rexp(10))
+ }
```

To use this function, we type the name and parentheses

```
> f()
[1] 0.7301
```

This function is named `f`. The keyword `function()` creates a function and assigns it to `f`. The body of the function is enclosed in braces: {}. The return value is the last line evaluated. In this case, only one line is evaluated—the one finding the `mean()`. (As with for loops, in this case the braces are optional.) In the next example we will discuss how to input arguments into a function.

If we define a function to find a single observation from the sampling distribution, then our simulation can be done with generic commands such as these:

```
> res = c()
> for(i in 1:500) res[i] = f()
```

6.4.1 Editing a function

An advantage of using functions to do our work is that they can be edited. The entire function needn't be retyped when changes are desired. Basic editing can be done with either the `fix()` function or the `edit()` function. For example, the command `fix(f)` will open an editor (in Windows this defaults to notepad) to the definition of your function `f`. You make the desired changes to your function then exit the editor. The changes are assigned to `f` which can be used as desired.

The `edit()` function works similarly, but you must assign its return value, as in

```
> f = edit(f)
```

6.4.2 Function arguments

A function usually has a different answer depending on the value of its arguments. Passing arguments to R functions is quite flexible. We can do this by name or position. As well, as writers of functions, we can create reasonable defaults.

Let's look at our function f, which finds the mean of ten exponentials. If we edit its definition to be

```
f = function(n = 10) {
  mean(rexp(n))
}
```

then we can pass in the size of the sample, n, as a parameter. We can call this function in several ways: `f()`, `f(10)`, and `f(n=10)` are all the same and use $n = 10$. This command uses $n = 100$: `f(100)`. The first argument to f is named n and is given a *default value* of 10 by the `n = 10` specification in the definition. Calling f by `f()` uses the defaults values. Calling f by `f(100)` uses the position of the argument to assign the values inside the function. In this case, the 100 is assigned to the only argument, `n=`. When we call f with `f(n=100)` we use a named argument. With this style there is no doubt what value n is being set to.

With f defined, simulating 200 samples of $\bar{X}$ for $n = 50$ can be done as follows:

```
> res = c()
> for(i in 1:200) res[i] = f(n=50)
```

Better still, we might want to pass in a parameter to the exponential. The rate of the exponential is 1 over its mean. So changing f to

```
f = function(n = 10, rate = 1) {
  mean(rexp(n, rate = rate))
}
```

sets the first argument of f to n with a default of 10 and the second to rate with a default of 1. This allows us to change the size and rate as in `f(50,2)`, which would take 50 X_i's each with rate 2 or mean 1/2. Alternately, we could do `f(rate=1/2)`, which would use the default of 10 for n and use the value of 1/2 for rate. (Note that `f(1/2)` will not do this, as the `1/2` would match the position for n and not that of rate.)

The arguments of a function are returned by the `args()` command. This can help you sort out the available arguments and their names as a quick alternative to the more informative help page. When consulting the help pages of R's built-in functions, the `...` argument appears frequently. This argument allows the function writer to pass along arbitrarily named arguments to function calls inside the body of a function.

6.4.3 The function body

The function body is a block of commands enclosed in braces. As mentioned, the braces are optional if there is a single command. The return value for a function is the last command executed. The function `return()` will force the return of a function, with its argument becoming the return value.

Some commands executed during a function behave differently from when they are executed at the command line—in particular, printing and assignment.

During interactive usage, typing the name of an R object causes it to be "printed." This shows the contents in a nice way, and varies by the type of object. For example, factors and data vectors print differently. Inside a function, nothing is printed unless you ask it to be.* The function `print()` will display an object as though it were typed on the command line. The function `cat()` can be used to concatenate values together. Unlike `print()`, the `cat()` function will not print a new line character, nor the element numbers, such as `[1]`. A new line can be printed by including `"\n"` in the `cat()` command. When a function is called, the return value will print unless it is assigned to some object. If you don't want this, such as when producing a graphic, the function `invisible()` will suppress the printing.

Assignment inside a function block is a little different. Within a block, assignment to a variable masks any variable outside the block. This example defines `x` to be 5 outside the block, but then assigns `x` to be 6 inside the block. When `x` is printed inside the block the value of 6 shows; however, `x` has not changed once outside the block.

```
> x = 5
> f = function() {
+ x = 6
+ x
+ }
> f()
[1] 6
> x
[1] 5
```

If you really want to force `x` to change inside the block, the global assignment operator `<<-` can be used, as can the function `assign()`. Consult the help pages `?"<<-"` and `?assign` for more detail.

In the example above, the value of `x` used inside the block is the one assigned inside the block. If none had been assigned, R would have looked for a definition outside the block. For example:

```
> x = 5
> f = function() print(x)
> f()
[1] 5
> rm(x)
> f()
Error: Object "x" not found
```

When no variable named `x` is be found, an error message is issued.

* In Windows you may need to call `flush.console()()` to get the output. See the FAQ for details.

6.5 Investigating distributions

■ **Example 6.1: The sample median**

The sample median, M, is a measurement of central tendency like the sample mean. Does it, too, have an approximately normal distribution? How does the sampling distribution of M reflect the parent distribution of the sample? Will M converge to some parameter of the parent distribution as $\bar{X}$ converges to μ?

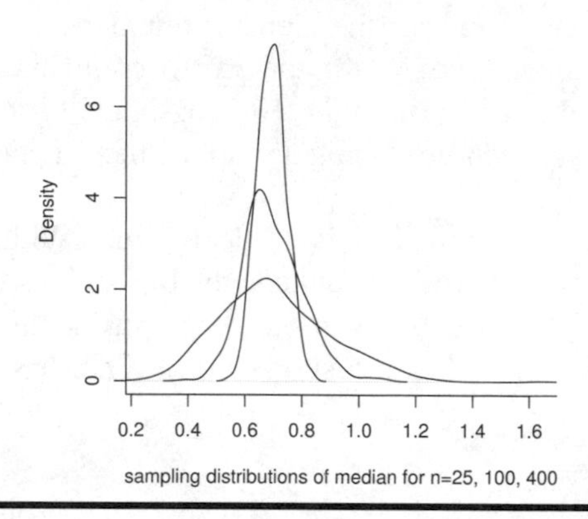

Figure 6.4 Density estimates for simulations of the sample median from exponential data. As n increases, the sampling distribution appears to become normally distributed and concentrates on the median of the parent population.

To investigate these questions, we will perform a simulation. Assume we have a random sample $X_1, X_2, \ldots, X_n$ taken from the $\mathsf{Exponential}(1)$ distribution. This distribution has mean 1 and median $\log(2) = .6931$. We perform a simulation for $n = 25$, 100, and 400. First we define a function to find the median from the sample:

```
> f = function(n) median(rexp(n))
```

Now we generate samples for different sizes of n.

```
> m = 500
> res.25 = c(); res.100 = c(); res.400 = c()
> for(i in 1:m) res.25[i]  = f(25)
> for(i in 1:m) res.100[i] = f(100)
> for(i in 1:m) res.400[i] = f(400)
> summary(res.25)
   Min. 1st Qu.  Median    Mean 3rd Qu.    Max.
  0.237   0.571   0.688   0.707   0.822   1.640
> summary(res.100)
   Min. 1st Qu.  Median    Mean 3rd Qu.    Max.
  0.393   0.629   0.682   0.699   0.764   1.090
> summary(res.400)
   Min. 1st Qu.  Median    Mean 3rd Qu.    Max.
  0.539   0.659   0.692   0.693   0.727   0.845
```

The `summary()` commands show that the mean and median are similar for each sample and appear to be centered around the median of the parent population.

We plot three density estimates to see the shape of the distribution in Figure 6.4. We plot the one for $n = 400$ first, so the y-axis is large enough to accommodate the three graphs.

```
> plot(density(res.400), xlim = range(res.25), type="l", main="",
+ xlab="sampling distributions of median for n=25, 100, 400")
)
> lines(density(res.100))
> lines(density(res.25))
```

As n gets large, the sampling distribution tends to a normal distribution which is centered on the median of the parent population. ■

■ **Example 6.2: Comparing measurements of spread** Simulations can help guide us in our choices. For example, we can use either the standard deviation or the IQR to measure spread. Why would we choose one over the other? One good reason would be if the sampling variation of one is significantly smaller than that of the other.

Let's compare the spread of the sampling distribution for both statistics using boxplots. First, define two functions `f()` and `g()` as

```
> f = function(n) sd(rnorm(n))
> g = function(n) IQR(rnorm(n))
```

Then we can simulate with

```
> res.sd = c(); res.iqr = c()
> for(i in 1:200) {
+ res.sd[i] = f(100)
+ res.iqr[i] = g(100)
+ }
> boxplot(list(sd=res.sd, iqr=res.iqr))
```

Figure 6.5 shows side-by-side boxplots illustrating that the spread of the IQR is wider than that of the mean. For normal data, the standard deviation is a better measure of spread. ■

The standard deviation isn't always a better measure of spread. We will repeat the simulation with exponential data and investigate. Before doing so, we look at script files, which save a sequence of commands to be executed.

6.5.1 Script files and `source()`

`R` can "read" the contents of a file and execute the commands as though they were typed in at the command line. The command to do this is `source()`, as in `source(file="filename")`. (Most of the GUIs have this ability.)

For example, if a file named "sim.R" contains these commands

```
## file sim.R
f = function(n) sd(rexp(n))
g = function(n) IQR(rexp(n))
```

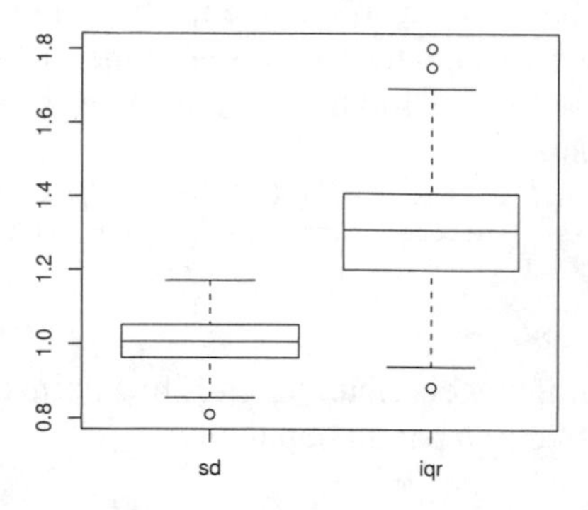

Figure 6.5 Boxplot of standard deviation and IQR for normal data

```
res.sd = c(); res.iqr = c()
for(i in 1:200) {
  res.sd[i] = f(100)
  res.iqr[i] = g(100)
}
boxplot(list(sd=res.sd, iqr = res.iqr))
```

then the command

```
> source("sim.R")
```

will read and evaluate these commands producing a graph similar to Figure 6.6. With exponential data, the spread of each statistic is similar. The more skewed or long-tailed the data is, the wider the spread of the standard deviation compared to the IQR.

By using a separate file to type our commands, we can more easily make changes than with the command line, and we can save our work for later use.

6.5.2 The geometric distribution

In a sequence of *i.i.d.* Bernoulli trials, there is a time of the first success. This can happen on the first trial, the second trial, or at any other point. Let X be the time of the first success. Then X is a random variable with distribution on the positive integers. The distribution of X is called the geometric distribution and is

$$f(k) = \mathrm{P}(X = k) = (1-p)^{k-1}p.$$

Let's simulate the random variable X to investigate its mean. To find a single sample from the distribution of X we can toss coins until we have a success. A `while()` loop is ideal for this type of situation, as we don't know in advance how many times we will need to toss the coin.

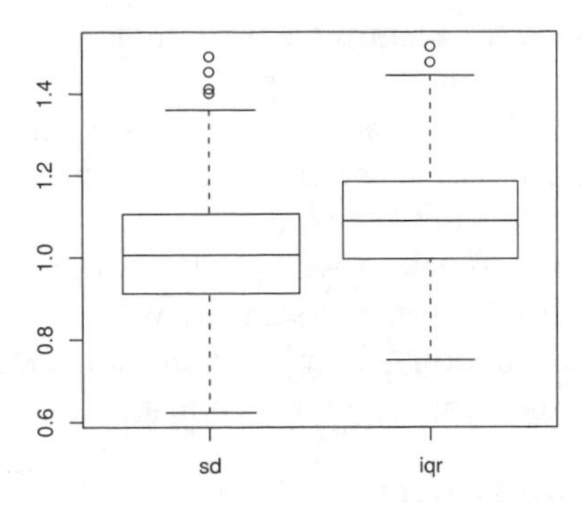

Figure 6.6 Simulation of standard deviation and IQR for Exponential(1) **data**

```
first.success = function(p) {
  k = 0;
  success = FALSE
  while(success == FALSE) {
    k = k + 1
    if(rbinom(1,1,p) == 1) success = TRUE
  }
  k
}
```

The `while` loop repeats a block of commands until its condition is met. In this case, the value of `success` is not `FALSE`. Inside the while loop, an `if` statement is used. When the `if` statement is true, it sets the value of `success` to `TRUE`, effectively terminating the `while` loop. The command `rbinom(1,1,p)` is one sample from the Binomial$(1,p)$ distribution—basically a coin toss.

We should expect that if the success probability is small, then the time of the first success should be large. We compare the distribution with $p = .5$ and with $p = .05$ using `summary()`.

```
> res.5 = c();res.05 = c()
> for(i in 1:500) {
+ res.5[i] = first.success(0.5)
+ res.05[i] = first.success(0.05)
+ }
> summary(res.5)
   Min. 1st Qu.  Median    Mean 3rd Qu.    Max.
   1.00    1.00    1.00    2.01    2.00   11.00
> summary(res.05)
   Min. 1st Qu.  Median    Mean 3rd Qu.    Max.
    1.0     6.0    13.0    20.1    28.0   120.0
```

From the output of `summary()` it appears that the sampling distribution has mean $2 = 1/0.5$ and $20 = 1/0.05$ respectively. For any p in $[0,1]$ the mean of the geometric distribution is $1/p$.

6.6 Bootstrap samples

The basic idea of a **bootstrap sample** is to sample *with replacement* from the data, thereby creating a new random sample of the same size as the original. For this random sample the value of the statistic is computed. Call this a replicate. This process is repeated to get the sampling distribution of the replicates. From this, inferences are made about the unknown parameters.

For example, we can estimate μ with the bootstrap. Let the replicate, $\bar{x}_i^*$, be the sample mean of the ith bootstrap sample. We estimate μ with the sample mean of these replicates. In doing so, we get an estimate for the population parameter and a sense of the variation in the estimate.

■ **Example 6.3: Albatross bycatch**

The bycatch (UsingR) data set[†] contains the number of albatross incidentally caught by squid fishers for 897 hauls of fishing nets, as measured by an observer program.

We wish to investigate the number of albatross caught. We can summarize this with the sample mean, but to get an idea of the underlying distribution of the sample mean, we generate 1,000 bootstrap samples and look at their means.

First, the data in bycatch (UsingR) is summarized for compactness. We expand it to include all 897 hauls.

```
> data(bycatch)
> hauls = with(bycatch, rep(no.albatross,no.hauls))
> n = length(hauls)
```

Now n is 897, and hauls is a data vector containing the number of albatross caught on each of the 897 hauls. A histogram shows a skewed distribution. Usually, none are caught, but occasionally many are. As the data is skewed, we know the sample mean can be a poor predictor of the center. So we create 1,000 bootstrap samples as follows, using sample().

```
> xbarstar = c()
> for(i in 1:1000) {
+ boot.samp = sample(hauls, n, replace=TRUE)
+ xbarstar[i] = mean(boot.samp)
+ }
```

For each bootstrap sample we find the replicate $\bar{x}_i^*$. The data vector xbarstar contains 1,000 realizations. Estimates for the population mean and variance are

```
> mean(xbarstar)
[1] 0.2789
> sd(xbarstar)
[1] 0.04001
```

Furthermore, we can find out where the sample mean usually is with the quantile() function:

```
> quantile(xbarstar,c(0.05,0.95))
    5%    95%
0.2107 0.3467
```

[†] From Hilborn and Mangel, *The Ecological Detective*

Which says that 90% of the time it is in $[.2107, .3467]$. ■

6.7 Alternates to for loops

Although for loops are a good way to approach a problem of repeating something, they are not necessarily the preferred approach to a problem in R. For practical reasons, alternatives to for loops can be faster. For aesthetic reasons, a vectorized approach may be desired. In this approach, we use a function on multiple values at once, rather than one at a time as in a for loop.

The speed issue can often be overcome by using a matrix. We illustrate by using a matrix to simulate the sample mean. We create a matrix with $m = 100$ rows and $n = 10$ columns of random numbers using `matrix()` as follows:

```
> m = 100; n =10
> tmp = matrix(rnorm(m*n), nrow=m)
```

The `rnorm(m*n)` command returns 1,000 normal samples, which are placed in a matrix with 100 rows. This is specified by the argument `nrow=m`. We want to find the mean of each row. We do so with `apply()`:

```
> xbar = apply(tmp,1, mean)
```

We did simulations by repeatedly calling a function to generate a realization of our statistic. Conceptually, we can picture this as applying the function to each value in the vector of values looped over in the for loop. The `sapply()` function also does just that, so we can use `sapply()` to replace the for loop in a simulation.

For example, this command will generate ten random samples of the mean of 25 random numbers:

```
> m = 10; n = 25
> sapply(1:m, function(x) mean(rnorm(n)))
 [1] -0.06627  0.09835 -0.29290 -0.42287  0.47297 -0.26416
 [7] -0.39610 -0.04068 -0.08084  0.20296
```

The `sim()` function in the `UsingR` package uses `sapply()` to produce simulations of the sampling distribution of many statistics for user-specified populations. The above simulation would be done as

```
> library(UsingR)
> sim(n=25, m=10, statistic="mean", family="norm", mean=0, sd=1)
```

The argument `statistic=` is the name of the desired statistic and `family=` the family name of the R function that produces the desired random sample.

6.8 Problems

6.1 Do simulations of the binomial for $n = 100$ and $p = 0.02$ and for $n = 100$ and $p = 0.2$. Do both distributions appear to be approximately normal? Discuss.

6.2 The data set `lawsuits (UsingR)` is very long tailed. However, the central limit theorem will apply for the sampling distribution of $\bar{X}$. To see how big n needs to be for approximate normality, repeat the following simulation for different values of n until the sampling distribution of $\bar{X}$ appears normal.

```
> data(lawsuits)
> res = c()
> n = 5
> for(i in 1:300) res[i] = mean(sample(lawsuits,n,replace=TRUE))
> plot(density(scale(res)))
```

The `scale()` command finds the z-scores for each value in the data vector. After scaling, compare the shapes of distributions with different means and standard deviations. This way you can layer subsequent density estimates with the command `lines(density(scale(res)))` and look for normality.

How big should n be to get a bell-shaped sampling distribution for $\bar{X}$?

6.3 For what value of n does $\bar{X}$ look approximately normal when each X_i is $\mathsf{Uniform}(0,1)$? (Run several simulations for different n's and decide where the switch to normality begins.)

6.4 For what value of n does $\bar{X}$ look approximately normal when each X_i is ($\mathsf{Exponential}(1)$ (`rexp(n,1)`)?

6.5 For what value of n does $\bar{X}$ look approximately normal when each X_i has a t-distribution with 3 degrees of freedom (`rt(n,3)`)?

6.6 Compare the distributions of the sample mean and sample median when the X_i have the t distribution with 3 degrees of freedom and $n = 10$. Which has a bigger spread?

6.7 The χ^2 distribution arises when we add a number of independent, squared standard normals. Instead of using `rchisq()` to generate samples, we can simulate by adding normally distributed numbers. For example, we can simulate a χ^2 distribution with 4 degrees of freedom with

```
> res = c()
> for(i in 1:500) res[i] = sum(rnorm(4)^2)
> qqnorm(res)
```

Repeat the above for 10, 25, and 50 degrees of freedom. Does the data ever appear approximately normal? Why would you expect that?

6.8 The correlation between $\bar{x}$ and s^2 depends on the parent distribution. For a normal parent distribution the two are actually independent. For other distributions, this isn't so.

To investigate, we can simulate both statistics from a sample of size 10 and observe their correlation with a scatterplot and the `cor()` function.

```
> xbar = c();std = c()
> for(i in 1:500) {
```

```
+ sam = rnorm(10)
+ xbar[i]=mean(sam); std[i] = sd(sam)
+ }
> plot(xbar,std)
> cor(xbar,std)
[1] 0.09986
```

The scatterplot (not shown) and small correlation is consistent with known independence of the variables.

Repeat the above with the t-distribution with 3 degrees of freedom (a long-tailed symmetric distribution) and the exponential distribution with rate 1 (a skewed distribution). Are there differences? Explain.

6.9 For a normal population the statistic $Z = (\bar{X} - \mu)/(\sigma/\sqrt{n})$ has a normal distribution. Let

$$T = \frac{\bar{x} - \mu}{s/\sqrt{n}}.$$

That is, σ is replaced by s, the sample standard deviation. The sampling distribution of T is different from that of Z.

To see that the sampling distribution is not a normal distribution, perform a simulation for $n = 3, 10, 25, 50$, and 100. Compare normality with a q-q plot. For $n = 3$ this is done as follows:

```
> n = 3; m = 200;
> res = c()
> for(i in 1:m) {
+ x = rnorm(n)                         # mu = 0, sigma = 1
+ res[i] = (mean(x) - 0)/ (sd(x)/sqrt(n))
+ }
> qqnorm(res)
```

For which values of n is the sampling distribution long tailed? For which values is it approximately normal?

6.10 In the previous exercise it is seen that for a normal population the sampling distribution of

$$T = \frac{\bar{x} - \mu}{s/\sqrt{n}}$$

is not the normal distribution. Rather, it is the t-distribution with $n-1$ degrees of freedom. Investigate this with a q-q plot using a random sample from the t-distribution to compare with the simulation of T. For $n = 3$ this is done as follows:

```
> n = 3; m = 1000;
> res = c()
> for(i in 1:m) {
+ x = rnorm(n)                         # mu = 0, sigma = 1
+ res[i] = (mean(x) - 0)/ (sd(x)/sqrt(n))
```

```
+ }
> qqplot(res,rt(m, df=n-1))
```

Verify this graphically for $n = 3, 10, 25, 50$, and 100.

6.11 In the previous exercise, the sampling distribution of

$$T = \frac{\bar{x} - \mu}{s/\sqrt{n}}$$

was seen to be the t-distribution when the sample comes from a normally distributed population. What about other populations? Repeat the above with the following three mean-zero distributions: the t-distribution with 15 degrees of freedom (symmetric, longish tails), the t-distribution with 2 degrees of freedom (symmetric with long tails), and exponential with rate 1 minus 1 (`rexp(10)-1`), which is skewed right with mean 0. Do all three populations result in T having an approximate t-distribution? Do any?

6.12 We can use probability to estimate a value of π. How? The idea dates to 1777 and Georges Buffon. Pick a point at random from a square of side length 2 centered at the origin. Inside the square draw a circle of radius 1. The probability that the point is inside the circle is equal to the area of the circle divided by the area of the square: $(\pi \cdot 1^2)/(2^2) = \pi/4$. We can simulate the probability and then multiply by 4 to estimate π.

This function will do the simulation and make the plot:

```
simpi <- function(n = 1000) {
  ## draw box, circle plot points, and return no inside
  plot(0,0,pch=" ",xlim=c(-1,1),ylim=c(-1,1))
  polygon(c(-1,-1,1,1,-1),c(-1,1,1,-1,-1)) # square
  theta=seq(0,2*pi,length=100)
  polygon(cos(theta),sin(theta))        # circle

  x = runif(n,min=-1,max=1)
  y = runif(n,min=-1,max=1)

  inorout = x^2 + y^2 < 1
  points(x,y,pch=as.numeric(inorout))

  return(sum(inorout))
}
```

The simulation could be done with just one line:

```
> n = 1000; x=runif(n,-1,1);y=runif(n,-1,1);sum(x^2 + y^2<1)/n
```

Do a simulation to estimate π. What do you get? Use the binomial model and the known value of π to find the standard deviation of the random variable you estimated.

Chapter 7

Confidence intervals

In this chapter we use probability models to make statistical inferences about the parent distribution of a sample. A motivating example is the way in which a public-opinion poll is used to make inferences about the unknown opinions of a population.

7.1 Confidence interval ideas

■ **Example 7.1: How old is the universe?** The `age.universe (UsingR)` data set contains estimates for the age of the universe, dating back to some early estimates based on the age of the earth. As of 2003, the best estimate for the age of the universe is 13.7 billion years old, as computed by the Wilkinson microwave anisotropy probe (`http://map.gsfc.nasa.gov`). This is reported to have a margin of error of 1% with 95% confidence. That is, the age is estimated to be in the interval $(13.56, 13.84)$ with high probability. Figure 7.1 shows other such intervals given by various people over time. Most, but not all, of the modern estimates contain the value of 13.7 billion years. This does not mean any of the estimates were calculated incorrectly. There is no guarantee, only a high probability, that a confidence interval will always contain the unknown parameter. ■

7.1.1 Finding confidence intervals using simulation

To explore the main concepts of a confidence interval, let's consider the example of a simple survey. We'll assume the following scenario. A population exists of 10,000 people; each has a strong opinion for or against some proposition. We

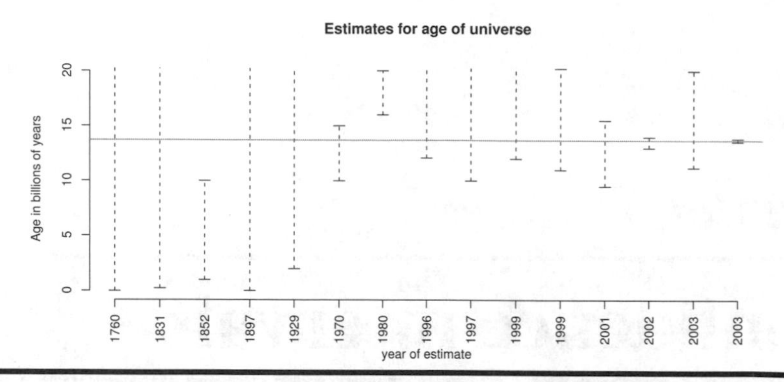

Figure 7.1 Various estimates for the age of universe, some historic, some modern. Ranges are represented by dotted lines. When the estimate is a lower bound, only a bottom bar is drawn. The current best estimate of 13.7 billion years old is drawn with a horizontal line. This estimate has a margin of error of 1%.

wish to know the true proportion of the population that is for the proposition. We can't afford to ask all 10,000 people, but we can survey 100 at random. If our sample proportion is called $\widehat{p}$, and the true proportion is p, what can we infer about the unknown p based on $\widehat{p}$?

Sometimes it helps to change the question to one involving simple objects. In this case, imagine we have 10,000 balls inside a box with a certain proportion, p, labeled with a "1," and the remaining labeled with a "0." We then select 100 of the balls at random and count the 1's.

In order to gain some insight, we will do a simulation for which we know the answer (the true value of p). Suppose the true proportion for the population is $p = 0.56$. That is 5,600 are for the proposition and 4,400 are against. To sample *without replacement* we use the `sample()` command. A single sample is found as follows:

```
> pop = rep(0:1,c(10000 - 5600,5600))
> phat = mean(sample(pop,100))
> phat
[1] 0.59
```

In this example, using the `mean()` function is equivalent to finding the proportion. If we simulate this sampling 1,000 times we can get a good understanding of the sampling distribution of $\widehat{p}$ from the repeated values of `phat`. The following will do so and store the results in a data vector called `res`.

```
> res = c()
> for(i in 1:1000) res[i] = mean(sample(pop,100))
```

From the values in `res` we can discern intervals where we are pretty confident $\widehat{p}$ will be. In particular, using the `quantile()` function, we have these intervals for 80%, 90%, and 95% of the data (see Figure 7.2):

```
> quantile(res,c(0.1,0.9))        # 80% of the time
 10%  90%
0.50 0.63
> quantile(res,c(0.05,0.95))      # 90% of the time
  5%  95%
0.48 0.64
> quantile(res,c(0.025,0.975))  # 95% of the time
 2.5% 97.5%
 0.47  0.66
```

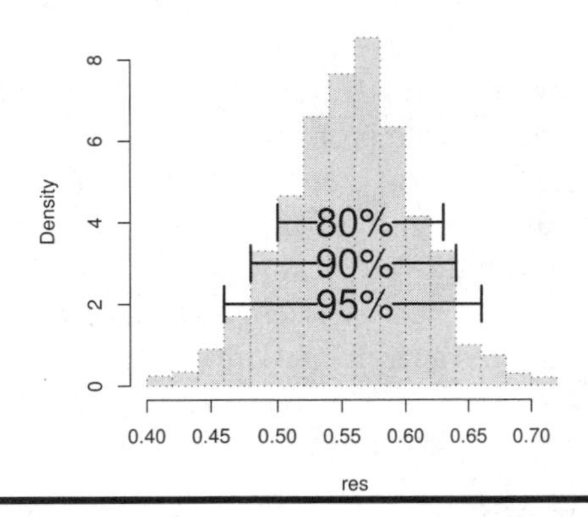

Figure 7.2 Percent of $\widehat{p}$ values in the indicated intervals

These suggest the following probability statements:

$$P(0.50 \leq \widehat{p} \leq 0.63) = 0.80,$$

$$P(0.48 \leq \widehat{p} \leq 0.64) = 0.90, \qquad \text{and}$$

$$P(0.47 \leq \widehat{p} \leq 0.66) = 0.95.$$

We interpret these by envisioning picking one of the 1,000 samples at random and asking the probability that it is in that range. This should be close to the true probability that $\widehat{p}$ is in the range, as we have generated many realizations of $\widehat{p}$.

In this example, we know that $p = 0.56$. Rewriting $0.50 = p - 0.06$, for example, the first one becomes

$$P(p - 0.06 \leq \widehat{p} \leq p + 0.07) = 0.80,$$

which, when we subtract p from all sides, becomes

$$P(-0.06 \leq \widehat{p} - p \leq 0.07) = 0.80.$$

This says that the distance between p and $\widehat{p}$ is less than 0.07 with 80% probability. We could have done this for any percentage, not just 80%. So we get this relationship: If we fix a probability, we can find an amount whereby the distance between p and $\widehat{p}$ is less than this amount with the given probability.

Continuing, we turn this around by *subtracting* $\widehat{p}$ from all sides (and flipping the signs) to get

$$\mathsf{P}(\widehat{p} - 0.07 \leq p \leq \widehat{p} + 0.06) = 0.80.$$

That is, from a single randomly chosen value of $\widehat{p}$ we can find an interval, $(\widehat{p} - 0.07, \widehat{p} + 0.06)$, that contains p with probability 0.80.

As before, something similar is true for other probabilities. If we specify a probability, then we can find an interval around a randomly chosen sample value, $\widehat{p}$, that contains p with the specified probability. This interval is called a **confidence interval**, as we have a certain confidence (given by the probability) that the parameter, p, is in this random interval.

7.2 Confidence intervals for a population proportion, p

In the previous example, the exact distribution of $\widehat{p}$ is hypergeometric, as we sampled without replacement. If we sampled with replacement, then the distribution of $\widehat{p}$ would be $\mathsf{Binomial}(100, p)$ divided by 100. Unless p is very close to 1 or 0, the distribution of $\widehat{p}$ should be normal by the normal approximation to the binomial. Thus, we should expect that $\widehat{p}$ is approximately normal, as the differences between sampling with or without replacement should be slight when there are 10,000 in the population and only 100 chosen.

If we know that the distribution of $\widehat{p}$ is approximately normal, we can find confidence intervals that use this known distribution rather than a simulation. This will allow us to make general statements about the relationship between the confidence probability and the size of the interval.

In order to use the normal approximation we need to make some assumptions about our sampling. First, we either sample with replacement or sample from a population that is so much bigger in size than the size of the sample that it is irrelevant. Next, we need to assume that np and $n(1-p)$ are both bigger than 5, even though p is unknown.

Assuming the binomial model applies, we can derive that the mean of $\widehat{p}$ is p and $\mathsf{SD}(\widehat{p})$ is $\sqrt{p(1-p)/n}$. Thus, if we let $1-\alpha$ be our confidence probability, then we can find from the normal distribution a corresponding z^* for which

$$\mathsf{P}(-z^* \leq Z \leq z^*) = 1 - \alpha.$$

As $\widehat{p}$ is approximately normal, we standardize it to get this relationship:

$$P(-z^* \leq \frac{\widehat{p}-p}{\mathsf{SD}(\widehat{p})} \leq z^*) \approx 1-\alpha. \quad (7.1)$$

That is, with probability $1-\alpha$, p is in the interval $(\widehat{p}-z^*\mathsf{SD}(\widehat{p}), \widehat{p}+z^*\mathsf{SD}(\widehat{p}))$.

This almost specifies a confidence interval, except that $\mathsf{SD}(\widehat{p})$ involves the unknown value of p. There are two ways around this problem. In this case, we can actually solve the equations and get an interval for p in terms of $\widehat{p}$ alone. However, for instructive purposes, we will make another assumption to simplify the math. Let's assume that the value of $\mathsf{SD}(\widehat{p}) = \sqrt{p(1-p)/n}$ is approximately $\mathsf{SE}(\widehat{p}) = \sqrt{\widehat{p}(1-\widehat{p})/n}$. The central limit still applies with this divisor. Consequently, for n large enough

$$P(-z^* \leq \frac{\widehat{p}-p}{\mathsf{SE}(\widehat{p})} \leq z^*) \approx 1-\alpha.$$

The value $\mathsf{SE}(\widehat{p})$ is called the **standard error** of $\widehat{p}$. It is known from the sample and is found by replacing the unknown population parameter in the standard deviation with the known statistic. This assumption is good provided n is large enough.

Confidence intervals for p

Assume n is large enough so that

$$\frac{\widehat{p}-p}{\mathsf{SE}(\widehat{p})}$$

is approximately normal where

$$\mathsf{SE}(\widehat{p}) = \sqrt{\widehat{p}(1-\widehat{p})/n}.$$

Let α and z^* be related by the distribution of a standard normal random variable through

$$P(-z^* \leq Z \leq z^*) = 1-\alpha.$$

Then the interval $(\widehat{p}-z^*\mathsf{SE}(\widehat{p}), \widehat{p}+z^*\mathsf{SE}(\widehat{p}))$ contains p with approximate probability $1-\alpha$. The interval is referred to as a $(1-\alpha)100\%$ **confidence interval** and is often abbreviated $\widehat{p} \pm z^*\mathsf{SE}(\widehat{p})$. The probability is called the **level of confidence** and the value $z^*\mathsf{SE}(\widehat{p})$ the **margin of error**.

The `prop.test()` function can be used to compute confidence intervals of proportions.

Finding z^* from α. From Figure 7.3 we see that z^* (also called $z_{\alpha/2}$ in other books) is related to $\alpha/2$. In particular, either

$$\alpha/2 = P(Z \leq -z^*) \qquad \text{or, similarly} \qquad 1-\alpha/2 = P(Z \leq z^*).$$

In R this becomes one of

```
> zstar = -qnorm(alpha/2)         # left tail
> zstar = qnorm(1-alpha/2)        # right tail
```

The inverse relationship would be found by

```
> alpha = 2*pnorm(-zstar)
```

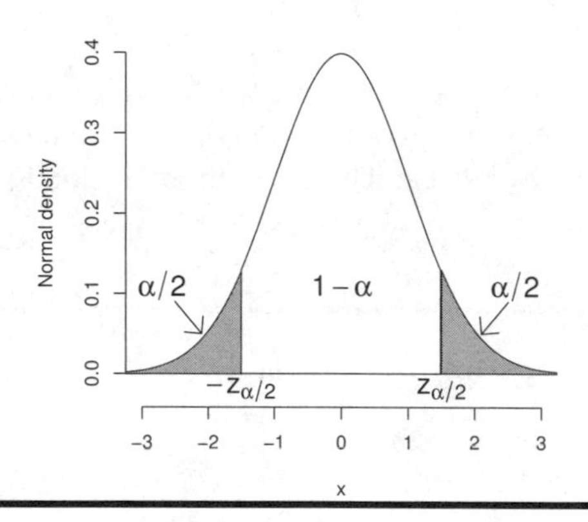

Figure 7.3 The relationship between z^*, or $z_{\alpha/2}$, and α

■ **Example 7.2: Presidential job performance** A Zogby America poll involved 1,013 likely voters selected randomly from throughout the 48 contiguous United States using listed residential telephone numbers.The surveyers found that 466 voters rated the president's job performance as "good" or "excellent." Find a 95% confidence interval for the true proportion.

This type of polling is about as close to a random sample as can be gotten with limited resources, though there are several sources of possible bias to consider. For example, not everyone in the United States has a listed residential telephone number, so the sample is only from households that do. Additionally, nonresponse can be very high in such surveys, introducing another potential bias. For simplicity, we'll assume the sample is a random sample from the population of likely voters and that $n = 1013$ is large enough so that the normal approximation applies.

As $\widehat{p} = 466/1013$, a 95% confidence interval for p would be $\widehat{p} \pm z^* \mathsf{SE}(\widehat{p})$. To find this we have

```
> n = 1013
> phat = 466/n
> SE = sqrt(phat*(1-phat) /n)
> alpha = .05
> zstar = -qnorm(alpha/2)
> zstar                          # nearly 2 if doing by hand
[1] 1.96
> c(phat - zstar * SE, phat + zstar*SE)
[1] 0.4293 0.4907
```

The confidence interval does not include $p = .5$.

The last line matches the formulas, but it takes less typing to use

```
> phat + c(-1,1)*zstar*SE
[1] 0.4293 0.4907
```

■

■ **Example 7.3: The missing confidence level** In United States newspapers the results of a survey are often printed with the sample proportion, the sample size, and a margin of error. The confidence level is almost always missing but can be inferred from the three pieces of information. If a survey has $\widehat{p} = .57$, $n = 1,000$, and a margin of error of 3 percentage points, what is α?

Assuming the survey is done with a random sample, we are given that $z^* \mathsf{SE}(\widehat{p}) = .03$ and $\mathsf{SE} = \sqrt{.57 \cdot (1 - .57)/1000}$. Solve for z^* and then $1 - \alpha$ as follows:

```
> zstar = 0.03 / sqrt(.57*(1-.57)/1000)
> zstar
[1] 1.916
> alpha = 2* pnorm(-zstar)
> alpha
[1] 0.05533
> 1- alpha
[1] 0.9447
```

There is an implied 95% confidence level. ■

7.2.1 Using `prop.test()` to find confidence intervals

The above examples were done the hard way. R provides a built-in function, `prop.test()`, to compute confidence intervals of proportions. A template for usage is

```
prop.test(x, n, conf.level=0.95, conf.int=TRUE)
```

The frequency, given by `x`; the sample size, given by `n`; and a confidence level, set with `conf.level=`, need to be specified. The default confidence level is 0.95, and by default a confidence interval will be returned.

For instance, in the example of the Zogby poll, $n = 1013, x = n\widehat{p} = 466$, and $1 - \alpha = 0.95$. The confidence interval is found with

```
> prop.test(466,1013,conf.level=0.95)

        1-sample proportions test with continuity correction

data:  466 out of 1013, null probability 0.5
X-squared = 6.318, df = 1, p-value = 0.01195
alternative hypothesis: true p is not equal to 0.5
95 percent confidence interval:
 0.4290 0.4913
sample estimates:
   p
0.46
```

The output contains more than we needed. We see, though, that just after the line `95 percent confidence interval:` are two values, 0.4290 and 0.4913, which are the endpoints of our interval. These are slightly different from our previously found endpoints, as the formula used by `prop.test()` is found by solving Equation 7.1 exactly rather than by using the standard error to approximate the answer.

The extra argument `conf.level=0.95` sets the confidence level. The default is 0.95, so in this case it could have been left off.

binom.test()

The function `binom.test()` will also find confidence intervals. In this case, it uses the binomial distribution in place of the normal approximation.

7.2.2 Problems

7.1 Find an example in the media in which the results of a poll are presented. Identify the population, the size of the sample, the confidence interval, the margin of error, and the confidence level.

7.2 In Example 7.2 a random sample from the United States population is taken by using listed residential phone numbers. Which segments of the population would be missed by this sampling method?

7.3 The web site `http://www.cnn.com` conducts daily polls. Explain why the site's disclaimer reads:

> This QuickVote is not scientific and reflects the opinions of only those Internet users who have chosen to participate. The results cannot be assumed to represent the opinions of Internet users in general, nor the public as a whole.

7.4 Suppose a Zogby poll with 1,013 randomly selected participants and the `http://www.cnn.com` poll (see the previous problem) with 80,000 respondents ask the same question. If there is a discrepancy in the sample proportion, which would you believe is closer to the unknown population parameter?

7.5 Find 80% and 90% confidence intervals for a survey with $n = 100$ and $\widehat{p} = 0.45$.

7.6 A student wishes to find the proportion of left-handed people. She surveys 100 fellow students and finds that only 5 are left-handed. Does a 95% confidence interval contain the value of $p = 1/10$?

7.7 Of the last ten times you've dropped your toast, it has landed sticky-side down nine times. If these are a random sample from the Bernoulli(p) distribution, find an 80% confidence interval for p, the probability of the sticky side landing down.

7.8 A *New York Times* article from October 9, 2003, contains this explanation about an exit survey for a California recall election:

> In theory, in 19 cases out of 20, the results from such polls should differ by no more than plus or minus two percentage points from what would have been obtained by seeking to interview everyone who cast a ballot in the recall election.

Assume a simple random sample and $\widehat{p} = .54$. How big was n?

7.9 An erstwhile commercial claimed that "Four out of five dentists surveyed would recommend Trident for their patients who chew gum."

Assume the results were based on a random sample from the population of all dentists. Find a 90% confidence interval for the true proportion if the sample size was $n = 5$. Repeat with $n = 100$ and $n = 1,000$.

7.10 A survey is taken of 250 students, and a $\widehat{p}$ of 0.45 is found. The same survey is repeated with 1,000 students, and the same $\widehat{p}$ value is found. Compare the two 95% confidence intervals. What is the relationship? Is the margin of error for the second one four times smaller? How much smaller is it?

7.11 How big a survey is needed to be certain that a 95% confidence interval has a margin of error no bigger than 0.01? How does this change if you are asked for an 80% confidence interval?

7.12 The phrasing, "The true value, p, is in the confidence interval with 95% probability" requires some care. Either p is or isn't in a given interval. What it means is that if we *repeated* the sampling, there is a 95% chance the true value is in the random interval. We can investigate this with a simulation. The commands below will find several confidence intervals at once.

```
> m = 50; n=20; p = .5;           # toss 20 coins 50 times,
> alpha = 0.10;zstar = qnorm(1-alpha/2)
> phat = rbinom(m,n,p)/n           # divide by n for proportions
> SE = sqrt(phat*(1-phat)/n)       # compute SE
```

We can find the proportion that contains p using

```
> sum(phat - zstar*SE < p & p < phat + zstar * SE)/m
```

and draw a nice graphic with

```
> matplot(rbind(phat - zstar*SE, phat + zstar*SE),
+         rbind(1:m,1:m),type="l",lty=1)
> abline(v=p)                      # indicate parameter value
```

Do the simulation above. What percentage of the 50 confidence intervals contain $p = 0.5$?

7.3 Confidence intervals for the population mean, μ

The success of finding a confidence interval for p in terms of $\widehat{p}$ depended on knowing the sampling distribution of $\widehat{p}$ once we standardized it. We can use the same approach to find a confidence interval for μ, the population mean, from the sample mean $\bar{X}$.

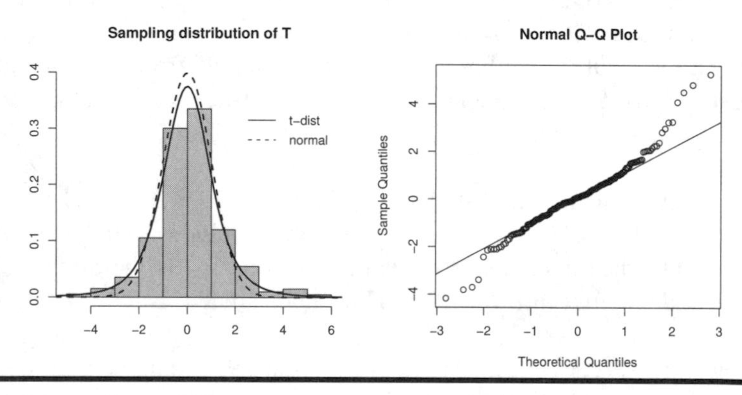

Figure 7.4 Simulation of sampling distribution of T with $n = 5$. Densities of normal distribution and t-distribution are drawn on the histogram to illustrate that the sampling distribution of T has longer tails than the normal distribution.

For a random sample $X_1, X_2, \ldots, X_n$, the central limit theorem and the formulas for the mean and standard deviation of $\bar{X}$ tell us that for large n

$$Z = \frac{\bar{X} - \mu}{\sigma/\sqrt{n}} = \frac{\bar{X} - \mu}{\mathsf{SD}(\bar{X})}$$

will have an approximately normal distribution. This implies, for example, that roughly 95% of the time Z is no larger than 2 in absolute value. In terms of intervals, this can be used to say that μ is in the random interval $\bar{X} \pm 2 * \sigma/\sqrt{n}$ with probability 0.95.

However, σ is usually not known. The standard error, $\mathsf{SE}(\bar{X}) = s/\sqrt{n}$, replaces the unknown σ by the known sample standard deviation, s. Consider

$$T = \frac{\bar{X} - \mu}{s/\sqrt{n}} = \frac{\bar{X} - \mu}{\mathsf{SE}(\bar{X})}.$$

Again, as the central limit theorem still applies, T has a sampling distribution that is approximately normal when n is large enough. This fact can be used to construct confidence intervals such as a 95% confidence interval of $\bar{X} \pm 1.96s/\sqrt{n}$.

When n is not large, T will also be of value when the population for the random sample is normally distributed. In this case, the sampling distribution of T is the t-distribution with $n-1$ degrees of freedom. The t-distribution is a symmetric, bell-shaped distribution that asymptotically approaches the standard normal distribution but for small n has fatter tails. The degrees of freedom, $n-1$, is a parameter for this distribution the way the mean and standard deviation are for the normal distribution. Figure 7.4 shows the results of a simulation of T for $n = 5$. The figures show that T, with 5 degrees of freedom, is long tailed compared to the normal distribution.

Confidence intervals for the mean

Let $X_1, X_2, \ldots, X_n$ be a random sample from a population with mean μ and variance σ^2. Let $\bar{X}$ be the sample mean, and $\mathsf{SE}(\bar{X}) = s/\sqrt{n}$.

If n is small and the population is $\mathsf{Normal}(\mu, \sigma)$, then a $(1-\alpha)100\%$ confidence interval for μ is given by

$$\bar{X} \pm t^* \mathsf{SE}(\bar{X}),$$

where t^* is related to α through the t-distribution with $n-1$ degrees of freedom by

$$\mathsf{P}(-t^* \leq T_{n-1} \leq t^*) = 1 - \alpha.$$

For unsummarized data, the function `t.test()` will compute the confidence intervals. A template for its usage is

```
t.test(x, conf.level=0.95)
```

The data is stored in a data vector (named `x` above) and the confidence level is specified with `conf.level=`.

If n is large enough for the central limit theorem to apply to the sampling distribution of T, then a $(1-\alpha)100\%$ confidence interval for μ is given by

$$\bar{X} \pm z^* \mathsf{SE}(\bar{X}),$$

where z^* is related to α by

$$\mathsf{P}(-z^* \leq Z \leq z^*) = 1-\alpha.$$

Finding t^* with R Computing the value of t^* (also called $t_{\alpha/2,k}$) for a given α and vice versa is done in a manner similar to finding z^*, except that a different density is used. As R is a consistent language, changing to a new density requires nothing more than using the proper family name—`t`, for the t-distribution, and `norm` for the normal—and specifying the parameter values. In particular, if n is the sample size, then the two are related as follows:

```
> tstar = qt(1 - alpha/2,df=n-1)
> alpha = 2*pt(-tstar, df=n-1)
```

By way of contrast, for z^* the corresponding commands are

```
> zstar = qnorm(1 - alpha/2)
> alpha = 2*pnorm(-zstar)
```

■ **Example 7.4: Average height** Students in a class of 30 have an average height of 66 inches, with a standard deviation of 4 inches. Assume that these heights are normally distributed, and that the class can be considered a random sample from the entire college population. What is an 80% confidence interval for the mean height of all the college students?

Our assumptions allow us to apply the confidence interval for the mean, so the answer is $\bar{X} \pm t^* \mathsf{SE}(\bar{X})$. Computing gives

```
> xbar = 66; s = 4; n = 30
> alpha = 0.2
> tstar = qt(1 - alpha/2, df = n-1)
> tstar
[1] 1.311
```

```
> SE = s/sqrt(n)
> c(xbar - tstar*SE, xbar + tstar*SE)
[1] 65.04 66.96
```

■

■ **Example 7.5: Making coffee** A barista at "t-test espresso" has been trained to set the bean grinder so that a 25-second espresso shot results in 2 ounces of espresso. Knowing that variations are the norm, he pours eight shots and measures the amounts to be 1.95, 1.80, 2.10, 1.82, 1.75, 2.01, 1.83, and 1.90 ounces. Find a 90% confidence interval for the mean shot size. Does it include 2.0?

As we have the data, we can use `t.test()` directly. We enter in the data, verify normality (with a quantile-quantile plot), and then use `t.test()`:

```
> ozs = c(1.95, 1.80, 2.10, 1.82, 1.75, 2.01, 1.83, 1.90)
> qqnorm(ozs)                        # approximately linear
> t.test(ozs,conf.level=0.80)
        One Sample t-test
data:  ozs
t = 45.25, df = 7, p-value = 6.724e-10
alternative hypothesis: true mean is not equal to 0
80 percent confidence interval:
 1.836 1.954
sample estimates:
mean of x
    1.895
```

Finding the confidence interval to be $(1.836, 1.954)$, the barista sees that 2.0 is not in the interval. The barista adjusts the grind to be less fine and switches to decaf. ■

The T-statistic is robust The confidence interval for the mean relies on the fact that the sampling distribution of $T = (\bar{X} - \mu)/\mathsf{SE}(\bar{X})$ is the t-distribution with $n-1$ degrees of freedom. This is true when the X_i are *i.i.d.* normal. What if the X_i are not normal?

If n is small, we can do simulations to see that the distribution of T is still approximately the t-distribution if the parent distribution of the X_i is not too far from normal. That is, the tails can't be too long, or the skew can't be too great. When n is large, the central limit theorem applies. A statistic whose sampling distribution doesn't change dramatically for moderate changes in the population distribution is called a **robust statistic**.

7.3.1 One-sided confidence intervals

When finding a confidence interval for the mean for a given α, we found t^* so that $\mathsf{P}(-t^* \leq T_{n-1} \leq t^*) = 1 - \alpha$. This method returns symmetric confidence intervals. The basic idea is that the area under the density of the sampling distribution that lies outside the confidence interval is evenly split on each side. This leaves $\alpha/2$ area in each tail. This is similar to Figure 7.3, in which the normal density is shown with equal allocation of the area to the tails. This approach is not the only one. This extra area can be allocated in any proportion to the left or right of the confidence interval. One-sided confidence intervals put the area all on one side or the other. For confidence intervals for the mean, based on the T statistic, these would be found for a given α by finding t^* such that $\mathsf{P}(t^* \leq T_{n-1}) = 1-\alpha$ or $\mathsf{P}(T_{n-1} \leq t^*) = 1-\alpha$.

In R, the `prop.test()`, `binom.test()`, and `t.test()` functions can return one-sided confidence intervals. When the argument `alt="less"` is used, an interval of the type $(-\infty, b]$ is printed. Similarly, when `alt="greater"` is used, an interval of the type $[b, \infty)$ is printed.

■ **Example 7.6: Serving coffee** The barista at "t-test espresso" is told that the optimal serving temperature for coffee is 180 °F. Five temperatures are taken of the served coffee: 175, 185, 170, 184, and 175 degrees. Find a 90% confidence interval of the form $(-\infty, b]$ for the mean temperature.

Using `t.test()` with `alt="less"` will give this type of one-sided confidence interval:

```
> x = c(175, 185, 170, 184, 175)
> t.test(x,conf.level = 0.90, alt="less")

        One Sample t-test

data:  x
t = 61.57, df = 4, p-value = 1
alternative hypothesis: true mean is less than 0
90 percent confidence interval:
  -Inf 182.2
sample estimates:
mean of x
    177.8
```

The confidence interval contains the value of 180 degrees. ■

7.3.2 Problems

7.13 A hard-drive manufacturer would like to ensure that the mean time between failures (MTBF) for its new hard drive is 1 million hours. A stress test is designed that can simulate the workload at a much faster pace. The testers assume that a test lasting 10 days correlates with the failure time exceeding the 1-million-hour mark. In stress tests of 15 hard drives they found an average of 9.5 days, with a standard deviation of 1 day. Does a 90% confidence level include 10 days?

7.14 The stud.recs (UsingR) data set contains math SAT scores in the variable sat.m. Find a 90% confidence interval for the mean math SAT score for this data.

7.15 For the homedata (UsingR) data set find 90% confidence intervals for both variables y1970 and y2000. Use t.test(), but first discuss whether it is appropriate.

7.16 The variable weight in the kid.weights (UsingR) data set contains the weights of a random sample of children. Find a 90% confidence interval for the weight of 5-year-olds. You'll need to isolate just the 5-year-olds' data first. Here's one way:

```
> attach(kid.weights)
> ind = age < (5+1)*12 & age >= 5*12
> weight[ind]                    # just five-year olds
> detach(kid.weights)
```

7.17 The brightness (UsingR) data set contains information on the brightness of stars in a sector of the sky. Find a 90% confidence interval for the mean.

7.18 The data set normtemp (UsingR) contains measurements of 130 healthy, randomly selected individuals. The variable temperature contains normal body temperature. Does the data appear to come from a normal distribution? Is so, find a 90% confidence interval for the mean normal body temperature. Does it include 98.6 °F?

7.19 The t-distribution is also called the Student t-distribution. (A Guinness Brewery employee, William Gosset, derived the distribution of T to handle small samples. As Guinness did not allow publication of research results at the time, Gosset chose to publish under the pseudonym Student.)

Gosset applied his research to a data set containing height and left-middle-finger measurements of 3,000 criminals. These values were written on cards and randomly sorted into 750 samples, each containing four criminals. (This is how simulations were done previously.)

Suppose the first sample of four had an average height of 67.5 inches, with a standard deviation of 2.54. From this sample, find a 95% confidence interval for the mean height of the 3,000 data points.

7.20 We can investigate how robust the T statistic is to changes in the underlying parent population from normality. In particular, we can verify that if the parent population is not too skewed or is symmetric without too heavy a tail then the T statistic will still have the t-distribution for its sampling distribution.

A simulation of the T statistic when X_i are $\mathsf{Normal}(0, 1)$ may be done as follows:

```
> n = 10; m = 250; df = n - 1
> res = c()
> for(i in 1:m) {
+ x = rnorm(n)                    # change this line only
+ res[i] = (mean(x) - 0)/(sd(x)/sqrt(n))
+ }
> qqplot(res, rt(m,df=df))
```

The quantile-quantile plot compares the distribution of the sample with a sample from the t-distribution. If you type this in you will see that the points are close to linear, as the sampling distribution is the t-distribution.

To test different parent populations you can change the line `x = rnorm(n)` to some other distributions with mean 0. For example, try a short-tailed distribution with `x = runif(n)-1/2`; a symmetric, long-tailed distribution with `x = rt(n,3)`; a not so long-tailed, symmetric distribution with `x = rt(n,30)`; and a skewed distribution with `x = rexp(n) - 1`.

7.21 We can compare the relationship of the t-distribution with $n-1$ degrees of freedom with the normal distribution in several ways. As n gets large, the t-distribution converges to the standard normal. But what happens when n is "small," and what do we mean by "large"?

A few comparative graphs can give us an idea. For $n = 10$ we can use boxplots of simulated data to examine the tails, or we can compare plots of theoretical quantiles or densities. These plots are created as follows:

```
> n = 10
> boxplot(rt(1000,df=n-1),rnorm(1000))
> x = seq(0,1,length=150)
> plot(qt(x,df=n-1), qnorm(x));abline(0,1)
> curve(dnorm(x),-3.5,3.5)
> curve(dt(x,df=n-1), lty=2, add=TRUE)
```

Repeat the above for $n = 3$, 25, 50, and 100. What value of n seems "large" enough to say that the two distributions are essentially the same?

7.22 When the parent population is $\mathsf{Normal}(\mu, \sigma)$ with known σ, then confidence intervals of the type

$$\bar{X} \pm z^* \mathsf{SD}(\bar{X}) \text{ and } \bar{X} \pm t^* \mathsf{SE}(\bar{X})$$

are both applicable. We have that far enough in the tail, $z^* < t^*$, but sometimes $s < \sigma$, so there is no clear winner as to which confidence interval is smaller.

Run a simulation 200 times in which the margin of error is calculated both ways for a sample of size 10 with $\sigma = 2$ and $\mu = 0$. Use a 90% confidence level. What percent of the time was the confidence interval using $\mathsf{SD}(\bar{X})$ smaller?

7.4 Other confidence intervals

To form confidence intervals, we have used the key fact that certain statistics,

$$\frac{\widehat{p}-p}{\sqrt{\widehat{p}(1-\widehat{p})/n}} \quad \text{and} \quad \frac{\bar{X}-\mu}{s/\sqrt{n}},$$

have known sampling distributions that do not involve any population parameters. From this, we could then solve for confidence intervals for the parameter in terms of known quantities.

In general, such a statistic is called a **pivotal quantity** and can be used to generate a number of confidence intervals in various situations.

7.4.1 Confidence interval for σ^2

For example, if the X_i are *i.i.d.* normals, then the distribution of

$$\frac{(n-1)s^2}{\sigma^2}$$

is known to be the χ^2-distribution (chi-squared) with $n-1$ degrees of freedom. This allows us to solve for confidence intervals for σ^2 in terms of the sample variance s^2.

In particular, a $(1-\alpha)100\%$ confidence interval can be found as follows. For a given α, let l^* and r^* solve

$$\mathsf{P}(l^* \leq \chi^2_{n-1} \leq r^*) = 1-\alpha.$$

If we choose l^* and r^* to yield equal areas in the tails, we can find them with

```
> lstar = qchisq(alpha/2, df=n-1)
> rstar = qchisq(1-alpha/2, df=n-1)
```

Then

$$\mathsf{P}(l^* \leq \frac{(n-1)s^2}{\sigma^2} \leq r^*) = 1-\alpha$$

can be rewritten as

$$\mathsf{P}(\frac{(n-1)s^2}{r^*} \leq \sigma^2 \leq \frac{(n-1)s^2}{l^*}) = 1-\alpha.$$

In other words, the interval $((n-1)s^2/r^*, (n-1)s^2/l^*)$ gives a $(1-\alpha)100\%$ confidence interval for σ^2.

■ **Example 7.7: How long is your commute?** A commuter believes her commuting times are independent and vary according to a normal distribution, with unknown mean and variance. She would like to estimate the variance to get an idea of the spread of times.

To compute the variance, she records her commute time on a daily basis. Over 10 commutes she reports a mean commute time of 25 minutes, with a variance of 12 minutes. What is a 95% confidence interval for the variance?

We are given $s^2 = 12$ and $n = 10$, and we assume each X_i is normal and *i.i.d.* From this we find

```
> s2 = 12; n = 10
> alpha = .05
> lstar = qchisq(alpha/2, df = n-1)
> rstar = qchisq(1 - alpha/2, df = n-1)
> (n-1)*s2 * c(1/rstar,1/lstar)           # CI for sigma squared
[1]  5.677 39.994
> sqrt((n-1)*s2 * c(1/rstar,1/lstar))     # CI for sigma
[1] 2.383 6.324
```

After taking the square roots, we get a 95% confidence interval for σ, which is $(2.324, 6.324)$. ■

7.4.2 Problems

7.23 Let $X_1, X_2, \ldots, X_n$ and $Y_1, Y_2, \ldots, Y_m$ be two *i.i.d.* samples with sample variances s_x and s_y respectively. A confidence interval for the equivalence of sample variances can be given from the following statistic:

$$F = \frac{(s_x^2/\sigma_x^2)}{(s_y^2/\sigma_y^2)}.$$

If the underlying X_i and Y_i are normally distributed, then the distribution of F is known to be the F-distribution with $n-1$ and $m-1$ degrees of freedom. That is, F is a pivotal quantity, so probability statements such as $\mathsf{P}(a \leq (s_x^2/\sigma_x^2)/(s_y^2/\sigma_y^2) \leq b)$ can be answered with the known quantiles of the F distribution. For example,

```
> n = 11; m = 16
> alpha = 0.10
> qf(c(alpha/2, 1- alpha/2),df1=n-1,df2=m-1)
[1] 0.3515 2.5437
```

says that $\mathsf{P}(0.3515 \leq (s_x^2/s_y^2)/(\sigma_x^2/\sigma_y^2) \leq 2.5437) = 0.9$ when $n = 11$ and $m = 16$. That is,

$$\frac{1}{2.5437}\frac{s_x^2}{s_y^2} < \frac{\sigma_x^2}{\sigma_y^2} < \frac{1}{.3515}\frac{s_x^2}{s_y^2}$$

with 90% confidence.

Suppose $n = 10, m = 20$, $s_x = 2.3$, and $s_y = 2.8$. Find an 80% confidence interval for the ratio of σ_x/σ_y.

7.24 Assume our data, $X_1, X_2, \ldots, X_n$ is uniform on the interval $[0, \theta]$ (θ is an unknown parameter). Set $\mathsf{max}(X)$ to the be maximum value in the data set. Then the quantity $\mathsf{max}(X)/\theta$ is pivotal with distribution

$$\mathsf{P}(\frac{\mathsf{max}(X)}{\theta} < x) = x^n, \quad 0 \leq x \leq 1.$$

Thus $\mathsf{P}(\mathsf{max}(X)/x < \theta) = x^n$. As θ is always bigger than $\mathsf{max}(X)$, we can solve for $x^n = \alpha$ and get that θ is in the interval $[\max(X), \max(X)/x]$ with probability $1 - \alpha$.

Use this fact to find a 90% confidence interval for the number of entries in the 2002 New York City Marathon. The `place` variable from the data set `nyc.2002` `(UsingR)` contains the place of the runner in the sample and is randomly sampled from all the possible places.

7.5 Confidence intervals for differences

When we have two samples, we might ask whether the two came from the same population. For example, Figure 7.5 shows results for several polls on presidential approval rating from early 2001 to early 2004.* The rating varies over time, but for any given time period the polls are all pretty much in agreement. This is to be expected, as the polls are tracking the same population proportion for a given time period. However, how can we tell if the differences between polls for different time periods are due to a change in the underlying population proportion or merely an artifact of sampling variation?

7.5.1 Difference of proportions

We compare two proportions when assessing the results of surveys, as with the approval ratings, but we could do the same to compare other proportions, such as market shares.

To see if a difference in the proportions is explainable by sampling error, we look at $\widehat{p}_1 - \widehat{p}_2$ and find a confidence interval for $p_1 - p_2$. This can be done, as

* A similar figure appeared in a February 9, 2004, edition of *Salon* (`http://www.salon.com`). The data is in the data set `BushApproval (UsingR)`, and the graphic can be produced by running `example(BushApproval)`.

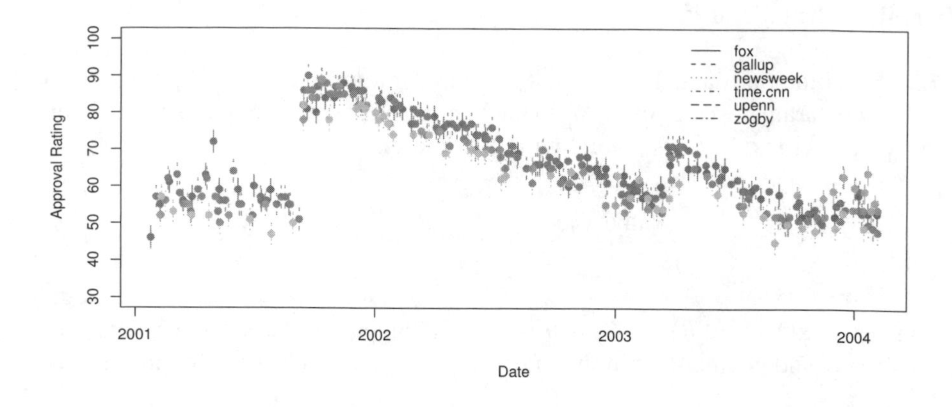

Figure 7.5 Presidential approval rating in United States from spring 2001 to spring 2004

the statistic

$$Z = \frac{(\widehat{p}_1 - \widehat{p}_2) - (p_1 - p_2)}{\mathsf{SE}(\widehat{p}_1 - \widehat{p}_2)}$$

is a pivotal quantity with standard normal distribution when n_1 and n_2 are large enough. The standard error is

$$\mathsf{SE}(\widehat{p}_1 - \widehat{p}_2) = \sqrt{\frac{\widehat{p}_1(1-\widehat{p}_1)}{n_1} + \frac{\widehat{p}_2(1-\widehat{p}_2)}{n_2}}.$$

Z has an asymptotic normal distribution, as it may be viewed as a sample average minus its expectation divided by its standard error. The central limit theorem then applies.

The function `prop.test()` can do the calculations for us. We use it as

```
prop.test(x, n, conf.level=0.95)
```

The data is specified in terms of counts, `x`, and sample sizes, `n`, using data vectors containing two entries. The results will differ slightly from the above description, as `prop.test()` uses a continuity correction.

■ **Example 7.8: Comparing poll results** In a span of two weeks the same poll is taken. The first time, 1,000 people are interviewed, and 560 agree; the second time, 1,200 are interviewed, and 570 agree. Find a 95% confidence interval for the difference of proportions.

Rather than do the work by hand, we let prop.test() find a confidence interval.

```
> prop.test(x=c(560,570), n=c(1000,1200), conf.level=0.95)

        2-sample test for equality of proportions with
        continuity correction

data:  c(560, 570) out of c(1000, 1200)
X-squared = 15.44, df = 1, p-value = 8.53e-05
alternative hypothesis: two.sided
95 percent confidence interval:
 0.04231 0.12769
sample estimates:
prop 1 prop 2
 0.560  0.475
```

We see that a 95% confidence interval is $(0.04231, 0.12769)$, which just misses including 0. We conclude that there appears to be a real difference in the population parameters. ■

7.5.2 Difference of means

Many problems involve comparing independent samples to see whether they come from identical parent populations. A teacher could compare two sections of the same class to look for differences; a pharmaceutical company could compare the effects of two drugs; or a manufacturer could compare two samples taken at different times to monitor quality control.

Let $X_1, X_2, \ldots, X_{n_x}$ and $Y_1, Y_2, \ldots, Y_{n_y}$ be the two samples with sample means $\bar{X}$ and $\bar{Y}$ and sample variances s_x^2 and s_y^2. Assume the populations for each sample are normally distributed. The sampling distribution of $\bar{X} - \bar{Y}$ is asymptotically normal, as each is asymptotically normal. Consequently, the standardized statistic

$$T = \frac{(\bar{X} - \bar{Y}) - (\mu_x - \mu_y)}{\mathsf{SE}(\bar{X} - \bar{Y})} \tag{7.2}$$

will have an approximately normal distribution, with mean 0 and variance 1 for large n_x and n_y. For small n_x and n_y, T will have the t-distribution.

The standard error of $\bar{X} - \bar{Y}$ is computed differently depending on the assumptions. For independent random variables, the variance of a sum is the sum of a variance. This is used to show that the variance of $\bar{X} - \bar{Y}$ is $\sigma_x^2/n_x + \sigma_y^2/n_y$.

When the two population variances are equal, the data can be pooled to give an estimate of the common variance σ^2. Let s_p^2 be the pooled estimate. It is defined by

$$s_p^2 = \frac{(n_x - 1)s_x^2 + (n_y - 1)s_y^2}{n_x + n_y - 2}. \tag{7.3}$$

When the population variances are not equal, the sample standard deviations are used to estimate the respective population standard deviations.

The standard error is then

$$\mathsf{SE}(\bar{X} - \bar{Y}) = \begin{cases} \sqrt{s_p^2(1/n_x + 1/n_y)} & \text{if } \sigma_x = \sigma_y, \\ \sqrt{s_x^2/n_x + s_y^2/n_y} & \text{if } \sigma_x \neq \sigma_y. \end{cases} \tag{7.4}$$

The statistic T will have a sampling distribution given by the t-distribution. When the variances are equal, the sampling variation of s_p is smaller, as all the data is used to estimate σ. This is reflected in a larger value of the degrees of freedom. The values used are

$$\text{degrees of freedom} = \begin{cases} n_x + n_y - 2 & \text{if } \sigma_x = \sigma_y, \\ \left(\frac{s_x^2}{n_x} + \frac{s_y^2}{n_y}\right)^2 \cdot \left(\frac{(s_x^2/n_x)^2}{n_x - 1} + \frac{(s_y^2/n_y)^2}{n_y - 1}\right)^{-1} & \text{if } \sigma_x \neq \sigma_y. \end{cases} \tag{7.5}$$

(The latter value is the Welch approximation.)

Given this, the T statistic is pivotal, allowing for the following confidence intervals.

Confidence intervals for difference of means for two independent samples

Let $X_1, X_2, \ldots, X_{n_x}$ and $Y_1, Y_2, \ldots, Y_{n_y}$ be two independent samples with distribution $\mathsf{Normal}(\mu_i, \sigma_i)$, $i = x$ or y. A $(1 - \alpha) \cdot 100\%$ confidence interval of the form

$$(\bar{X} - \bar{Y}) \pm t^* \mathsf{SE}(\bar{X} - \bar{Y})$$

can be found where t^* is given by the t-distribution. This is based on the sampling distribution of T given in Equation 7.2.

This distribution is the t-distribution. The standard error and degrees of freedom differ depending on whether or not the variances are equal. The standard error is given by Equation 7.4 and the degrees of freedom by Equation 7.5.

If the unsummarized data is available, the `t.test()` function can be used to compute the confidence interval for the difference of means. It is used as

```
t.test(x, y, var.equal=FALSE, conf.level=0.95)
```

The data is contained in two data vectors, `x` and `y`. The assumption on the equality of variances is specified by the argument `var.equal=` with default of `FALSE`.

■ **Example 7.9: Comparing independent samples** In a clinical trial, a weight-loss drug is tested against a placebo to see whether the drug is effective. The amount of weight lost for each group is given by the stem-and-leaf plot in Table 3.6. Find a 90% confidence interval for the difference in mean weight loss.

From inspection of the boxplots of the data in Figure 7.6, the assumption of equal variances is reasonable, prompting the use of `t.test()` with the argument `var.equal=TRUE`.

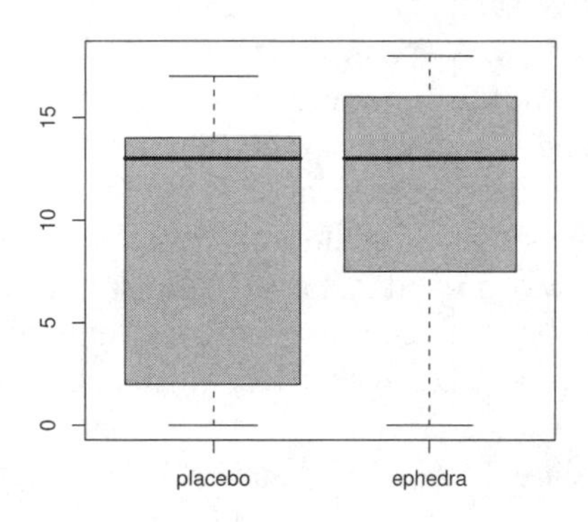

Figure 7.6 Boxplots used to investigate assumption of equal variances

```
> x = c(0,0,0,2,4,5,13,14,14,14,15,17,17)
> y = c(0,6,7,8,11,13,16,16,16,17,18)
> boxplot(list(placebo=x, ephedra=y), col="gray") # compare spreads
> t.test(x,y, var.equal=TRUE)

        Two Sample t-test

data:  x and y
t = -1.054, df = 22, p-value = 0.3032
alternative hypothesis: true difference in means is not equal to 0
```

```
95 percent confidence interval:
 -8.279  2.699
sample estimates:
mean of x mean of y
    8.846    11.636
```

By comparison, if we do not assume equal variances,

```
> t.test(x,y)
...
t = -1.072, df = 21.99, p-value = 0.2953
...
```

When we assume `var.equal=TRUE`, we have $13+11-2=22$ degrees of freedom. In this example, the approximate degrees of freedom in the unequal variance case is found to be 21.99: essentially identical. The default 95% confidence interval is $(-8.279, 2.699)$, so the difference of 0 is still in the confidence interval, even though the sample means differ quite a bit at first glance (8.846 versus 11.636). ■

7.5.3 Matched samples

Sometimes we have two samples that are not independent. They may be paired or matched up in some way. A classic example in statistics involves the measurement of shoe wear. If we wish to measure shoe wear, we might give five kids one type of sneaker and five others another type and let them play for a while. Afterward, we could measure shoe wear and compare. The only problem is that variation in the way the kids play could mask small variations in the wear due to shoe differences. One way around this is to put mismatched pairs of shoes on all ten kids and let them play. Then, for each kid, the amount of wear on each shoe is related, but the difference should be solely attributable to the differences in the shoes.

If the two samples are $X_1, X_2, \ldots, X_n$ and $Y_1, Y_2, \ldots, Y_n$, then the statistic

$$T = \frac{(\bar{X} - \bar{Y}) - (\mu_x - \mu_y)}{\mathsf{SE}(\bar{X} - \bar{Y})}$$

is pivotal with a t-distribution. What is the standard error? As the samples are not independent, the standard error for the two-sample T statistic is not applicable. Rather, it is just the standard error for the single sample $X_i - Y_i$.

Comparison of means for paired samples

Let $X_1, X_2, \ldots, X_n$ and $Y_1, Y_2, \ldots Y_n$ be two samples. If the sequence of differences, $X_i - Y_i$, is an *i.i.d.* sample from a $\mathsf{Normal}(\mu, \sigma)$ distribution, then a $(1-\alpha)\cdot 100\%$ confidence interval for the difference of means, $\mu_x - \mu_y$, is given by

$$(\bar{X} - \bar{Y}) \pm t^* s/\sqrt{n},$$

where s is the sample standard deviation of the $X_i - Y_i$ and t^* is found from the t-distribution with $n-1$ degrees of freedom.

The `t.test()` function can compute the confidence interval. If x and y store the data, the function may be called as either

`t.test(x, y, paired=TRUE)` or `t.test( x - y )`

We use the argument `conf.level=...` to specify the confidence level.

■ **Example 7.10: Comparing shoes** The `shoes` (MASS) data set contains shoe wear for ten children each wearing two different shoes. By comparing the differences, we can tell whether the two types of shoes have different mean wear amounts.

```
> library(MASS)                          # load data set
> names(shoes)
[1] "A" "B"
> with(shoes, t.test(A-B,conf.level = 0.9))

        One Sample t-test

data:  A - B
t = -3.349, df = 9, p-value = 0.008539
alternative hypothesis: true mean is not equal to 0
90 percent confidence interval:
 -0.6344 -0.1856
sample estimates:
mean of x
    -0.41
### Alternately:
> with(shoes, t.test(A,B,conf.level = 0.9,paired=TRUE))
...
```

Both approaches produce the same 90% confidence interval. In this case, it does not include 0, indicating that there may be a difference in the means. ■

7.5.4 Problems

7.25 Two different AIDS-treatment "cocktails" are compared. For each, the time it takes (in years) to fail is measured for seven randomly assigned patients. The data is in Table 7.1. Find an 80% confidence interval for the difference of means. What assumptions are you making on the data?

Table 7.1 Time to fail for AIDS cocktails, in years

Type	1	2	3	4	5	6	7	$\bar{x}$	s
Cocktail 1:	3.1	3.3	1.7	1.2	0.7	2.3	2.9	2.24	0.99
Cocktail 2:	1.8	2.3	2.2	3.5	1.7	1.6	1.4	2.13	0.69

7.26 In determining the recommended dosage of AZT for AIDS patients, tests were done comparing efficacy for various dosages. If a low dosage is effective, then that would be recommended, as it would be less expensive and would have fewer potential side effects.

A test to decide whether a dosage of 1,200 mg is similar to one of 400 mg is performed on two random samples of AIDS patients. A numeric measurement of a patient's health is made, and the before-and-after differences are recorded after treatment in Table 7.2. Find a 90% confidence interval for the differences of the means. What do you assume about the data?

Table 7.2 Health measurements after AZT treatment

400 mg group	7	0	8	1	10	12	2	9	5	2
1200 mg group	2	1	5	1	5	7	-1	8	7	3

7.27 The data in Table 7.3 is from IQ tests for pairs of twins that were separated at birth. One twin was raised by the biological parents, the other by adoptive parents. Find a 90% confidence interval for the differences of mean. What do you assume about the data? In particular, are the two samples independent?

Table 7.3 IQ scores for identical twins

Foster	80	88	75	113	95	82	97	94	132	108
Biological	90	91	79	97	97	82	87	94	131	115

7.28 For the `babies (UsingR)` data set, the variable `age` contains the mother's age and the variable dage contains the father's age for several babies. Find a 95% confidence interval for the difference in mean age. Does it contain 0? What do you assume about the data?

7.6 Confidence intervals for the median

The confidence intervals for the mean are based on the fact that the distribution of the statistic

$$T = \frac{\bar{X} - \mu}{\mathsf{SE}(\bar{X})}$$

is known. This is true when the sample is an *i.i.d.* sample from a normal population or one close to normal. However, many data sets, especially long-tailed skewed ones, are not like this. For these situations, **nonparametric** methods are preferred. That is, no parametric assumptions on the population distribution for the sample are made, although assumptions on its shape may apply.

7.6.1 Confidence intervals based on the binomial

The binomial distribution can be used to find a confidence interval for the median for any continuous parent population. The key is to look at whether a data point is more or less than the median. As the median splits the area in half, the probability that a data point is more than the median is exactly 1/2. (We need a continuous distribution with positive density over its range to say "exactly" here.) Let T count the number of data points more than the median in a sample of size n. T is a $\mathsf{Binomial}(n, 1/2)$ random variable.

Let $X_{(1)}, X_{(2)}, \ldots, X_{(n)}$ be the sample after sorting from smallest to largest. A $(1-\alpha)\cdot 100\%$ confidence interval is constructed by finding the largest $j \geq 1$ so that $\mathsf{P}(X_{(j)} \leq M \leq X_{(n+1-j)}) \geq 1-\alpha$. In terms of T, this becomes the largest j so that $\mathsf{P}(j \leq T \leq n-j) \geq 1-\alpha$, which in turn becomes a search for the largest j with $\mathsf{P}(T < j) < \alpha/2$. We can find this in the data after sorting.

A concrete example can clarify.

■ **Example 7.11: CEO compensation in 2000** The following data is compensation in \$10,000s of a random sampling from the top 200 CEOs in America for the year 2000:†

```
110 12 2.5 98 1017 540 54 4.3 150 432
```

Find a 90% confidence interval for the median based on the sign test.

† See `http://www.aflcio.org/corporateamerica/paywatch/ceou/database.cfm` for such data.

We enter in the data and then look at the binomial probabilities for this size sample:

```
> x = c(110, 12, 2.5, 98, 1017, 540, 54, 4.3, 150, 432)
> n = length(x)
> pbinom(0:n,n,1/2)                   # P(T <= k) not P(T < k)
 [1] 0.0009766 0.0107422 0.0546875 0.1718750 0.3769531 0.6230469
 [7] 0.8281250 0.9453125 0.9892578 0.9990234 1.0000000
```

For a 90% confidence interval, $\alpha/2 = 0.05$. Thus, j is 2, as $\mathsf{P}(T < 2) = 0.0107422$, but $\mathsf{P}(T < 3) = 0.0546875$. Sorting the data we get

```
> sort(x)
 [1]    2.5    4.3   12.0   54.0   98.0  110.0  150.0  432.0
 [9]  540.0 1017.0
```

Our 90% confidence interval is then $[4.3, 540.0]$.

The $\alpha/2$-quantile for the binomial returns the smallest k with $\mathsf{P}(X \leq k) \geq \alpha/2$. This is just our j in a different context. So we could automate the above with

```
> j = qbinom(0.05, n, 1/2)
> sort(x)[c(j,n+1-j)]
[1]   4.3 540.0
```

■

7.6.2 Confidence intervals based on signed-rank statistic

The Wilcoxon signed-rank statistic allows for an improvement on the confidence interval given by counting the number of data points above the median. Its usage is valid when the X_i are assumed to be *symmetric about their median*. If this is so, then a data point is equally likely to be on the left or right of the median, and the distance from the median is independent of what side of the median the data point is on. If we know the median then we can rank the data by distance to the median. Add up the ranks for the values where the data is more than the median. The distribution of this statistic, under the assumption, can be computed and used to give confidence intervals. It is available in R under the family name `signrank`. In particular, `qsignrank()` will return the quantiles.

This procedure is implemented in the `wilcox.test()` function. Unlike with `prop.test()` and `t.test()`, to return a confidence interval when using `wilcox.test()` we need to specify that a confidence interval is desired with the argument `conf.int=TRUE`.

■ **Example 7.12: CEO confidence interval** The data on CEOs is too skewed to apply this test, but after taking a log transform we will see a symmetric data set (Figure 7.7).

```
> boxplot(scale(x),scale(log(x)),names=c("CEO","log.CEO"))
> title("Boxplot of CEO data and its logarithm")
```

Using `scale()` makes a data set have mean 0 and variance 1, so the shape is all that is seen and comparisons of shapes are possible.

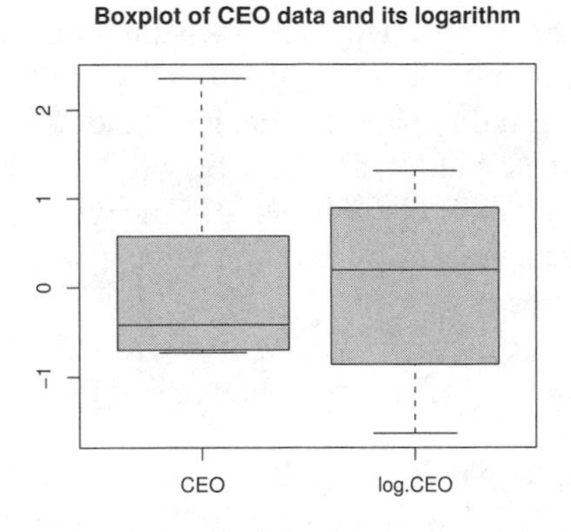

Figure 7.7 Comparison of CEO data and its logarithm on the same scale

Thus we can apply the Wilcoxon method to the log-transformed data, and then transform back.

```
> wilcox.test(log(x), conf.int=TRUE, conf.level=0.9)

        Wilcoxon signed rank test

data:  log(x)
V = 55, p-value = 0.001953
alternative hypothesis: true mu is not equal to 0
90 percent confidence interval:
 2.963 5.540
sample estimates:
(pseudo)median
         4.345

> exp(c(2.863,5.540))          # inverse of log.
[1]  17.51 254.68
```

Compare the interval $(17.51, 254.68)$ to that found previously: $(4.3, 540.0)$. ■

7.6.3 Confidence intervals based on the rank-sum statistic

The t-test to compare samples is robust to nonnormality in the parent distribution but is still not appropriate when the underlying distributions are decidedly nonnormal. However, if the two distributions are the same up to a possible shift of center, then a confidence interval based on a nonparametric statistic can be given.

Let $f(x)$ be a density for a mean-zero distribution, and suppose we have two independent random samples: the first, $X_1, X_2, \ldots, X_{n_x}$, from a population with density $f(x-\mu_x)$, and the second, $Y_1, Y_2, \ldots, Y_{n_y}$, from a population with density $f(x-\mu_y)$. The basic statistic, called the **rank-sum statistic**, looks at all possible pairs of the data and counts the number of times the X value is greater than or equal to the Y value. If the population mean for the X values is larger than the population mean for the Y values, this statistic will likely be large. If the mean is smaller, then the statistic will likely be small. The distribution of this statistic is given by R with the `wilcox` family and is used to give a confidence interval for the difference of the means.

The command `wilcox.test(x,y,conf.int=TRUE)`. function will find a confidence interval for the difference in medians of the two data sets.

■ **Example 7.13: CEO pay in 2002** In Example 7.12, the compensation for a sampling of the top 200 CEOs in the year 2000 was given. For the year 2002, a similar sampling was performed and gave this data:

`312 316 175 200 92 201 428 51 289 1126 822`

From these two samples, can we tell if there is a difference in the center of the distribution of CEO pay?

Figure 7.8 shows two data sets that are quite skewed, so confidence intervals based on the T statistic would be inappropriate. Rather, as the two data sets have a similar shape, we find the confidence interval returned by `wilcox.test()`. As before, we need to specify that a confidence interval is desired. To answer our question, we'll look at a 90% confidence interval and see if it contains 0.

```
> pay.02 = c(312, 316, 175, 200, 92, 201, 428, 51, 289, 1126, 822)
> pay.00 = c(110, 12, 2.5, 98, 1017, 540, 54, 4.3, 150, 432)
> plot(density(pay.02),main="densities of y2000, y2002")
> lines(density(pay.00),lty=2)
> wilcox.test(pay.02, pay.00, conf.int=TRUE, conf.level=0.9)

        Wilcoxon rank sum test

data:  pay.02 and pay.00
W = 75, p-value = 0.1734
alternative hypothesis: true mu is not equal to 0
90 percent confidence interval:
 -18 282
sample estimates:
difference in location
                 146.5
```

The 90% confidence interval, $[-18, 282]$, contains a value of 0.

This example would be improved if we had matched or paired data—that is, the salaries for the same set of CEOs in the year 2000 and 2002—as then

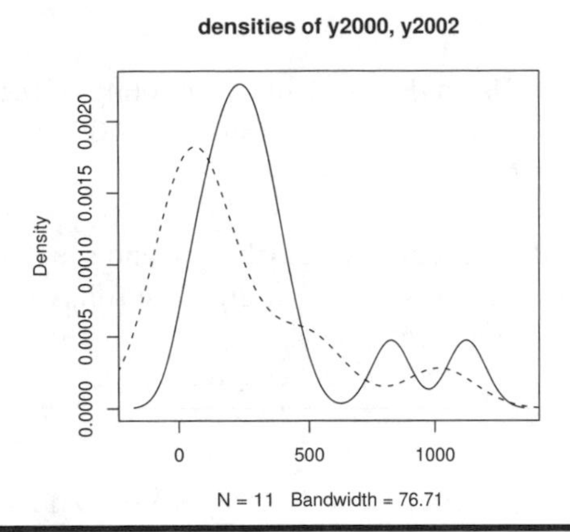

Figure 7.8 Densities of 2000 and 2002 CEO compensations indicating similarly shaped distributions with possible shift

differences in the sampling would be minimized. If that case is appropriate, then adding the argument `paired=TRUE` to `wilcox.test()` computes a confidence interval based on the signed-rank statistic. ■

7.6.4 Problems

7.29 The commuter revisited: the commuter records 20 commute times to gauge her median travel time. The data has a sample median of 24 and is summarized in this stem-and-leaf diagram:

```
> stem(commutes)
2 | 1112233444444
2 | 5569
3 | 113
```

If the data is appropriate for `t.test()`, use that to give a 90% confidence interval for the mean. Otherwise use `wilcox.test()` (perhaps after a transform) to give a confidence interval for the median.

7.30 The data set `cabinet (UsingR)` contains information on the amount of money each member of President Bush's cabinet saved due to the passing of a 2003 tax bill. This is stored in the variable named `est.tax.savings`. Find a confidence interval for the median amount.

7.31 The data set `u2 (UsingR)` contains song lengths for several albums by the band U2. How might you interpret the lengths as a sample from a larger population? Use `wilcox.test()` to construct a 95% confidence interval for the difference of population means between the album *October* and the album *The*

Joshua Tree.

7.32 The data set `cfb (UsingR)` contains a sampling of the data contained in the Survey of Consumer Finances. For the variables `AGE` and `INCOME` find 95% confidence intervals for the median.

7.33 We can simulate the signed-rank distribution and see that it applies for any symmetric distribution regardless of tail length. The following will simulate the distribution for $n = 20$ using normal data.

```
> n = 20;m=250                          # 250 samples
> res = c()                             # the results
> for(i in 1:m) {
+ x = rnorm(n)
+ res[i]=sum(rank(abs(x))[x>0]) # only add positive values
+ }
```

This can be plotted with

```
> hist(res,prob=TRUE)
> x = 40:140
> lines(x,dsignrank(x,n))         # density-like, but discrete.
```

If you change the line `x = rnorm(x)` to `x = rt(n,df=2)`, the underlying distribution will be long tailed, and short tailed if you change it to `x=runif(n,-1,1)`. Do both, and then compare all three samples. Are they different or the same? What happens if you use skewed data, such as `x=rexp(n)-1`?

Chapter 8

Significance tests

Finding a confidence interval for a parameter is one form of statistical inference. A significance test, or test of hypothesis, is another. Rather than specify a range of values for a population parameter, a significance test assumes a value for the population parameter and then computes a probability based on a sample given that assumption.

■ **Example 8.1: A criminal trial** The ideas behind a significance test can be illustrated by analogy to a criminal trial in the United States—as seen on TV. Imagine the following simplified scenario: a defendant is charged with a crime and must stand trial. During the trial, a prosecutor and defense lawyer respectively try to convince the jury that the defendant is either guilty or innocent. The jury is supposed to be unbiased. When deciding the defendant's fate, the jurors are instructed to *assume* that the defendant is innocent unless *proven* guilty beyond *a shadow of a doubt*. At the end of the trial the jurors decide the guilt or innocence of the defendant based on the strength of their belief in the assumption of his innocence given the evidence. If the jurors believe it very unlikely that an innocent person could have evidence to the contrary, they will find the defendant "guilty." If it is not so unlikely, they will rule "not guilty."

The system is not foolproof. A guilty man can go free if he is found not guilty, and an innocent man can be erroneously convicted. The frequency with which these errors occur depends on the threshold used to find guilt. In a criminal trial, to decrease the chance of a erroneous guilty verdict, the stringent *shadow of a doubt* criterion is used. In a civil trial, this phrasing is relaxed to *a preponderance of the evidence*. The latter makes it easier to err with a truly innocent person but harder to err with a truly guilty one. ■

Let's rephrase the above example in terms of significance tests. The assumption of innocence is replaced with the **null hypothesis**, H_0. This stands in contrast

Table 8.1 Level of significance for range of p-values

p-value range	significance stars	common description
$[0, .001]$	***	extremely significant
$(.001, .01]$	**	highly significant
$(.01, .05]$	*	statistically significant
$(.05, .1]$	.	could be significant
$(.1, 1]$		not significant

to the **alternative hypothesis**, H_A. This would be an assumption of guilt in the trial analogy. In a trial, this alternative is not used as an assumption; it only gives a direction to the interpretation of the evidence. The determination of guilt by a jury is not proof of the alternative, only a failure of the assumption of innocence to explain the evidence well enough. A guilty verdict is more accurately called a verdict of "not innocent." The performer of a significance test seeks to determine whether the null hypothesis is reasonable given the available data. The evidence is replaced by an experiment that produces a **test statistic**. The probability that the test statistic is the observed value *or is more extreme* as implied by the alternative hypothesis is calculated *using the assumptions of the null hypothesis*. This is called the ***p*-value**. This is like the weighing of the evidence—the jury calculating the likelihood that the evidence agrees with the assumption of innocence.

The calculation of the p-value is called a **significance test**. The p-value is based on both the sampling distribution of the test statistic under H_0 and the single observed value of it during the trial. In words, we have

$$p\text{-value} = \mathsf{P}(\text{test statistic is the observed value or is more extreme} \mid H_0).$$

The p-value helps us decide whether differences in the test statistic from the null hypothesis are attributable to chance or sampling variation, or to a failure of the null hypothesis. If a p-value is small, the test is called **statistically significant**, as it indicates that the null hypothesis is unlikely to produce more extreme values than the observed one. Small p-values cast doubt on the null hypothesis; large ones do not.

What is "large" or "small" depends on the area of application, but there are some standard levels that are used. Some `R` functions will mark p-values with significance stars, as described in Table 8.1. Although these are useful for quickly identifying significance, the cutoffs are arbitrary, settled on more for ease of calculation than actual relevance.

In some instances, as with a criminal trial, a decision is made based on the p-value. A juror is instructed that a defendant, to be found guilty, must be thought guilty beyond a shadow of a doubt. A significance test is less vague, as a **significance level** is specified that the p-value is measured against. A typical significance level is 0.05. If the p-value is less than the significance level, then the null hypothesis is said to be *rejected*, or viewed as false. If the p-value is larger than the significance level, then the null hypothesis is *accepted*.

The words "reject" and "accept" are perhaps more absolute than they should be. When rejecting the null, we don't actually prove the null to be false or the alternative to be true. All that is shown is that the null hypothesis is unlikely to produce values more extreme than the observed value. When accepting the null we don't prove it is true, we just find that the evidence is not too unlikely if the null hypothesis is true.

By specifying a significance level, we indirectly find values of the test statistic that will lead to rejection. This allows us to specify a **rejection region** consisting of all values for the observed test statistic that produce p-values smaller than the significance level. The boundaries between the acceptance and rejection regions are called **critical values**. The use of a rejection region avoids the computation of a p-value: reject if the observed value is in the rejection region and accept otherwise. We prefer, though, to find and report the p-value rather than issue a simple verdict of "accept" or "reject."

This decision framework has been used historically in scientific endeavors. Researchers may be investigating whether a specific treatment has an effect. They might construct a significance test with a null hypothesis of the treatment having no effect, against the alternative hypothesis of some effect. (In this case, the alternative hypothesis is known as the *research hypothesis*.) The significance test then determines the reasonableness of the assumption of no effect. If this is rejected, there has been no proof of the research hypothesis, only that the null hypothesis is not supported by the data.

As with a juried trial, the system is not foolproof. When a decision is made based on the p-value, mistakes can happen. If the null hypothesis is falsely rejected, it is a **type-I error** (innocent man is found guilty). If the null hypothesis is false, it may be falsely accepted (guilty man found not guilty). This is a **type-II error**.

A simple example can illustrate the process.

■ **Example 8.2: Which mean?** Imagine we have a widget-producing machine that sometimes goes out of calibration. The calibration is measured in terms of a mean for the widgets. How can we tell if the machine is out of calibration by looking at the output of a single widget?

Assume, for simplicity, that the widgets produced are random numbers that usually come from a normal distribution with mean 0 and variance 1. When the machine slips out of calibration, the random numbers come from normal distribution with mean 1 and variance 1. Based on the value of a single one of these random numbers, how can we decide whether the machine is in calibration or not?

This question can be approached as a significance test. We might assume that the machine is in calibration (has mean 0), unless we see otherwise based on the value of the observed number.

Let X be the random number. The hypotheses become

$$H_0 : X \text{ is } \mathsf{Normal}(0,1), \quad H_A : X \text{ is } \mathsf{Normal}(1,1).$$

We usually write this as

$$H_0 : \mu = 0, \quad H_A : \mu = 1,$$

where the assumption on the normal distribution and a variance of 1 are implicit.

Suppose we observe a value 0.7 from the machine. Is this evidence that the machine is out of calibration?

The p-value in this case is the probability that a $\mathsf{Normal}(0,1)$ random variable produces a 0.7 or *more*. This is `1 - pnorm(0.7,0,1)`, or 0.2420. Why this probability? The calculation is done under the null hypothesis, so a normal distribution with mean 0 and variance 1 is used. The observed value of the test statistic is 0.7. Larger values than this are more extreme, given the alternative hypothesis. This p-value is not very small, and there is no evidence that the null hypothesis is false. It may be, if the alternative were true, that a value of 0.7 or less is `pnorm(.7,1,1)`, or 0.3821, so it, too, is not unlikely. (See Figure 8.1.)

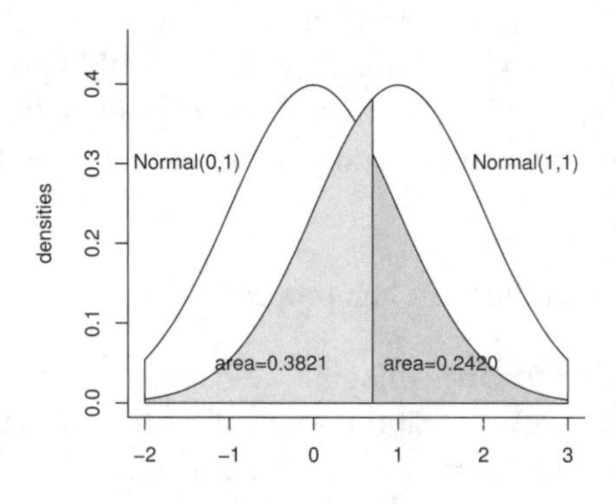

Figure 8.1 The p-value of 0.2420 is the probability of 0.7 or more for the $\mathsf{Normal}(0,1)$ distribution

Even though 0.7 is closer to the mean of 1 than the mean of 0, it is really not conclusive evidence that the null hypothesis (the assumption of calibration) is incorrect. The problem is that the two distributions are so "close" together. It would be much easier to decide between the two if the means were 10 units apart instead of just 1 (with the same variance). Alternatively, if the standard deviations were smaller, the same would be true. This can be achieved by taking averages, as we know that the standard deviation of an average is $\sigma/\sqrt{n}$, or smaller than the population standard deviation by the divisor $\sqrt{n}$.

With this in mind, suppose our test statistic is now the sample mean of a random sample of ten widgets. How does our thinking change if the sample mean is now 0.7?

The p-value looks similar, $\mathsf{P}(\bar{X} \geq 0.7 \mid H_0)$, but when we compute, we use the sampling distribution of $\bar{X}$ under H_0, which is $\mathsf{Normal}(0, 1/\sqrt{10})$. The p-value is 0.0134, as found by `1 - pnorm(0.7,0,1/sqrt(10))`. This is illustrated in Figure 8.2. Now the evidence is more convincing that the machine is out of calibration. ■

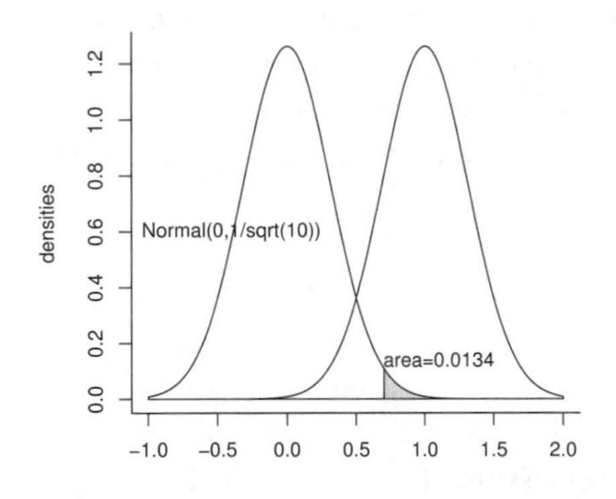

Figure 8.2 ***p*-value calculated for $n = 10$ when observed value of $\bar{X}$ is 0.7**

The above example illustrates the steps involved in finding the p-value:

1. Identify H_0 and H_A, the null and alternative hypotheses.
2. Specify a test statistic that discriminates between the two hypotheses, collect data, then find the observed value of the test statistic.
3. Using H_A, specify values of the test statistic that are "extreme" under H_0 in the direction of H_A. The p-value will be the probability of the event that the test statistic is the observed value or more extreme.
4. Calculate the p-value under the null hypothesis. The smaller the p-value, the stronger the evidence is against the null hypothesis.

8.1 Significance test for a population proportion

A researcher may wish to know whether a politician's approval ratings are falling, or whether unemployment rate is rising, or whether the poverty rate is changing. In many cases, a known proportion exists. What is asked for is a comparison against this known proportion. A test of proportion can be used to help answer these questions.

Assume p_0 reflects the historically true proportion of some variable of interest. A researcher may wish to test whether the current unknown proportion, p, is different from p_0. A test of proportion would check the null hypothesis,

$$H_0 : p = p_0,$$

versus an alternative hypothesis on p. Possible alternatives are

$$H_A : p > p_0, \; H_A : p < p_0, \text{ or } H_A : p \neq p_0.$$

If the survey is a random sample from the target population, the number of successes, x, is binomially distributed, $\widehat{p} = x/n$, and $\widehat{p}$ will be approximately normal for large enough values of n. We might use $\widehat{p}$ directly as a test statistic, but it is more common to standardize $\widehat{p}$, yielding the following test statistic:

$$Z = \frac{\widehat{p} - \mathsf{E}(\widehat{p} \mid H_0)}{\mathsf{SD}(\widehat{p} \mid H_0)} = \frac{\widehat{p} - p_0}{\sqrt{p_0(1-p_0)/n}}.$$

We use the notation $\mathsf{E}(\widehat{p} \mid H_0)$ to remind ourselves that we use the null hypothesis when calculating this expected value. In this case, under H_0, p_0 is the expected value of $\widehat{p}$ and is assumed to be known. Thus, we can use $\mathsf{SD}(\widehat{p} \mid H_0) = \sqrt{(p_0(1-p_0))/n}$ in our test statistic, in contrast to use of $\mathsf{SE}(\widehat{p})$ when we found confidence intervals for p.

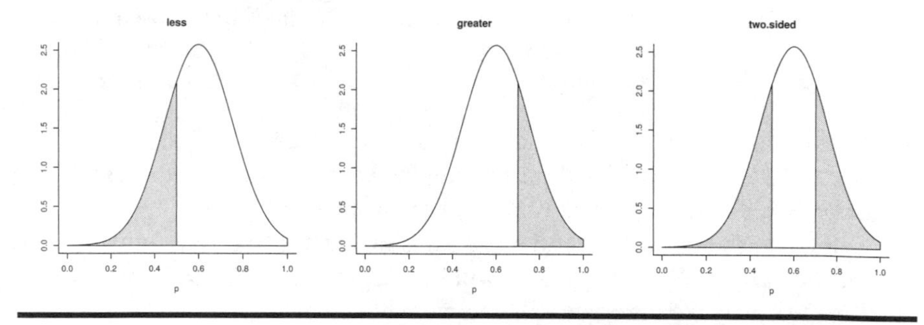

Figure 8.3 Illustration of the three alternative hypotheses. In `R`**,** `less` **is** $H_A : p < p_0$**,** `greater` **is** $H_A : p > p_0$**,** `two.sided` **is** $H_A : p \neq p_0$**.**

The p-value varies based on the alternative hypothesis. This is because what is meant by "more extreme" for the value of the test statistic depends on H_A. In this instance there are three cases:

$$p\text{-value} = \begin{cases} \mathrm{P}(\widehat{p} \leq \text{observed value} \mid H_0) & \text{if } H_A : p < p_0, \\ \mathrm{P}(\widehat{p} \geq \text{observed value} \mid H_0) & \text{if } H_A : p > p_0, \\ \mathrm{P}(|\widehat{p} - p_0| \geq |\text{observed value} - p_0| \mid H_0) & \text{if } H_A : p \neq p_0. \end{cases} \quad (8.1)$$

The first two cases are "one-sided" or "one-tailed," the last "two-sided" or "two-tailed." The absolute values in the third case can be confusing but are there to say that large differences in either direction of p_0 are "more extreme." Figure 8.3 illustrates the areas.

Significance test for a population proportion

A significance test for an unknown proportion between

$$H_0 : p = p_0, \quad H_A : p < p_0,\ p > p_0,\ \text{ or } p \neq p_0$$

can be performed with test statistic

$$Z = \frac{\widehat{p} - p_0}{\sqrt{p_0(1 - p_0)/n}}.$$

If $\widehat{p}$ is based on a simple random sample and n is large enough, Z has a standard normal distribution under the null hypothesis. The p-values can be computed from (8.1).

In `R` the function `prop.test()` will perform this significance test.

■ **Example 8.3: Poverty-rate increase** In the United States, the poverty rate rose from 11.3 percent in 2000 to 11.7 percent in 2001 to 12.1 percent in 2002, as reported by the United States Census Bureau. A national census takes place every decade. The year-2000 number comes from a census. For the Census Bureau to decide the 2001 and 2002 figures, a random sampling scheme is employed. Assume that the numbers come from a simple random sample (they don't), so that we can use the binomial model for the problem, and that the sample sizes are 50,000 for 2001 and 60,000 for 2002.

We investigate whether the 11.7% figure for the year 2001 shows an increase from the year-2000 figure. The null hypothesis is that it is the same as the 11.3%

amount of 2000; the alternative is that the new figure is greater than the old:

$$H_0: p = 0.113, \quad H_A: p > 0.113.$$

A test statistic is based on the proportion living in poverty of a random sample of size 50,000. In the sample, 5,850, or a proportion of .117, were found to be in poverty. Is this difference significant?

The direction of the alternative would be that the rate is .117 *or higher*, as larger values support this one-sided alternative. The *p*-value is $P(\widehat{p} \geq .117 \mid H_0)$, which is found in R with

```
> p0 = .113; n = 50000; SD = sqrt(p0*(1-p0)/n)
> pnorm(.117,mean=p0, sd=SD, lower.tail=FALSE)
[1] 0.002363
```

Thus the *p*-value is 0.002363 and is "very significant." This data casts much doubt on the null hypothesis of no change. We would think sampling variation alone does not explain the difference. ■

8.1.1 Using `prop.test()` to compute p-values

The calculation above is done by hand. The pre-loaded `stats` package in R has many built-in functions to do significance tests more directly. The R function for the above test of proportion is `prop.test()`. This was also used to find confidence intervals under the same assumptions.

The `prop.test()` function needs a few arguments. A template for usage to perform a significance test is

```
prop.test(x, n, p=..., alternative="two.sided")
```

The value for `x` is the sample frequency; in our case, $5,850 = 0.117 \cdot 50,000$. The value of `n` is the sample size $50,000$. These are the same as when we used this function to find confidence intervals.

To perform a significance test, the null and alternative hypotheses must be specified. The null is done with the p= argument; for our example `p = .113`. The alternative hypothesis is specified with the argument `alternative=`, which we abbreviate to `alt=`. This argument has one of these values: `"less"`, `"greater"`, or `"two.sided"`. The default is `two.sided`. As $H_A: p > 0.11$, we will use `"greater"`. This argument is common to many of the functions in R that perform significance tests.

To illustrate, the above calculation is done with

```
> prop.test(x=5850, n=50000, p=.113, alt="greater")

        1-sample proportions test with continuity correction

data:  5850 out of 50000, null probability 0.113
X-squared = 7.942, df = 1, p-value = 0.002415
```

```
alternative hypothesis: true p is greater than 0.113
95 percent confidence interval:
 0.1146 1.0000
sample estimates:
    p
0.117
```

The p-value, 0.002415, and the null and alternative hypotheses are repeated in the output. In addition, a confidence interval is given, as is a sample estimate that we term $\widehat{p}$. The p-value is slightly different from above, as a continuity correction is used by R.

It isn't any more difficult to test the alternative hypothesis, that the rate has changed, or $H_A : p \neq p_0$. This is done by specifying the alternative as `two.sided` (just the differences shown):

```
> prop.test(x=5850, n = 50000, p = .113, alt = "two.sided")
...
X-squared = 7.942, df = 1, p-value = 0.004831
...
```

The p-value is different—it is twice as big—as we would guess by looking at the symmetry in Figure 8.3.

8.1.2 Problems

8.1 United States federal law on dietary supplements requires that the Food and Drug Administration (FDA) prove a supplement harmful in order to ban its sale. In contrast, for a new prescription drug, a pharmaceutical company must prove the product is safe.

Write null and alternative hypotheses for a hypothetical significance test by the FDA when testing a dietary supplement. Do you think the same standard should be used for both dietary supplement and new prescription drugs?

8.2 The `samhda (UsingR)` data set contains information on marijuana usage among children as collected at the the Substance Abuse and Mental Health Data Archive. The variable `marijuana` indicates whether the individual has ever tried marijuana. A 1 means yes, a 2 no. If it used to be that 50% of the target population had tried marijuana, does this data indicate an increase in marijuana usage? Do a significance test of proportions to decide.

8.3 A new drug therapy is tested. Of 50 patients in the study, 40 had no recurrence in their illness after 18 months. With no drug therapy, the expected percentage of no recurrence would have been 75%. Does the data support the hypothesis that this percentage has increased? What is the p-value?

8.4 In the United States in 1998, the proportion of adults age 21-24 who had no medical insurance was 34.4 percent, according to the Employee Benefit Research

Institute. A survey of 75 recent college graduates in this age range finds that 40 are without insurance. Does this support a difference from the nationwide proportion? Perform a test of significance and report the p-value. Is it significant?

8.5 On a number of highways a toll is collected for permission to travel on the roadway. To lessen the burden on drivers, electronic toll-collection systems are often used. An engineer wishes to check the validity of one such system. She arranges to survey a collection unit for single day, finding that of 5,760 transactions, the system accurately read 5,731. Perform a one-sided significance test to see if this is consistent with a 99.9% accuracy rating at the 0.05 significance level. (Do you have any doubts that the normal approximation to the binomial distribution should apply here?)

8.6 In Example 8.3 a count of 5,850 in the survey produced a p-value of 0.002363. What range of counts would have produced a p-value less than 0.05? (Start by asking what observed proportions in the survey would have such a p-value.)

8.7 Historically, a car from a given company has a 10% chance of having a significant mechanical problem during its warranty period. A new model of the car is being sold. Of the first 25,000 sold, 2,700 have had an issue. Perform a test of significance to see whether the proportion of these new cars that will have a problem is more than 10%. What is the p-value?

8.8 A poll taken in 2003 of 200 Europeans found that only 16% favored the policies of the United States. Do a test of significance to see whether this is significantly different from the 50% proportion of Americans in favor of these policies.

8.2 Significance test for the mean (t-tests)

Significance tests can also be constructed for the unknown mean of a parent population. The hypotheses take the form

$$H_0 : \mu = \mu_0, \quad H_A : \mu < \mu_0, \ \mu > \mu_0, \ \text{or } \mu \neq \mu_0.$$

For many populations, a useful test statistic is

$$T = \frac{\bar{X} - \mathsf{E}(\bar{X} \mid H_0)}{\mathsf{SE}(\bar{X} \mid H_0)} = \frac{\bar{X} - \mu_0}{s/\sqrt{n}}.$$

T takes the form of "observed" minus "expected," divided by the standard error, where the expected value and the standard error are found under the null hypothesis.

In the case of normally distributed initial data, the sampling distribution of T under the null hypothesis is known to be the t-distribution with $n-1$ degrees of freedom. If n is large enough, the sampling distribution of T is a standard normal by the central limit theorem. As both the t distribution and normal distribution are similar for large n, the following applies to both assumptions.

Test of significance for the population mean

If the data $X_1, X_2, \ldots, X_n$ is an *i.i.d.* sequence from a $\mathsf{Normal}(\mu, \sigma)$ distribution, or n is large enough for the central limit theorem to apply, a test of significance for

$$H_0 : \mu = \mu_0, \quad H_A : \mu < \mu_0,\ \mu > \mu_0, \text{ or } \mu \neq \mu_0$$

can be performed with test statistic

$$T = \frac{\bar{X} - \mu_0}{s/\sqrt{n}}.$$

For a normally distributed population, T has the t-distribution with $n-1$ degrees of freedom under H_0. For large n, T has the standard normal distribution. Let $t = (\bar{x} - \mu_0)/(s/\sqrt{n})$ be the observed value of the test statistic. The p-value is computed by

$$p\text{-value} = \begin{cases} \mathsf{P}(T \leq t \mid H_0) & H_A : \mu < \mu_0 \\ \mathsf{P}(T \geq t \mid H_0) & H_A : \mu > \mu_0 \\ \mathsf{P}(|T - \mu_0| \geq |t - \mu_0| \mid H_0) & H_A : \mu \neq \mu_0. \end{cases}$$

In R, the function `t.test()` can be used to compute the p-value with unsummarized data, as in

```
t.test(x, mu=..., alt="two.sided")
```

The null hypothesis is specified by a value for the argument `mu=`. The alternative is specified as appropriate by `alt="less"`, `alt="greater"`, or `alt="two.sided"` (the default).

■ **Example 8.4: SUV gas mileage** A consumer group wishes to see whether the actual mileage of a new SUV matches the advertised 17 miles per gallon. The group suspects it is lower. To test the claim, the group fills the SUV's tank

Table 8.2 SUV gas mileage

stem	leaf
11	4
12	
13	1
14	77
15	0569
16	08

and records the mileage. This is repeated ten times. The results are presented in a stem-and-leaf diagram in Table 8.2.

Does this data support the null hypothesis that the mileage is 17 or the alternative, that it is less?

The data is assumed to be normal, and the stem-and-leaf plot shows no reason to doubt this. The null and alternative hypotheses are

$$H_0 : \mu = 17, \quad H_A : \mu < 17.$$

This is a one-sided test. The p-value will be computed from those values of the test statistic less than the observed value, as these are more extreme given the alternative hypothesis.

```
> mpg = c(11.4,13.1,14.7,14.7,15.0,15.5,15.6,15.9,16.0,16.8)
> xbar = mean(mpg)
> s = sd(mpg)
> n = length(mpg)
> c(xbar, s, n)
[1] 14.870  1.572 10.000
> SE = s/sqrt(n)
> (xbar - 17)/SE
[1] -4.285
> pt(-4.285, df = 9, lower.tail = T)
[1] 0.001017
```

The p-value is very small and discredits the claim of 17 miles per gallon, as the difference of $\bar{X}$ from 17 is not well explained by sampling variation.

The above calculations could be done using `t.test()` as follows:

```
> t.test(mpg, mu = 17, alt="less")

        One Sample t-test

data:  mpg
t = -4.285, df = 9, p-value = 0.001018
alternative hypothesis: true mean is less than 17
95 percent confidence interval:
  -Inf 15.78
```

```
sample estimates:
mean of x
    14.87
```

The output contains the same p-value (up to rounding), plus a bit more information, including the observed value of the test statistic, a one-sided confidence interval, and $\bar{x}$ (the estimate for μ). ■

It is easy to overlook the entire null hypothesis. We assume not only that $\mu = \mu_0$, but also that the random sample comes from a normally distributed population with unspecified variance. With these assumptions, the test statistic has a known sampling distribution. The t-statistic is robust to small differences in the assumed normality of the population, but a really skewed population distribution would still be a poor candidate for this significance test unless n is large. It is recommended that you plot the data prior to doing any analysis, to ensure that it is appropriate.

■ **Example 8.5: Rising textbook costs?** A college bookstore claims that, on average, a college student will pay $101.75 per class for textbooks. A student group investigates this claim by randomly selecting ten courses from the course catalog and finding the textbook costs for each. The data collected is

```
140 125 150 124 143 170 125  94 127  53
```

Do a test of significance of $H_0 : \mu = 101.75$ against the alternative hypothesis $H_A : \mu > 101.75$.

We assume independence and normality of the data. Once the data is entered, we can use `t.test()`, with `"greater"` for the alternative. This gives

```
> x = c(140, 125, 150, 124, 143, 170, 125, 94, 127, 53)
> qqnorm(x)                      # check normality, OK
> t.test(x, mu = 101.75, alt="greater")

        One Sample t-test

data:  x
t = 2.291, df = 9, p-value = 0.02385
alternative hypothesis: true mean is greater than 101.8
95 percent confidence interval:
 106.4   Inf
sample estimates:
mean of x
    125.1
```

The p-value is small, indicating that the true amount per class may be more than that indicated under the null hypothesis. ■

8.2.1 Problems

8.9 A study of the average salaries of New York City residents was conducted for 770 different jobs. It was found that massage therapists average $58,260 in yearly income. Suppose the study surveyed 25 massage therapists and had a standard deviation of $3,250. Perform a significance test of the null hypothesis that the average massage therapist makes $55,000 per year against the one-sided alternative that it is more. Assume the data is normally distributed.

8.10 The United States Department of Energy conducts weekly phone surveys on the price of gasoline sold in the United States. Suppose one week the sample average was $2.03, the sample standard deviation was $0.22, and the sample size was 800. Perform a one-sided significance test of $H_0 : \mu = 2.00$ against the alternative $H_A : \mu > 2.00$.

8.11 The variable `sat.m` in the data set `stud.recs (UsingR)` contains math SAT scores for a group of students. Test the null hypothesis that the mean score is 500 against a two-sided alternative. Would you accept or reject at a 0.05 significance level?

8.12 In the `babies (UsingR)` data set, the variable `dht` contains the father's height. Do a significance test of the null hypothesis that the mean height is 68 inches against an alternative that it is taller. Remove the values of 99 from the data, as these indicate missing data.

8.13 A consumer-reports group is testing whether a gasoline additive changes a car's gas mileage. A test of seven cars finds an average improvement of 0.5 miles per gallon with a standard deviation of 3.77. Is this difference significantly greater than 0? Assume the values are normally distributed.

8.14 The data set `OBP (UsingR)` contains on-base percentages for the 2002 major league baseball season. Do a significance test to see whether the mean on-base percentage is 0.330 against a two-sided alternative.

8.15 The data set `normtemp (UsingR)` contains measurements of 130 healthy, randomly selected individuals. The variable `temperature` contains normal body temperature. Does the data appear to come from a normal distribution? If so, perform a t-test to see if the commonly assumed value of 98.6 °F is correct. (A recent study suggests that 98.2 degrees is actually correct.)

8.16 We can perform simulations to see how robust the t-test is to changes in the parent distribution. For a normal population we can run a simulation with the commands:

```
> m = 250; n = 10                # m simulations with sample size n
> res = c();                     # store values here
> for(i in 1:m) res[i] = t.test(rnorm(n), mu = 0, df = n-1)$p.value
> sum(res < 0.05)/length(res)    # proportion of "rejections"
```

```
[1] 0.052
```

(The `$p.value` after `t.test` extracts just the p-value from the output.) This example shows that 5.2% of the time we rejected at the $\alpha = 0.05$ significance level, as expected. Repeat with exponential data (`rexp(n)`, and `mu=1`), uniform data (`runif(n)` and `mu=1/2`), and t-distributed data (`rt(n,df=4)` and `mu=0`).

8.3 Significance tests and confidence intervals

You may have noticed that the R functions for performing a significance test for a population proportion or mean are the same functions used to compute confidence intervals. This is no coincidence, as performing a significance test and constructing a confidence interval both make use of the same test statistic, although in different ways.

Suppose we have a random sample from a normally distributed population with mean μ and variance σ^2. We can use the sample to find a confidence interval for μ, or we could use the sample to do a significance test of

$$H_0 : \mu = \mu_0, \quad H_A : \mu \neq \mu_0.$$

In either case, the T statistic

$$T = \frac{\bar{X} - \mu}{\mathsf{SE}(\bar{X})}$$

is used to make the statistical inference. The two approaches are related by the following: a significance test with significance level α will be rejected if and only if the $(1 - \alpha) \cdot 100\%$ confidence interval around $\bar{X}$ does not contain μ_0.

To see why, suppose α is given. The confidence interval uses t^* found from

$$\mathsf{P}(-t^* \leq T \leq t^*) = 1 - \alpha.$$

From this, the confidence interval will not contain μ_0 if the value of T is more than t^* or less than $-t^*$. This same relationship is used to find the critical values defining the boundaries of the rejection region. If the observed value of T is more than t^* or less than $-t^*$, then the observed value is in the rejection region, and the null hypothesis is rejected. This is illustrated in Figure 8.4.

Many people prefer the confidence interval to the p-value of the significance test for good reasons. If the null hypothesis is that the mean is 16, but the true mean is just a bit different, then the probability that a significance test will fail can be made arbitrarily close to 1 just by making n large enough. The confidence interval, on the other hand, would show much better that the value of the mean is likely close to 16. The language of significance tests, however, is more flexible

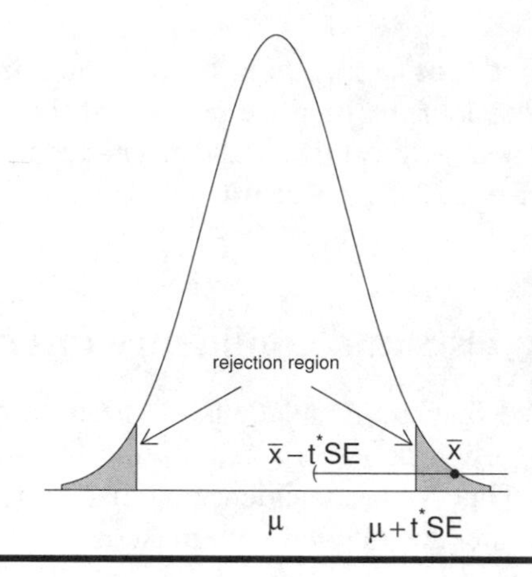

Figure 8.4 If $\bar{x}$ is in the two-sided rejection region, then a confidence interval around $\bar{x}$ does not contain μ

and allows us to consider more types of problems. Both approaches are useful to have.

R is agnostic: it can return both the confidence interval and the p-value when asked, although the defaults for the functions usually return just the confidence interval.

8.4 Significance tests for the median

The significance test for the mean relies on a large sample, or on an assumption that the parent distribution is normally (or nearly normally) distributed. In the situation where this isn't the case, we can use test statistics similar to the ones used to find confidence intervals for the median. Significance tests based on these test statistics are nonparametric tests, as they do not make assumptions about the population parameters to calculate the test statistic (though there may be assumptions about the shape of the distribution).

8.4.1 The sign test

The sign test is a simple test for the median of a distribution that has no assumptions on the parent distribution except that it is continuous with positive density. Let H_0 suppose that the median is m. If we count the number of data points higher than the median, we get a number that will have a $\text{Binomial}(n, 1/2)$ distribution, as under H_0, a data point is equally likely to be more or less than the median.

This leads to the following test.

Sign test for the median

Assume $X_1, X_2, \cdots X_n$ are from a continuous distribution with positive density. A significance test of the hypotheses

$$H_0: \text{ median } = m, \quad H_A: \text{ median } < m, \text{ median } > m, \text{ or median } \neq m,$$

can be performed with test statistic

$$T = \text{the number of } X_i \text{ with } X_i > m.$$

If the data has values equal to m, then delete those values from the data set. Under H_0, T has a $\mathsf{Binomial}(n, 1/2)$ distribution. Large values of T support the alternative that the median is greater than M; small values of T support the alternative that the median is smaller than M. For two-sided alternatives, large or small values of T support H_A.
The p-value is calculated by

$$p\text{-value} = \begin{cases} \mathsf{P}(T \geq \text{observed} \mid H_0) & H_A: \text{median} > m \\ \mathsf{P}(T \geq n - \text{observed} \mid H_0) & H_A: \text{median} < m \\ \mathsf{P}(T \geq \mathsf{max}(\text{observed}, n - \text{observed}) \mid H_0) & H_A: \text{median} \neq m. \end{cases}$$

In `R`, the test statistic can be computed using `sum()`. The p-values are found using `pbinom(k)`. However, as $\mathsf{P}(T \geq k) = 1 - \mathsf{P}(T \leq k-1)$, the p-value is is found with `1 - pbinom(k - 1, n, 1/2)`.

■ **Example 8.6: Length of cell-phone calls** Suppose a cell-phone bill contains this data for the number of minutes per call:

```
2  1  3  3  3  3  1  3 16  2  2 12 20  3  1
```

Is this consistent with an assumption that the median call length is 5 minutes, or does it indicate that the median length is less than 5?

The hypothesis test is

$$H_0: \text{ the median } = 5, \quad H_A: \text{ the median } < 5.$$

The data is clearly nonnormal, so a t-test is inappropriate. A sign test can be used. Here, small values of T support the alternative.

```
> calls = c(2, 1, 3, 3, 3, 3, 1, 3, 16, 2, 2, 12, 20, 3, 1)
> obs = sum(calls > 5)              # find observed value of T
> n = length(calls)
> n - obs
[1] 12
> 1 - pbinom(11,n,1/2)              # we want P(T >= 12)
[1] 0.01758
```

We get a p-value of 0.0176, which leads us to believe that the median is less than 5.

For illustration, the p-value for the two-sided alternative can be computed as follows:

```
> k = max(obs, n - obs)
> k
[1] 12
> 2*(1 - pbinom(k-1 , n, 1/2))
[1] 0.03516
```

■

8.4.2 The signed-rank test

The signed-rank test is an improvement to the sign test when the population is symmetric, but not close enough to normal to use a t-test. Assume H_0 : median $= m$. If X_i are from a continuous distribution with density $f()$ that is symmetric about m, then not only is X_i equally likely to be on either side of m, but the distance from m is independent of the side. Thus, if we rank all the data by its distance to m, the sum corresponding to the values larger than m may be viewed as a random sample of a certain size from the numbers 1 through n. The distribution of this sum can be characterized, so the sum of the ranks can be an effective test statistic.

The Wilcoxon signed-rank test for the median

If the data, $X_1, X_2, \ldots, X_n$, is an *i.i.d.* sample from a continuous, symmetric distribution, then a significance test of the hypotheses

$$H_0: \text{ the median} = m, \quad H_A: \text{ median } < m, \text{ median } > m, \text{ or median } \neq m$$

can be performed with test statistic

$$T = \sum_{i:X_i > m} \mathsf{rank}(|X_i - m|).$$

Under H_0, the distribution of T can be calculated. Large values of T support the alternative hypothsis H_A : median $> m$.

In R, the function `wilcox.test()` performs the test as

```
wilcox.text(x, mu=..., alt="two.sided")
```

The data is contained in `x`, the null hypothesis is specified by the argument `mu=`, and the alternative is specified with the argument `alt=`. This argument takes a value of `"less"`, `"greater"`, or `"two.sided"` (the default value). If desired, the distribution of T is given by the function `psignrank()`.

A typical application of the signed-rank test is to use it after transforming the data to make it look like the parent distribution is symmetric.

■ **Example 8.7: Number of recruits** In salmon populations, there is a relationship between the number of spawners and the subsequent number of "recruits" that they produce. A common model involves two parameters, which describe how many recruits there are when there are few spawners and how many there are when there are many spawners. The data set `salmon.rate (UsingR)` contains simulated data on one of the parameters. A plot of the data shows that a normal population assumption is not correct; rather, the population appears to be lognormal.

Perform a significance test of

$$H_0 : \text{median} = .005, \quad H_A : \text{median} > .005.$$

After taking logs, we can see that the data is symmetric, so the signed-rank test can apply to the log-transformed data. The significance test of this data is

$$H_0 : \text{median} = \log(.005), \quad H_A : \text{median} > \log(.005).$$

```
> wilcox.test(log(salmon.rate),mu = log(.005),alt="greater")

        Wilcoxon signed rank test with continuity correction

data:  log(salmon.rate)
V = 2077, p-value = 0.065
alternative hypothesis: true mu is greater than -5.298
```

A small p-value is found.

To contrast, the p-value for the sign test is found with these commands:

```
> T = sum(salmon.rate > .005); n = length(salmon.rate)
> 1 - pbinom(T - 1, n, 1/2)
[1] 0.1361
```

■

8.4.3 Problems

8.17 The `exec.pay (UsingR)` data set contains data on the salaries of CEOs at 199 top companies in the United States. The amounts are in \$10,000s. The data is not symmetric. Do a sign test to determine whether the median pay is more than \$220,000.

8.18 Repeat the previous exercise, using the signed-rank test on the log-transformed data. Do you reach the same conclusion?

8.19 The `babies (UsingR)` data set contains data covering many births. Information included is the gestation period, and a factor indicating whether the mother was a smoker. Extracting the gestation times for mothers who smoked during pregnancy can be done with these commands:

```
> attach(babies)
> smokers = gestation[smoke == 1 & gestation != 999]
> detach(babies)
```

Perform a significance test of the null hypothesis that the average gestation period is 40 weeks against a two-sided alternative. Explain what test you used, and why you chose that one.

8.20 If the sign test has fewer assumptions on the population, why wouldn't we always use that instead of a t-test? The answer lies in the power of the sign test to detect when the null hypothesis is false. The sign test will not reject a false null as often as the t-test. The following commands will perform a simulation comparing the two tests on data that has a $\mathsf{Normal}(1,2)$ distribution. The significance tests performed are both

$$H_0 : \mu = 0, \quad H_A : \mu > 0.$$

Run the simulation. Is there a big difference between the two tests?

```
> m = 200; n = 10
> res.t = rep(0,m);res.sign=rep(0,m)
> for(i in 1:m) {
+   x = rnorm(n, mean=1, sd=2)
+   if(t.test(x,mu=0,alt = "greater")$p.value < 0.05)
+     res.t[i] = 1
+   T = sum(x>0)
+   if (1-pbinom(T-1,n,1/2) < .05)
+     res.sign[i]=1
+ }
> sum(res.t)/m                      # proportion rejected by t-test
> sum(res.sign)/m                   # proportion rejected by sign-test
```

(The notation `$p.value` appended to the output of `t.test()` retrieves just the p-value from the test results.)

8.5 Two-sample tests of proportion

In the previous sections our tests of significance compared a sample to some assumed value of the population and determined whether the sample supported the null hypothesis. This assumes some specific knowledge about the population parameters. In many situations, we'd like to compare two parameters.

In this section we consider how to compare two population proportions. This can be useful in many different contexts: comparing polling results taken over different periods of time, surveying results after an intervention such as an advertising campaign, or comparing attitudes among different ethnic groups.

In Example 8.3, we compared the 2001 poverty rate, which was found by a sample, with the 2000 poverty rate which was known from a census. To compare the 2002 rate to the 2001 rate, we would compare two samples. How do we handle this with a significance test?

Let $\widehat{p}_1$ be the estimated 2001 poverty rate and $\widehat{p}_2$ be the estimated 2002 poverty rate. We wish to perform a significance test of

$$H_0: p_1 = p_2, \quad H_A: p_1 < p_2$$

using the values of $\widehat{p}_1$ and $\widehat{p}_2$ in the test statistic. If we think of the test as one of differences, we can rephrase it as

$$H_0: p_1 - p_2 = 0, \quad H_A: p_1 - p_2 < 0.$$

A natural test statistic would be

$$Z = \frac{(\widehat{p}_1 - \widehat{p}_2) - E(\widehat{p}_1 - \widehat{p}_2 \mid H_0)}{\mathrm{SE}(\widehat{p}_1 - \widehat{p}_2 \mid H_0)}.$$

We assume that the surveys were a simple random sample from the population, so that the number responding favorably, x_i, has a binomial distribution with $n = n_i$ and $p = p_i$ for $i = 1,2$. (So $\widehat{p}_i = x_i/n_i$.) Thus, the expectation in Z is simply $p_1 - p_2 = 0$ under the null hypothesis. The standard error is found from the standard deviation under the null hypothesis

$$\mathrm{SD}(\widehat{p}_1 - \widehat{p}_2 \mid H_0) = \sqrt{\frac{p_1(1-p_1)}{n_1} + \frac{p_2(1-p_2)}{n_2}} = \sqrt{p(1-p)(\frac{1}{n_1} + \frac{1}{n_2})},$$

where $p = p_1 = p_2$ under the null hypothesis. The value of p is not assumed in H_0, so we estimate it and use the standard error instead. To estimate p it makes sense to use the entire sample:

$$\widehat{p} = \frac{\text{total who are favorable}}{\text{total size of both samples}} = \frac{x_1 + x_2}{n_1 + n_2} = \frac{n_1\widehat{p}_1 + n_2\widehat{p}_2}{n_1 + n_2}.$$

This leaves

$$Z = \frac{\widehat{p}_1 - \widehat{p}_2}{\sqrt{\widehat{p}(1-\widehat{p})(\frac{1}{n_1} + \frac{1}{n_2})}}. \tag{8.2}$$

Two-sample test of proportions

If we have sample proportions for two random samples, a significance test of

$$H_0: p_1 = p_2, \quad H_A: p_1 < p_2,\ p_1 > p_2, \text{ or } p_1 \neq p_2$$

can be carried out with test statistic Z given by (8.2). Under H_0, Z has a standard normal distribution if n_1 and n_2 are sufficiently large. Large values of Z support the alternative $p_1 > p_2$; small values support $p_1 < p_2$.

In R, the function `prop.test()` will perform a two-sample test of proportions:

```
prop.test(x, n, alt="two.sided")
```

The data is specified by a vector of values with `x` storing the counts and `n` the sample size. There is no need to specify a null hypothesis, as it is always the same. The alternative hypothesis is specified by one of `alt="less"`, `alt="greater"`, or `alt="two.sided"` (the default).

■ **Example 8.8: Poverty rate, continued** Assume the 2001 poverty rate of 11.7% was derived from a random sample of 50,000 people, and the 2002 poverty rate of 12.1% was derived from a simple random sample of 60,000. Is the difference between the proportions statistically significant?

Let $\widehat{p}_1 = 0.117$ and $\widehat{p}_2 = 0.121$ be the given sample proportions. Our null and alternative hypotheses are

$$H_0: p_1 = p_2, \quad H_A: p_1 < p_2.$$

We can use `prop.test()` using $n\widehat{p}$ to give the frequencies of those in the sample

```
> phat = c(.121, .117)          # the sample proportions
> n = c(50000, 60000)           # the sample sizes
> n*phat                        # the counts
[1] 5850 7260
```

```
> prop.test(n*phat,n,alt="less")

        2-sample test for equality of proportions with
        continuity correction

data:  n * phat out of n
X-squared = 4.119, df = 1, p-value = 0.02121
alternative hypothesis: less
95 percent confidence interval:
 -1.0000000 -0.0007589
sample estimates:
prop 1 prop 2
 0.117  0.121
```

The small p-value of 0.02107 indicates an increase in the rate.

If we were to do this by hand, rather than by using `prop.test()`, we would find:

```
> p = sum(n*phat)/sum(n)          # (n_1p_1 + n_2p_2)/(n_1 + n_2)
> obs = (phat[1]-phat[2])/sqrt(p*(1-p)*sum(1/n))
> obs
[1] -2.039
> pnorm(obs)
[1] 0.02073
```

This also gives a small p-value. The difference is due to a continuity correction used by `prop.test()`. ■

8.5.1 Problems

8.21 A cell-phone store has sold 150 phones of Brand A and had 14 returned as defective. Additionally, it has sold 125 phones of Brand B and had 15 phones returned as defective. Is there statistical evidence that Brand A has a smaller chance of being returned than Brand B?

8.22 In the year 2001, a poll of 600 people found that 250 supported gay marriage. A 2003 poll of 500 found 250 in support. Do a test of significance to see whether the difference in proportions is statistically significant.

8.23 There were two advance screenings of a new movie. The first audience was composed of a classical-music radio station's listeners, the second a rock-and-roll music station's listeners. Suppose the audience size was 350 for each screening. If 82% of the audience at the first screening rated the movie favorably, but only 70% of second audience did, is this difference statistically significant? Can you assume that each audience is a random sample from the population of the respective radio station listeners?

8.24 The HIP mammography study was one of the first and largest studies of the value of mammograms. The study began in New York in the 1960s and involved 60,000 women randomly assigned to two groups—one that received mammograms, and one that did not. The women were then observed for the next 18 years. Of the 30,000 who had mammograms, 153 died of breast cancer; of the 30,000 who did not, 196 died of breast cancer. Compare the two sample proportions to see whether there is a statistically significant difference between the death rates of the two groups. (There is debate about the validity of the experimental protocol.)

8.25 Ginkgo biloba extract is widely touted as a miracle cure for several ailments, including acute mountain sickness (AMS), which is common in mountaineering. A randomized study took 44 healthy subjects to the Himalayas; half received the extract (80 mg twice/day) and half received placebos. Each group was measured for AMS. The results of the study are given in Table 8.3. Compute a p-value for a significance test for the null hypothesis of equivalence of proportions against a two-sided alternative.

Table 8.3 Data on acute mountain sickness

Group	n	Number who suffered AMS
placebo	22	18
ginkgo biloba	22	3

source: *Aviation, Space, and Environmental Medicine* 67, 445-452, 1996

8.26 Immediately after a ban on using of hand-held cell phones while driving was implemented, compliance with the law was measured. A random sample of 1,250 found that 98.9% were in compliance. A year after the implementation, compliance was again measured. A sample of 1,100 drivers found 96.9% in compliance. Is the difference in proportions statistically significant?

8.27 The start of a May 5, 2004 *New York Times* article reads

> In the wake of huge tobacco tax increases and a ban on smoking in bars, the number of adult smokers in New York City fell 11 percent from 2002 to 2003, one of the steepest short-term declines ever measured, according to surveys commissioned by the city.

The article continues, saying that the surveys were conducted using methods—the questions and a random dialing approach—identical to those done annually by the federal Centers for Disease Control and Prevention. Each survey used a large sample of 10,000 people, giving a stated margin of error of 1 percentage point.

The estimated portion of the population that smoked in 2002 was 21.6% the estimated proportion in 2003 was 19.3%. Are these differences significant at the 0.01 level?

8.6 Two-sample tests of center

A physician may be interested in knowing whether the time to recover for one surgery is shorter than that for another surgery. A taxicab driver might wish to know whether the time to travel one route is faster than the time to travel another. A consumer group might wish to know whether gasoline prices are similar in two different cities. Or a government agency might want to know whether consumer spending is similar in two different states. All of these questions could be approached by taking random samples from the respective populations and comparing. We consider the situation when the question of issue can be boiled down to a comparison of the centers of the two populations. We can use a significance test to compare centers in the same manner as we compare two population proportions. However, as there are more possibilities for types of populations considered, there are more test statistics to consider.

Suppose $X_i, i = 1, \ldots, n_x$ and $Y_j, j - 1, \ldots, n_y$ arc random samples from the two populations of interest. A significance test to compare the centers of their parent distributions would use the hypotheses

$$H_0 : \mu_x = \mu_y, \quad H_A : \mu_x < \mu_y,\ \mu_x > \mu_y,\ \text{or}\ \mu_x \neq \mu_y. \tag{8.3}$$

A reasonable test statistic depends on the assumptions placed on the parent populations. If the populations are normally distributed or nearly so, and the samples are independent of each other, then a t-test can be used. If the populations are not normally distributed, then a nonparametric Wilcoxon test may be appropriate. If the samples are not independent but paired off in some way, then a paired test might be called for.

8.6.1 Two sample tests of center with normal populations

Suppose the two samples are independent with normally distributed populations. As $\bar{X}$ and $\bar{Y}$ estimate μ_x and μ_y respectively, the value of $\bar{X} - \bar{Y}$ should be a good estimate for $\mu_x - \mu_y$. We can use this to form a test statistic. Both sample means have normally distributed sampling distributions. A natural test statistic is then

$$T = \frac{(\bar{X} - \bar{Y}) - E(\bar{X} - \bar{Y} \mid H_0)}{\mathsf{SE}(\bar{X} - \bar{Y} \mid H_0)}.$$

Under H_0, the expected value of the difference is 0. The standard error is found from the formula for the standard deviation, which is based on the independence of the samples:

$$\mathsf{SD}(\bar{X} - \bar{Y} \mid H_0) = \sqrt{\frac{\sigma_x^2}{n_x} + \frac{\sigma_y^2}{n_y}}.$$

As with confidence intervals, the estimate used for the population variances depends on an assumption of equal variances.

If the two variances are assumed equal, the all the data is pooled to estimate $\sigma = \sigma_x = \sigma_y$ using

$$s_p = \sqrt{\frac{(n_x - 1)s_x^2 + (n_y - 1)s_y^2}{n_x + n_y - 2}}. \tag{8.4}$$

The standard error used is

$$\mathsf{SE}(\bar{X} - \bar{Y}) = s_p \sqrt{\frac{1}{n_x} + \frac{1}{n_y}}. \tag{8.5}$$

With this, T has a t-distribution with $n - 2$ degrees of freedom.

If the population variances are not assumed to be equal, then we estimate σ_x with s_x and σ_y with s_y to get

$$\mathsf{SE}(\mu_x - \mu_y) = \sqrt{\frac{s_x^2}{n_x} + \frac{s_y^2}{n_y}}. \tag{8.6}$$

Additionally, we use the Welch approximation for the degrees of freedom as described in Chapter 7. This again yields a test statistic that is described by the t-distribution under the null hypothesis.

---

t-tests for comparison of means of independent samples

Assume $X_1, X_2, \ldots, X_{n_x}$ and $Y_1, Y_2, \ldots, Y_{n_y}$ are independent random samples from $\mathsf{Normal}(\mu_i, \sigma_i)$ distributions, where $i = x$ or y. A significance test of

$$H_0: \mu_x = \mu_y, \quad H_A: \mu_x < \mu_y,\ \mu_x > \mu_y, \text{ or } \mu_x \neq \mu_y$$

can be done with test statistic T. T will have the t-distribution with a specified number of degrees of freedom under H_0. Larger values of T

support $H_A : \mu_x > \mu_y$.

If we assume that $\sigma_x^2 = \sigma_y^2$, then T has $n_x + n_y - 2$ degrees of freedom, and the standard error is given by (8.5).

If we assume that $\sigma_x^2 \neq \sigma_y^2$, then T has degrees of freedom given by the Welch approximation in Equation 7.5 and standard error given by (8.6).

In each case, the function `t.test()` will perform the significance test. It is used with the arguments

```
t.test(x, y, alt="two.sided", var.equal=FALSE)
```

The data is specified in two data vectors, `x` and `y`. There is no need to specify the null hypothesis, as it is always the same. The alternative is specified by `"less"`, `"greater"`, or `"two.sided"` (the default). The argument `var.equal=TRUE` is given to specify the equal-variance case. The default is to assume unequal variances.

■ **Example 8.9: Differing dosages of AZT**

AZT was the first FDA approved antiretroviral drug used in the care of HIV-infected individuals. The common dosage is 300 mg twice daily. Higher dosages cause more side effects. But are they more effective? A study done in 1990 compared dosages of 300 mg, 600 mg, and 1,500 mg (source `http://www.aids.org`). The study found higher toxicity with greater dosages, and, more importantly, that the lower dosage may be equally effective.

The p24 antigen can stimulate immune responses. The measurement of p24 levels for the 300 mg and 600 mg groups is given by the simulated data in Table 8.4. Perform a t-test to determine whether there is a difference in means.

Table 8.4 Levels of p24 in mg for two treatment groups

Amount	p24 level									
300 mg	284	279	289	292	287	295	285	279	306	298
600 mg	298	307	297	279	291	335	299	300	306	291

Let μ_x be the mean of the 300 mg group, and μ_y the mean of the 600 mg group. We can test the hypotheses

$$H_0 : \mu_x = \mu_y, \quad H_A : \mu_x \neq \mu_y$$

with a t-test. First, we check to see whether the assumption of a common variance and normality seems appropriate by looking at two densityplots:

```
> x = c(284, 279, 289, 292, 287, 295, 285, 279, 306, 298)
> y = c(298, 307, 297, 279, 291, 335, 299, 300, 306, 291)
> plot(density(x))
> lines(density(y), lty=2)
```

The graph (Figure 8.5) shows two density estimates that indicate normally distributed populations with similar spreads. As such, the t-test looks appropriate.

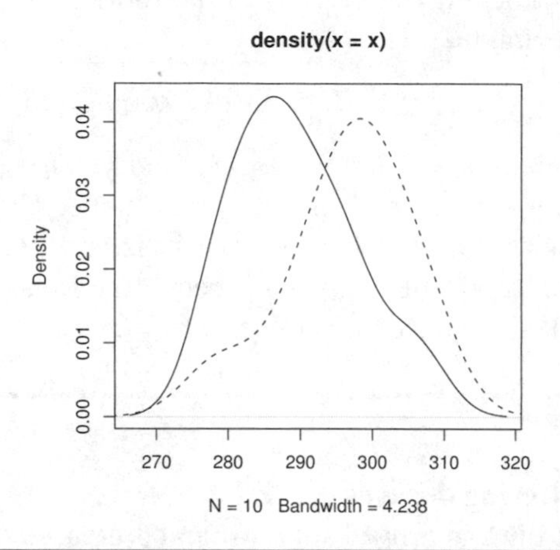

Figure 8.5 Densityplots to compare variances and shapes of the 300 mg dosage (solid) and the 600 mg dosage (dashed)

```
> t.test(x,y,var.equal=TRUE)

        Two Sample t-test

data:  x and y
t = -2.034, df = 18, p-value = 0.05696
alternative hypothesis: true difference in means is not equal to 0
...
```

The p-value is 0.05696 for the two-sided test. This suggests a difference in the mean values, but it is not statistically significant at the 0.05 significance level. A look at the reported confidence interval for the difference of the means shows a wide range of possible value for $\mu_x - \mu_y$. We conclude that this data is consistent with the assumption of no mean difference.

How would this change if we did not assume equal variances?

```
> t.test(x,y)

        Welch Two Sample t-test
```

```
data:  x and y
t = -2.034, df = 14.51, p-value = 0.06065
alternative hypothesis: true difference in means is not equal to 0
95 percent confidence interval:
 -22.3557   0.5557
sample estimates:
mean of x mean of y
    289.4     300.3
```

In this example, the same observed value of the test statistic (marked `t`) is found as in the equal-variance case, as (8.5) and (8.6) yield identical standard errors when the two sample sizes are the same. We get a larger p-value, though, as the degrees of freedom decrease. ■

8.6.2 Matched samples

There are times when two samples depend on each other in some way, for example, samples from twin studies, where identical or fraternal twins are used as pairs, so that genetic or environmental factors can be controlled. For this, the usual two-sample t-test is not applicable. We mention two examples.

■ **Example 8.10: Twin studies** An industry-sponsored clinical trial (`http://www.hairtoday.com/html/propeciatwins.cfm`) demonstrates that Finasteride inhibits male-pattern hair loss. How did the researchers show this? They used two treatment groups: one received a Finasteride treatment, the other a placebo. A randomized, double-blind study was performed. Hair loss was measured by photographs, hair counts, and questionnaires.

What was different about this study was the use of identical twins for the treatment groups. For each pair of twins, one was randomly assigned to the treatment group and the other to the control group. This allowed the researchers to "control" for genetic differences — differences that might be so great that the true effect of the Finasteride treatment could be hidden. The researchers stated

> As identical twins share the same genetic makeup, comparison between the responses of each subject in a twin pair, when one receives drug and the other receives placebo, allows for rigorous examination of the effects due to drug treatment in a limited number of subjects.

■

■ **Example 8.11: Pre- and post-tests** Outcomes assessment is an attempt to measure whether a certain action actually does what it is intended to do. For example, does a statistics book actually work for teaching R? Or, does a statistics class make you understand the difference between mere sampling variation and a true effect? One way to assess the effectiveness of something is with a pre-test and a post-test. If the scores are markedly better on the post-test, then we may be able to attribute the change to the teaching.

Imagine a class takes a pre-test and a post-test. Each student has two test scores, X_i for the first test and the matching Y_i for the second. How can we test whether there is a difference in the means? We might be tempted to use the t-test, but we should be careful, as the two samples are not independent. This assumption of independence was used implicitly when computing the standard error in the test statistic. Besides, what is really important is the change in the scores $X_i - Y_i$. ■

For paired data, even if there are large variations within the samples, we can still test a difference in the means by using a one-sample test applied to the data, $X_i - Y_i$.

Significance tests for paired samples

If the two sample $X_1, X_2, \ldots, X_n$ and $Y_1, Y_2, \ldots, Y_n$ are matched so that the differences $X_i - Y_i$ are an *i.i.d.* sample, then the significance test of hypotheses

$$H_0: \mu_x = \mu_y, \quad H_A: \mu_x < \mu_y,\ \mu_x \neq \mu_y,\ \text{or}\ \mu_x > \mu_y$$

becomes a significance test of

$$H_0: \mu = 0, \quad H_A: \mu < 0,\ \mu > 0,\ \text{or}\ \mu \neq 0.$$

If the differences have a normally distributed population, a t-test can be used. If the differences are from a symmetric distribution, the Wilcoxon signed-rank test can be used. Otherwise, the sign test can be used, where μ is interpreted as the difference of medians.

In R, both the `t.test()` and `wilcox.test()` functions have an argument `paired=TRUE` that will perform the paired tests.

■ **Example 8.12: Twin studies continued** For the Finasteride study, photographs are taken of each head. They are assessed using a standard methodology. This results in a score between 1 and 7: 1 indicating greatly decreased hair growth and 7 greatly increased. Simulated data, presented as pairs, is in Table 8.5.

We can assess the differences with a paired t-test as follows:

```
> Finasteride = c(5,3,5,6,4,4,7,4,3)
> placebo = c(2,3,2,4,2,2,3,4,2)
```

Table 8.5 Assessment for hair loss on 1–7 scale for twin study

Group	score								
Finasteride treatment	5	3	5	7	4	4	7	4	3
placebo	2	3	2	4	2	2	3	4	2

```
> t.test(Finasteride, placebo, paired=TRUE, alt="two.sided")

        Paired t-test

data:  Finasteride and placebo
t = 4.154, df = 8, p-value = 0.003192
alternative hypothesis: true difference in means is not equal to 0
95 percent confidence interval:
 0.8403 2.9375
sample estimates:
mean of the differences
                  1.889
```

We see a very small p-value, indicating that the result is significant. The null hypothesis of no effect is in doubt. ■

■ **Example 8.13: Pre- and post-tests, continued** To test whether a college course is working, a pre- and post-test is arranged for the students. The results are given in Table 8.6. Compare the scores with a t-test. First, assume that the scores are randomly selected from the two tests. Next, assume that they are pairs of scores for ten students.

Table 8.6 Pre- and post-test scores

Test	score									
Pre-test	77	56	64	60	57	53	72	62	65	66
Post-test	88	74	83	68	58	50	67	64	74	60

For each, we test the hypotheses that

$$H_0 : \mu_1 = \mu_2, \quad H_A : \mu_1 < \mu_2,$$

and we assume that the data is normally distributed.

If we assume that the scores are random samples from the two test populations, then the usual t-test is used. We first make a boxplot, to decide whether the variances are equal (not shown), and then we apply the test.

```
> pre = c(77, 56, 64, 60, 57, 53, 72, 62, 65, 66)
> post = c(88, 74, 83, 68, 58, 50, 67, 64, 74, 60)
> boxplot(pre,post)
> t.test(pre, post,var.equal=TRUE, alt="less")
...
t = -1.248, df = 18, p-value = 0.1139
...
```

The p-value is small but not significant.

If we assume these scores are paired off, then we focus on the differences. This gives a much smaller p-value:

```
> t.test(pre,post, paired=TRUE, alt="less")
...
t = -1.890, df = 9, p-value = 0.04564
...
```

This time, the difference is significant at the 0.05 level.

If small samples are to be used, it can often be advantageous to use paired samples, rather than independent samples. ■

8.6.3 The Wilcoxon rank-sum test for equality of center

The two-sample t-test tests whether two independent samples have the same center when both samples are drawn from a normal distribution. However, there are many situations in which the parent populations may be heavily skewed or have heavy tails. Then the t-test is not appropriate. However, if it is assumed that our two samples come from two distributions that are identical up a shift of center, then the Wilcoxon rank-sum test can be used to perform a significance test to test whether the centers are identical.

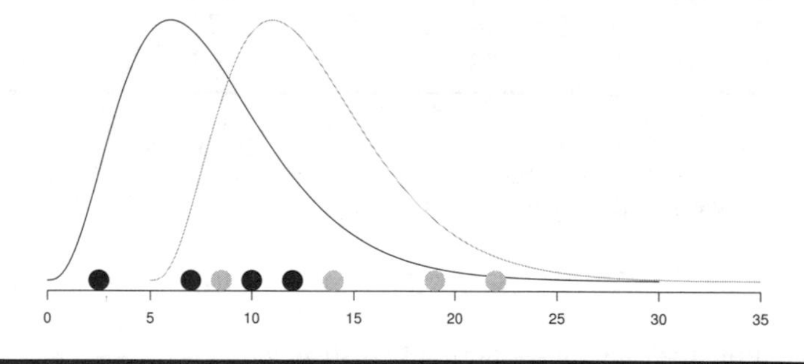

Figure 8.6 Two random samples from similar distributions indicated by points with different shades. The lower ranked ones come primarily from the distribution shifted to the left.

To be precise, suppose $f(x)$ is a density of a continuous distribution with mean 0. Further, assume that the X_i are a random sample from a population with density $f(x-\mu_1)$ (so it has mean μ_1), and that the Y_j are a random sample from a population with density $f(x-\mu_2)$. Figure 8.6 shows two samples where the μ's are different. The darker distribution is shifted to the left, and its sample, indicated with darker dots, has most of its values to the left of the other sample. This would not be expected if the two populations were identical. The rank-sum statistic quantifies this, allowing for a significance test.

Wilcoxon rank-sum test for two independent samples

Assume that the random sample, $X_1, X_2, \ldots X_{n_x}$, comes from a distribution with density $f(\cdot - \mu_x)$, and that $Y_1, Y_2, \ldots, Y_{n_y}$ from a distribution with density $f(\cdot - \mu_y)$ (same shaped density, but perhaps different centers). A test of significance of the hypotheses

$$H_0 : \mu_1 = \mu_2, \quad H_A : \mu_1 < \mu_2,\ \mu_1 > \mu_2, \text{ or } \mu_1 \neq \mu_2$$

can be performed with the rank-sum statistic.

To perform the significance test in R, the `wilcox.test()` function is used as

```
wilcox.test(x, y, alt="two.sided")
```

The variables `x` and `y` store the two data sets, and the alternative is specified as usual with one of `"less"`, `"greater"`, or `"two.sided"` (the default). The `wilcox.test()` function will also work in the case when there are ties in the data.

■ **Example 8.14: Comparing grocery checkers** A grocery store's management wishes to compare checkout personnel to see if there is a difference in their checkout times. A small sample of data comparing two checkers' times (in minutes) is given in Table 8.7. Compare the mean checkout times.

We use the `wilcox.test()` function after verifying that the assumptions are met.

```
> A = c(5.8, 1.0, 1.1, 2.1, 2.5, 1.1, 1.0, 1.2, 3.2, 2.7)
> B = c(1.5, 2.7, 6.6, 4.6, 1.1, 1.2, 5.7, 3.2, 1.2, 1.3)
> plot(density(A))
> lines(density(B))
```

The graph (not shown) suggests that the populations are skewed with long tails. As such, the t-test assumptions are not met. However, we also see that the sam-

Table 8.7 Ten checkout times for two grocery checkers

Checker	Times									
checker A	5.8	1.0	1.1	2.1	2.5	1.1	1.0	1.2	3.2	2.7
checker B	1.5	2.7	6.6	4.6	1.1	1.2	5.7	3.2	1.2	1.3

ples appear to have densities with the same shape, so the rank-sum test is available. A two-sided test can be done with

```
> wilcox.test(A,B)
        Wilcoxon rank sum test with continuity correction
data:  A and B
W = 34, p-value = 0.2394
alternative hypothesis: true mu is not equal to 0
```

The p-value is not significant. ■

8.6.4 Problems

8.28 A 2003 study at the Cleveland Clinic compared the effects of two cholesterol drugs, atorvastatin and pravastatin, on middle-aged heart-disease patients. It was found that the atorvastatin treatment group had an average LDL level of 79 after treatment, whereas the pravastatin group had an average LDL level of 110. Suppose the two groups contained 250 patients each, and the sample standard deviations were 25 for the atorvastatin group and 20 for the pravastatin. If the populations are assumed to be normally distributed, perform a two-sample test to compare whether the mean LDL levels for atorvastatin are lower than those for pravastatin, or whether the differences are explainable by chance variation.

8.29 A test to determine whether echinacea is beneficial in treating the common cold was set up as follows. If a child reported cold symptoms, then he was randomly assigned to be given either echinacea or a placebo. Recovery time was measured and is summarized in Table 8.8. Is this statistical evidence that children in the echinacea group had a quicker recovery?

Table 8.8 Recovery time for different treatment groups

group	n	$\bar{x}$	s
echinacea	200	5.3	2.5
placebo	207	5.4	2.5

8.30 For the `babies (UsingR)` data set, the variable `age` contains the mom's age and `dage` contains the dad's age for several babies. Do a significance test of the null hypothesis of equal ages against a one-sided alternative that the dads are older.

8.31 The data set `normtemp (UsingR)` contains body measurements for 130 healthy, randomly selected individuals. The variable `temperature` contains normal body temperature data and the variable `gender` contains gender information, with male coded as `1` and female as `2`. First split the data by `gender`, and then perform a two-sample test to see whether the population means are equivalent. Is the difference statistically significant?

8.32 Students wishing to graduate must achieve a specific score on a standardized test. Those failing must take a course and then attempt the test again. Suppose 12 students are enrolled in the extra course and their two test scores are given in Table 8.9. Do a t-test to see if there was any improvement in the students' mean scores following the class. If you assume equal variances or a paired test, explain why.

Table 8.9 Student scores on pre- and post-test

Student	scores											
Pre-test	17	12	20	12	20	21	23	10	15	17	18	18
Post-test	19	25	18	18	26	19	27	14	20	22	16	18

The p-value indicates that the null hypothesis of "no improvement" is not consistent with the data.

8.33 Water-quality researchers wish to measure biomass/chlorophyll ratio for phytoplankton (in milligrams per liter of water). There are two possible tests, one less expensive than the other. To see whether the two tests give the same results, ten water samples were taken and each was measured both ways, providing the data in Table 8.10. Do a t-test to see if there is a difference in the means of the measured amounts. If you assume equal variances or a paired test, explain why.

8.34 The `shoes` data set in the `MASS` package contains a famous data set on shoe wear. Ten boys wore two different shoes each, then measurements were taken on shoe wear. The wear amounts are stored in variables `A` and `B`. First make a scatterplot of the data, then compare the mean wear for the two types of shoes using the appropriate t-test.

Table 8.10 Measurements of biomass/chlorophyll in mg/L

Method	measurement									
method 1	45.9	57.6	54.9	38.7	35.7	39.2	45.9	43.2	45.4	54.8
method 2	48.2	64.2	56.8	47.2	43.7	45.7	53.0	52.0	45.1	57.5

8.35 The `galton (UsingR)` data set contains data collected by Francis Galton in 1885. Each data point contains a child's height and an average of his or her parents' heights. Do a t-test to see if there is a difference in the mean height. Assume the paired t-test is appropriate. What problems are there with this assumption?

8.36 The question of equal variances comes up when we perform a two sample t-test. We've answered this based on a graphical exploration. The F-test for equal variances of two normal populations can be used to test formally for equality. The test statistic is the ratio of the sample variances, which under the null hypothesis of equal variances has an F-distribution. This test is carried out by the function `var.test()`. A two-sided test ($H_A : \sigma_1^2 \neq \sigma_2^2$) is done with the command `var.test(x,y)`.

Do a two-sided test for equality of variance on the data in Example 8.9.

Chapter 9

Goodness of fit

In this chapter we return to problems involving categorical data. We previously summarized such data using tables. Here we discuss a significance test for the distribution of the values in a table. The test statistic will be based on how well the actual counts for each category fit the expected counts.

Such tests are called goodness-of-fit tests, as they measure how well the distribution of the data fits a probability model. In this chapter we will also discuss goodness-of-fit tests for continuous data. For example, we will learn a significance test for investigating whether a data set is normally distributed.

9.1 The chi-squared goodness-of-fit test

In a public-opinion poll, there are often more than two possible opinions. For example, suppose a survey of registered voters is taken to see which candidate is likely to be elected in an upcoming election. For simplicity, we assume there are two candidates, a Republican and a Democrat. A prospective voter may choose one of these or may be undecided. If 100 people are surveyed, and the results are 35 for the Republican, 40 for the Democrat, and 25 undecided, is the difference between the Republican and Democratic candidate significant?

9.1.1 The multinomial distribution

Before answering a question about significance, we need a probability model, so that calculations can be made. The above example is a bit different from the familiar polling model. When there are just two categories to choose from we use the binomial model as our probability model; in this case, with more categories, we generalize and use the multinomial model.

Assume we have k categories to choose from, labeled 1 through k. We pick one of the categories at random, with probabilities specified by $p_1, p_2, \ldots, p_k$; p_i gives the probability of selecting category i. We must have $p_1 + p_2 + \cdots + p_k = 1$. If all the p_i equal $1/k$, then each category is equally likely (like rolling a die). Picking a category with these probabilities produces a single random value; repeat this selection n times, with each pick being independent, to get n values. A table of values will report the frequencies. Call these table entries $Y_1, Y_2, \ldots, Y_k$. These k numbers sum to n. The joint distribution of these random variables is called the **multinomial distribution**.

We can create multinomial data in R with the `sample()` function. For example, an M&Ms bag is filled using colors drawn from a fixed ratio. A bag of 30 can be filled as follows:

```
> cols = c("blue","brown","green","orange","red","yellow","purple")
> prob = c(1,1,1,1,2,2,2)        # ratio of colors
> bagfull.mms = sample(cols,30,replace=TRUE, prob=prob)
> table(bagfull.mms)
bagfull.mms
  blue  brown  green orange purple    red yellow
     2      3      1      3      6     10      5
```

A formula for the multinomial distribution is similar to that for the binomial distribution except that more factors are involved, as there are more categories to choose from. The distribution can be specified as follows:

$$\mathsf{P}(Y_1 = y_1, \ldots, Y_k = y_k) = \binom{n}{y_1}\binom{n-y_1}{y_2}\cdots\binom{n-y_1-y_2-\cdots-y_{k-1}}{y_k} p_1^{y_i}\cdots p_k^{y_k}.$$

As an example, consider the voter survey. Suppose we expected the percentages to be 35% Republican, 35% Democrat, and 30% undecided. What is the probability in a survey of 100 that we see 35, 40, and 25 respectively? It is

$$\mathsf{P}(Y_1 = 35, Y_2 = 40, Y_3 = 25) = \binom{100}{35}\binom{65}{40}\binom{25}{25}(.35)^{35}(.35)^{40}(.3)^{25}.$$

This is found with

```
> choose(100,30)*choose(70,40) * .35^35 * .35^40 * .30^25
[1] 0.008794
```

(We skip the last coefficient, as $\binom{j}{j} = 1$ for any j.) This small value is the probability of the observed value, but it is not a p-value. A p-value also includes the probability of seeing more extreme values than the observed one. We still need to specify what that means.

9.1.2 Pearson's χ^2 statistic

Trying to use the multinomial distribution directly to answer a problem about the p-value is difficult, as the variables Y_i are correlated. If one is large the others are more likely to be small, so the meaning of "extreme" in calculating a p-value is not immediately clear. As an alternative, the problem of testing whether a given set of probabilities could have produced the data is done as before: by comparing the observed value with the expected value and then normalizing to get something with a known distribution.

Each Y_i is a random variable telling us how many of the n choices were in category i. If we focus on a single i, then Y_i is seen to be $\mathsf{Binomial}(n, p_i)$. Again, the Y_i are not independent but correlated, as one large one implies that the others are more likely smaller. However, we know that the expected number of Y_i is np_i. Based on this, a good statistic might be

$$\sum_{i=1}^{k} (Y_i - np_i)^2.$$

This gives the total discrepancy between the observed and the expected. We use the square as $\sum Y_i - np_i = 0$. This sum gets larger when a category is larger or smaller than expected. So a larger-than-expected value contributes, and any correlated smaller-than-expected values do, too. As usual, we scale this by the right amount to yield a test statistic with a known distribution. In this case, each term is divided by the expected amount, producing **Pearson's chi-squared statistic** (written using the Greek letter *chi*):

$$\chi^2 = \sum_{i=1}^{k} \frac{(Y_i - np_i)^2}{np_i} = \sum \frac{(\text{observed} - \text{expected})^2}{\text{expected}}. \tag{9.1}$$

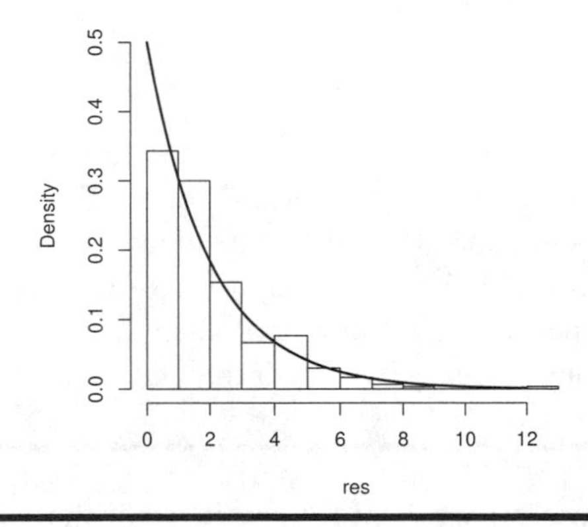

Figure 9.1 Simulation of χ^2 statistic with $n = 20$ and probabilities $3/12$, $4/12$, and $5/12$. The chi-squared density with 2 degrees of freedom is added.

If the multinomial model is correct, then the asymptotic distribution of Y_i is known to be the chi-squared distribution with $k - 1$ degrees of freedom. The number of degrees of freedom coincides with the number of free ways we can specify the values for p_i in the null hypothesis. We are free to choose $k-1$ of the values but not k, as the values must sum to 1.

The chi-squared distribution is a good fit if the expected cell counts are all five or more. Figure 9.1 shows a simulation and a histogram of the corresponding χ^2 statistic, along with a theoretical density.

Using this statistic as a test statistic allows us to construct a significance test. Larger values are now considered more extreme, as they imply more discrepancy from the predicted amount.

The chi-squared significance test for goodness of fit

Let $Y_1, Y_2, \ldots, Y_k$ be the observed cell counts in a table that arise from random sampling. Suppose their joint distribution is described by the multinomial model with probabilities $p_1, p_2, \ldots, p_k$. A significance test of

$$H_0: p_1 = \pi_1, \cdots, p_k = \pi_k, \quad H_A: p_i \neq \pi_i \text{ for at least } i$$

can be performed with the χ^2 statistic. The π_i are specified probabilities. Under H_0 the sampling distribution is asymptotically the chi-squared distribution with $k-1$ degrees of freedom. This is a good approximation, provided that the expected cell counts are all five or more. Large values of the statistic support the alternative.

This test is implemented by the `chisq.test()` function. The function is called with

```
chisq.test(x, p=...)
```

The data is given in tabulated form in x; the null hypothesis is specified with the argument p= as a vector of probabilities. The default is a uniform probability assumption. This should be given as a named argument, as it is not the second position in the list of arguments. The alternative hypothesis is not specified, as it does not change. A warning message will be returned if any category has fewer than five expected counts.

For example, suppose we wanted to know whether the voter data was generated according to the probabilities $p_1 = .35$, $p_2 = .35$, and $p_3 = .30$. To investigate, we can perform a significance test. This can be done directly with the `chisq.test()` function or "by hand." We illustrate both approaches, as we'll see soon that knowing how to do it the long way allows us to do more problems.

To do this by hand, we specify the counts in y and the probabilities in p, then form the test statistic:

```
> y = c(35,40,25)
> p = c(35,35,30)                # ratios
> p = p/sum(p)                   # proportions
> n = sum(y)
> chi2 = sum( (y - n*p)^2 / (n*p) )
> chi2
[1] 1.548
> pchisq(chi2, df = 3-1, lower.tail = F)
[1] 0.4613
```

In contrast, the above could have been done with

```
> chisq.test(y, p=p)

        Chi-squared test for given probabilities

data:  y
X-squared = 1.548, df = 2, p-value = 0.4613
```

The function returns the value of the test statistic (after `X-squared`), the degrees of freedom, and the *p*-value.

■ **Example 9.1: Teen smoking** The `samhda (UsingR)` data set contains information about health behavior for school-age children. For example, the variable `amt.smoke` measures how often a child smoked in the previous month. There are seven levels: a 1 means he smoked every day and a 7 means not at all. Values 98 and 99 indicate missing data. See `?samhda` for a description. We investigate whether the sample proportions agree with the probabilities:

$$p_1 = .15, p_2 = .05, p_3 = .05, p_4 = .05, p_5 = .10, p_6 = .20, p_7 = .40.$$

A test of significance can be constructed as follows:

```
> library(UsingR)
> attach(samhda)
> y = table(amt.smoke[amt.smoke < 98])
> y

  1   2   3   4   5   6   7
 32   7  13  10  14  43 105
> p = c(.15,.05,.05,.05,.10,.20,.40)
> chisq.test(y,p = p)

        Chi-squared test for given probabilities

data:  y
X-squared = 7.938, df = 6, p-value = 0.2427
> detach(samhda)                       # clean up
```

The *p*-value of 0.2427 is not significant. ■

Partially specified null hypotheses

In the example with voting data, we might be interested in knowing whether the Republican candidate is significantly trailing the Democrat or whether the differences are due to sampling variation. That is, we would want to test the hypotheses

$$H_0 : p_1 = p_2 \quad H_A : p_1 \neq p_2.$$

These, too, can be tested with the χ^2 statistic, but we need to specify what we mean by "expected," as under H_0 this is not fully specified.

To do so, we use any values completely specified by the null hypothesis; for those values that aren't, we estimate using the null hypothesis to pool our data as appropriate. For this problem, none of the p_i values are fully specified. To estimate $\widehat{p}_1 = \widehat{p}_2$, we use both of the cell counts through $(Y_1 + Y_2)/(2n)$. This leaves $\widehat{p}_3 = 1 - \widehat{p}_1 - \widehat{p}_2 = Y_3/n$. Then the χ^2 statistic in this case becomes

$$\chi^2 = \sum_{i=1}^{k} \frac{(Y_i - n\widehat{p}_i)^2}{n\widehat{p}_i}.$$

Again, if all the expected counts are large enough, this will have an approximately chi-squared distribution. There is only one degree of freedom in this problem, as only one thing is left to estimate, namely the value $p = p_1 = p_2$. Once we specify a value of p, then, by the assumptions in the null hypothesis, all the p_i are decided.

We get the p-value in our example as follows:

```
> y = c(35,40,25)
> phat = c(75/200,75/200,25/100)
> n = sum(y)
> sum( (y - n*phat)^2/(n*phat))
[1] 0.3333
> pchisq(.3333, df =1 , lower.tail=FALSE)
[1] 0.5637
```

The difference is not statistically significant.

In general, the χ^2 statistic can be used in significance tests where the null specifies some relationship among the multinomial probabilities. The asymptotic distribution of the test statistic under the null hypothesis will be chi-squared. The degrees of freedom will depend on the number of values that we are free to specify.

9.1.3 Problems

9.1 A die is rolled 100 times and yields the frequencies in Table 9.1. Is this a fair die? Answer using a significance test with $H_0 : p_i = 1/6$ for each i and

Table 9.1 100 rolls of a die

	value					
	1	2	3	4	5	6
count	13	17	9	17	18	26

$H_A : p_i \neq 1/6$ for at least one i.

9.2 Table 9.2 contains the results of a poll of 787 registered voters and the actual race results (in percentages of total votes) in the 2003 gubernatorial recall election in California.

Table 9.2 California gubernatorial recall election

Candidate	party	poll amount	actual
Schwarzenegger	Republican	315	48.6
Bustamante	Democrat	197	31.5
McClintock	Republican	141	12.5
Camejo	Green	39	2.8
Huffington	Independent	16	0.6
other	–	79	4.0

[a] Source `http://www.cnn.com`

Is the sample data consistent with the actual results? Answer this using a test of significance.

9.3 A package of M&M candies is filled from batches that contain a specified percentage of each of six colors. These percentages are given in the `mandms` `(UsingR)` data set. Assume a package of candies contains the following color distribution: 15 blue, 34 brown, 7 green, 19 orange, 29 red, and 24 yellow. Perform a chi-squared test with the null hypothesis that the candies are from a `milk chocolate` package. Repeat assuming the candies are from a `Peanut` package. Based on the p-values, which would you suspect is the true source of the candies?

9.4 The `pi2000 (UsingR)` data set contains the first 2,000 digits of π. Perform a chi-squared significance test to see if the digits appear with equal probability.

9.5 A simple trick for determining what language a document is written in is to compare the letter distributions (e.g., the number of z's) to the known proportions for a language. For these proportions, we use the familiar letter frequencies given in the `frequencies` variable of the `scrabble` (`UsingR`) data set. These are an okay approximation to those in the English language.

For simplicity (see `?scrabble` for more details), we focus on the vowel distribution of a paragraph from R's webpage appearing below. The counts and Scrabble frequencies are given in Table 9.3.

> R is a language and environment for statistical computing and graphics. It is a GNU project which is similar to the S language and environment which was developed at Bell Laboratories (formerly AT&T, now Lucent Technologies) by John Chambers and colleagues. R can be considered as a different implementation of S. There are some important differences, but much code written for S runs unaltered under R.

Table 9.3 Vowel distribution and Scrabble frequency

	a	e	i	o	u
count	28	39	23	22	11
Scrabble frequency	9	12	9	8	4

Perform a chi-squared goodness-of-fit test to see whether the distribution of vowels appears to be from English.

9.6 The names of common stars are typically Greek or Arab in derivation. The `bright.stars` (`UsingR`) data set contains 96 names of common stars. Perform a significance test on the letter distribution to see whether they could be mistaken for English words.

The letter distribution can be found with:

```
> all.names  = paste(bright.stars$name, sep="", collapse="")
> x = unlist(strsplit(tolower(all.names), ""))
> letter.dist = sapply(letters, function(i) sum(x == i))
```

The English-letter frequency is found using the `scrabble` (`UsingR`) data set with:

```
> p = scrabble$frequency[1:26];p=p/sum(p) # skip the blank
```

9.7 The number of murders by day of week in New Jersey during 2003 is shown in Table 9.4.

1. Perform a significance test to test the null hypothesis that a murder is equally likely to occur on any given day.

Table 9.4 Number of murders by day of week in New Jersey during 2003

Sunday	Monday	Tuesday	Wednesday	Thursday	Friday	Saturday
53	42	51	45	36	37	65

[a] Source: New Jersey State Police Uniform Crime Report `http://www.njsp.org`

2. Perform a significance test of the null hypothesis that murders happen on each weekday with equal probability; similarly on the weekends, but not necessarily with the same probability.

For each test, write down explicitly the null and alternative hypotheses.

9.8 A large bag of M&Ms is opened and some of the colors are counted: 41 brown, 48 orange, 105 yellow, and 58 green. Test the partially specified null hypothesis that the probability of brown is equal to the probability of orange. What do you conclude?

9.9 The data for Figure 9.1 was simulated using the following commands:

```
> n = 20; m = 250; k = 3
> f = factor(letters[1:k])
> p = c(3,4,5); p = p/sum(p)
> res=c()
> for(i in 1:m) {
+   x = sample(f,n,replace=TRUE,prob=p)
+   y = table(x)
+   res[i] = sum((y - n*p)^2/(n*p))
+ }
> hist(res,prob=T,col=gray(.8), ylim=c(0,.5)) # extend y limit
> curve(dchisq(x,df=k-1), add=TRUE)
```

The sampling distribution of χ^2 is well approximated by the chi-squared distribution, with $k-1$ degrees if the expected cell counts are all five or more. Do a simulation like the above, only with $n = 5$. Does the fit seem right? Repeat with $n = 20$ using the different probabilities `p=c(1,19,20)/40`.

9.10 When $k = 2$ you can algebraically simplify the χ^2 statistic. Show that it simplifies to

$$\chi^2 = \left(\frac{\widehat{p}_1 - p_1}{\sqrt{p_1(1-p_1)/n}} \right)^2 .$$

This is the square of the statistic used in the one-sample test of proportion and is asymptotically a single-squared normal or a chi-squared random variable with 1 degree of freedom. Thus, in this case, the chi-squared test is equivalent to the test of proportions.

9.2 The chi-squared test of independence

In a two-way contingency table we are interested in the relationship between the variables. In particular, we ask whether the levels of one variable affect the distribution of the other variable. That is, are they independent random variables in the same sense that we defined an independent sequence of random variables?

For example, in the seat-belt-usage data from Table 3.1 (reprinted in Table 9.5), does the fact that a parent has her seat belt buckled affect the chance that the child's seat belt will be buckled?

Table 9.5 Seat-belt usage in California

	Child	
Parent	buckled	unbuckled
buckled	56	8
unbuckled	2	16

The differences appear so dramatic that the answer seems to be obvious. We can set up a significance test to help decide, with a method that can be used when the data does not tell such a clear story.

To approach this question with a significance test, we need to state the null and alternative hypotheses, a test statistic, and a probability model.

First, our model for the sampling is that each observed car follows some specified probability that is recorded in any given cell. These probabilities don't change from observation to observation, and the outcome of one does not effect the distribution of another. That is, we have an *i.i.d.* sequence. Then a multinomial model applies. Fix some notation. Let n_r be the number of rows in the table (the number of levels of the row variable), n_c be the number of columns, and Y_{ij} be a random variable recording the frequency of the (i, j) cell. Let p_{ij} be the cell probability for the ith row and jth column. The marginal probabilities are denoted p^r_i and p^c_j where, for example, $p^r_i = p_{i1} + p_{i2} + \cdots + p_{in_j}$.

Our null hypothesis is that the column variable should be independent of the row variable. When stated in terms of the cell probabilities, p_{ij}, this says that $p_{ij} = p^r_i p^c_j$. This is consistent with the notion that independence means multiply.

Thus our hypotheses can be stated verbally as

$$H_0: \text{the variables are independent}, \quad H_A: \text{the variables are not independent}.$$

In terms of our notation, we can rewrite the null hypothesis as $H_0: p_{ij} = p^r_i p^c_j$.

The χ^2 statistic,

$$\chi^2 = \sum \frac{(\text{observed} - \text{expected})^2}{\text{expected}},$$

can still be used as a test statistic after we have estimated each p_{ij} in order to compute the "expected" counts. Again we use the data and the assumptions to estimate the p_{ij}. Basically, the data is used to estimate the marginal probabilities, and the assumption of independence allows us to estimate the p_{ij} from there.

Table 9.6 Seat-belt usage in California with marginal distributions

	Child		
Parent	buckled	unbuckled	marginal
buckled	56	8	64
unbuckled	2	16	18
marginal	58	24	82

The marginal probabilities are estimated by the marginal distributions of the data. For our example these are given in Table 9.6.

The estimate for $p_1^r = \text{P}(\text{parent is buckled})$ is $\widehat{p}_1^r = 64/82$, and for $p_2^r = \text{P}(\text{parent is unbuckled})$ it is $\widehat{p}_2^r = 18/82$. Similarly, for p_j^c we have $\widehat{p}_1^c = 58/82$ and $\widehat{p}_2^c = 24/82$. As usual, we've used a "hat" for estimated values.

With these estimates, we can use the relationship $p_{ij} = p_i^r p_j^c$ to find the estimate $\widehat{p}_{ij} = \widehat{p}_i^r \widehat{p}_j^c$. For our seat-belt data we have the estimates in Table 9.7. In order to show where the values comes from, the values have not been simplified.

Table 9.7 Seat-belt usage in California with estimates $\widehat{p}_{ij}$ for the corresponding p_{ij}

	Child		
Parent	buckled	unbuckled	marginal
buckled	$\frac{64}{82} \cdot \frac{58}{82}$	$\frac{64}{82} \cdot \frac{24}{82}$	$\frac{64}{82}$
unbuckled	$\frac{18}{82} \cdot \frac{58}{82}$	$\frac{18}{82} \cdot \frac{24}{82}$	$\frac{18}{82}$
marginal	$\frac{58}{82}$	$\frac{24}{82}$	1

With this table we can compute the expected amounts in the ijth cell with $n\widehat{p}_{ij}$. This is often written R_iC_j/n, where R_i is the row sum and C_j the column sum, as this simplifies computations by hand.

With the expected amounts now known, we form the χ^2 statistic as:

$$\chi^2 = \sum_{i=1}^{n_r}\sum_{j=1}^{n_c} \frac{(Y_{ij} - n\widehat{p}_{ij})^2}{n\widehat{p}_{ij}}. \tag{9.2}$$

Under the hypothesis of multinomial data and the independence of the variables, the sampling distribution of χ^2 will be the chi-squared distribution with $(n_r - 1)\cdot(n_c - 1)$ degrees of freedom. Why this many? In general, we subtract one degree of freedom from $n_r \cdot n_c - 1$ for each estimated parameter. As there are $n_r - 1 + n_c + 1$ estimated parameters, the value for the degrees of freedom is $n_r \cdot n_c - 1 - (n_r - 1 + n_c + 1) = n_r \cdot n_c - n_r - n_c + 1 = (n_r - 1)\cdot(n_c - 1)$.

We now have all the pieces to formulate the problem in the language of a significance test.

The chi-squared test for independence of two categorical variables

Let $Y_{ij}, i = 1, \ldots, n_r, j = 1, \ldots, n_c$ be the cell frequencies in a two-way contingency table for which the multinomial model applies. A significance test of

$$H_0: \text{the two variables are independent}$$

$$H_A: \text{the two variables are not independent}$$

can be performed using the chi-squared test statistic (9.2). Under the null hypothesis, this statistic has sampling distribution that is approximated by the chi-squared distribution with $(n_r - 1)(n_c - 1)$ degrees of freedom. The p-value is computed using $P(\chi^2 \geq \text{observed value} \mid H_0)$.

In R this test is performed by the `chisq.test()` function. If the data is summarized in a table or a matrix in the variable `x` the usage is

```
chisq.test(x)
```

If the data is unsummarized and is stored in two variables `x` and `y` where the ith entries match up, then the function can be used as

```
chisq.test(x,y).
```

Alternatively, the data could be summarized first using `table()`, as in `chisq.test(table(x,y))`.

For each usage, the null and alternative hypotheses are not specified, as they are the same each time the test is used.

The argument `simulate.p.value=TRUE` will return a p-value estimated using a Monte Carlo simulation. This is used if the expected counts in some cells are too small to use the chi-squared distribution to approximate the sampling distribution of χ^2.

To illustrate, the following will do the chi-squared test on the seat-belt data. This data is summarized, so we first need to make a table. We use `rbind()` to combine rows.

```
> seatbelt = rbind(c(56,8),c(2,16))
> seatbelt
     [,1] [,2]
[1,]   56    8
[2,]    2   16
> chisq.test(seatbelt)

        Pearson's Chi-squared test with Yates' continuity
        correction

data:  seatbelt
X-squared = 36.00, df = 1, p-value = 1.978e-09
```

The small p-value is consistent with our observation that the two variables are not independent.

■ **Example 9.2: Teen smoking and gender** The `samhda (UsingR)` data set contains survey data on 590 children. The variables `gender` and `amt.smoke` contain information about the gender of the participant and how often the participant smoked in the last month. Are the two variables independent? Is smoking dependent on gender?

We compute a p-value for the hypotheses

$$H_0 : \text{the two variables are independent}$$

$$H_A : \text{the two variables are not independent}$$

using the χ^2 statistic.

In this example we use `xtabs()` to make a table, then apply `chisq.test()`. The `xtabs()` function allows us to use the convenient `subset=` argument to eliminate the data for which the values are not applicable.

```
> tbl = xtabs( ~ gender + amt.smoke,      # no left side in formula
+ subset = amt.smoke < 98 & gender !=7,
+ data=samhda)
```

```
> tbl
        amt.smoke
gender  1  2  3  4  5  6  7
     1 16  3  5  6  7 24 64
     2 16  4  8  4  7 19 40
> chisq.test(tbl)

        Pearson's Chi-squared test

data:  tbl
X-squared = 4.147, df = 6, p-value = 0.6568

Warning message:
Chi-squared approximation may be incorrect in: chisq.test(tbl)
```

The significance test shows no reason to doubt the hypothesis that the two variables are independent.

The warning message is due to some expected counts being small. Could this significantly change the p-value reported? A p-value based on a simulation may be computed.

```
> chisq.test(tbl,simulate.p.value=TRUE)

        Pearson's Chi-squared test with simulated p-value (based
        on 2000 replicates)

data:  tbl
X-squared = 4.147, df = NA, p-value = 0.6612
```

The p-value is not changed significantly. ■

9.2.1 The chi-squared test of homogeneity

How can we assess the effectiveness of a drug treatment? Typically, there is a clinical trial, with each participant randomly allocated to either a treatment group or a placebo group. If the results are measured numerically, a t-test may be appropriate to investigate whether any differences in means are significant. When the results are categorical, we see next how to use the χ^2 statistic to test whether the distributions of the results are the same.

Stanford University Medical Center conducted a study to determine whether the antidepressant Celexa can help stop compulsive shopping. Twenty-four compulsive shoppers participated in the study: twelve were given a placebo and twelve a dosage of Celexa daily for seven weeks. After this time the individuals were surveyed to determine whether their desires to shop had been curtailed. Data simulated from a preliminary report is given in Table 9.8.

Does this indicate that the two samples have different distributions?

Table 9.8 Does Celexa treatment cut down on compulsive shopping?

	much worse	worse	same	much improved	very much improved
Celexa	0	2	3	5	2
placebo	0	2	8	2	0

We formulate this as a significance test using hypotheses:

$$H_0 : \text{the two distributions are the same}$$

$$H_A : \text{the two distributions are different.}$$

We use the χ^2 statistic. Again we need to determine the expected amounts, as they are not fully specified by H_0.

Let the random variable be the column variable, and the category that breaks up the data be the row variable in our table of data. For row i of the table, let p_{ij} be the probability that the random variable (the survey result) will be in the jth level of the random variable. We can rephrase the hypotheses as

$$H_0 : p_{ij} = p_j \text{ for all rows } i, \quad H_A : p_{ij} \neq p_j \text{ for some } i,\ j.$$

If we let n_i be the number of counts in each row (R_i before), then the expected amount in the (i, j) cell under H_0 should be $n_i p_j$. We don't specify the value of p_j in the null hypothesis, so it is estimated. Under H_0 all the data in the jth column of our table is binomial with n and p_j, so an estimator for p_j would be the column sum divided by n: C_j/n. Based on this, the expected number in the (i, j)-cell would be

$$e_{ij} = n_i \widehat{p}_j = \frac{R_i C_j}{n}.$$

This is the same formula as the chi-squared test of independence.

As the test statistic and its sampling distribution under H_0 are the same as with the test of independence, the chi-squared significance tests of homogeneity and independence are identical in implementation despite the differences in the hypotheses.

Before proceeding, let's combine the data so that there are three outcomes: "worse," "same," and "better."

```
> celexa = c(2,3,7); placebo = c(2,8,2)
> x = rbind(celexa,placebo)
> colnames(x) = c("worse","same","better")
> x
```

```
        worse same better
celexa      2    3      7
placebo     2    8      2
> chisq.test(x)

        Pearson's Chi-squared test

data:  x
X-squared = 5.05, df = 2, p-value = 0.08004

Warning message:
Chi-squared approximation may be incorrect in: chisq.test(x)
```

The warning notes that one or more of the expected cell counts is less than five, indicating a possible discrepancy with the asymptotic distribution used to find the p-value. We can use a simulation to find the p-value, instead of using the chi-squared distribution approximation, as follows:

```
> chisq.test(x, simulate.p.value=TRUE)

        Pearson's Chi-squared test with simulated p-value (based
        on 2000 replicates)

data:  x
X-squared = 5.05, df = NA, p-value = 0.1025
```

In both cases, the p-value is small but not tiny.

9.2.2 Problems

9.11 A number of drivers were surveyed to see whether they had been in an accident during the previous year, and, if so, whether it was a minor or major accident. The results are tabulated by age group in Table 9.9. Do a chi-squared hypothesis test of independence for the two variables.

Table 9.9 Type of accident by age

	Accident type		
Age	none	minor	major
under 18	67	10	5
18-25	42	6	5
26-40	75	8	4
40-65	56	4	6
over 65	57	15	1

9.12 Table 9.10 contains data on the severity of injuries sustained during car crashes. The data is tabulated by whether or not the passenger wore a seat belt. Are the two variables independent?

Table 9.10 Accidents by injury level and seat-belt usage

		Injury level			
		none	minimal	minor	major
Seat belt	yes	12,813	647	359	42
	no	65,963	4,000	2,642	303

9.13 The `airquality` data set contains measurements of air quality in New York City. We wish to see if ozone levels are independent of temperature. First we gather the data, using `complete.cases()` to remove missing data from our data set.

```
> aq = airquality[complete.cases(airquality),]
> attach(aq)
> te = cut(Temp, quantile(Temp))
> oz = cut(Ozone,quantile(Ozone))
```

Perform a chi-squared test of independence on the two variables `te` and `oz`. Does the data support an assumption of independence?

9.14 In an effort to increase student retention, many colleges have tried block programs. Assume that 100 students are broken into two groups of 50 at random. Fifty are in a block program; the others are not. The number of years each student attends the college is then measured. We wish to test whether the block program makes a difference in retention. The data is recorded in Table 9.11. Perform a chi-squared test of significance to investigate whether the distributions are homogeneous.

Table 9.11 Retention data by year and program

Program	1 year	2 year	3 year	4year	5+ years
nonblock	18	15	5	8	4
block	10	5	7	18	10

9.15 The data set `oral.lesion (UsingR)` contains data on location of an oral lesion for three geographic locations. This data set appears in an article by Mehta and Patel about differences in p-values in tests for independence when the exact or asymptotic distributions are used. Compare the p-values found by `chisq.test()` when the asymptotic distribution of the sampling distribution is used to find the p-value and when a simulated value is used. Are the p-values similar? If not, which do you think is more accurate? Why?

9.3 Goodness-of-fit tests for continuous distributions

When finding confidence intervals for a sample we were concerned about whether or not the data was sampled from a normal distribution. To investigate, we made a quantile plot or histogram and eyeballed the result. In this section, we see how to compare a continuous distribution with a theoretical one using a significance test.

The chi-squared test is used for categorical data. We can try to make it work for continuous data by "binning." That is, as in a construction of a histogram, we can choose some bins and count the number of data points in each. Now the data can be thought of as categorical and the test can be used for goodness of fit.

This is fine in theory but works poorly in practice. The Kolmogorov-Smirnov test will be a better alternative in the continuous distribution case.

9.3.1 Kolmogorov-Smirnov test

Suppose we have a random sample $X_1, X_2, \ldots X_n$ from some continuous distribution. (There should be no ties in the data.) Let $f(x)$ be the density and X some other random variable with this density. The cumulative distribution function for X is $F(x) = \mathsf{P}(X \leq x)$, or the area to the left of x under the density of X.

The c.d.f. can be defined the same way when X is discrete. In that case it is computed from the p.d.f. by summing: $\mathsf{P}(X \leq x) = \sum_{y \leq x} f(y)$.

For a sample, $X_1, X_2, \ldots, X_n$, the **empirical distribution** is the distribution generated by sampling from the data points. The probability that a number randomly picked from a sample is less than or equal to x is the number of data points in the sample less than or equal to x divided by n. We use the notation $F_n(x)$ for this:

$$F_n(x) = \frac{\#\{i : X_i \leq x\}}{n}.$$

$F_n(x)$ is referred to as the empirical cumulative distribution function, or e.c.d.f.

The function $F_n(x)$ can easily be plotted in R using the `ecdf()` function in the `stats` package.* This function is used in a manner similar to the `density()` function: the return value is plotted in a new figure using `plot()` or may be

* The `ecdf()` function from the useful `Hmisc` package can also be used to create these graphs. The `Hmisc` package needs to be installed separately. Use the menu bar or `install.packages("Hmisc")`.

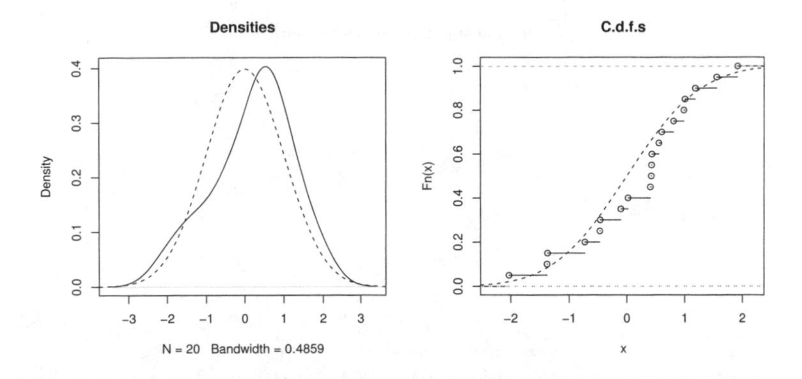

Figure 9.2 For a sample of size 20 from a normally distributed population, both sample and theoretical densities and cumulative distribution functions are drawn

added to the existing plot using `lines()`. The following commands produced Figure 9.2:

```
> y = rnorm(20)
> plot(density(y), main="Densities") # densities
> curve(dnorm(x), add=TRUE, lty=2)
> plot(ecdf(y), main="C.d.f.s")       # c.d.f.s
> curve(pnorm(x), add=TRUE, lty=2)
```

If the data is from the population with c.d.f. F, then we would expect that F_n is close to F is some way. But what does "close" mean? In this context, we have two different functions of x. Define the distance between them as the largest difference they have:

$$D = \text{maximum in } x \text{ of } |F_n(x) - F(x)|.$$

The surprising thing is that with only the assumption that F is continuous, D has a known sampling distribution called the Kolmogorov-Smirnov distribution. This is illustrated in Figure 9.3, where the sampling distribution of the statistic for $n = 25$ is simulated for several families of random data. In each case, we see the same distribution. This fact allows us to construct a significance test using the test statistic D. In addition, a similar test can be done to compare two independent samples.

The Kolmogorov-Smirnov goodness-of-fit test

Assume $X_1, X_2, \ldots, X_n$ is an *i.i.d.* sample from a continuous distribution with c.d.f. $F(x)$. Let $F_n(x)$ be the empirical c.d.f. A significance test of

$$H_0 : F(x) = F_0(x), \qquad H_A : F(x) \neq F_0(x)$$

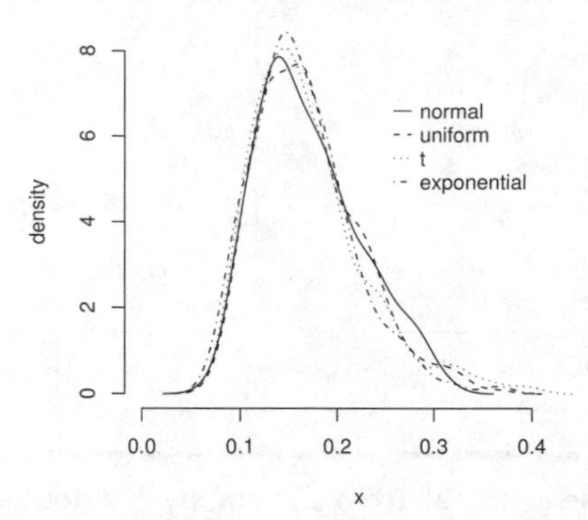

Figure 9.3 Density estimates for sampling distribution of the Kolmogorov-Smirnov statistic with $n = 25$ for normal, uniform, t, and exponential data

can be constructed with test statistic D. Large values of D support the alternative hypothesis.

In R, this test is implemented in the function `ks.test()`. Its usage follows this pattern:

```
ks.test(x, y="name", ...)
```

The variable `x` stores the data. The argument y= is used to set the family name of the distribution in H_0. It has a character value of `"name"` containing the "p" function that returns the c.d.f. for the family (e.g., `"pnorm"` or `"pt"`). The `...` argument allows the specification of the assumed parameter values. These depend on the family name and are specified as named arguments, as in `mean=1, sd=1`. The parameter values should not be estimated from the data, as this affects the sampling distribution of D.

If we have two *i.i.d.* independent samples $X_1, \ldots, X_n$ and $Y_1, \ldots, Y_m$, from two continuous distributions F^X and F^Y, then a significance test of

$$H_0 : F^X = F^Y, \qquad H_A : F^X \neq F^Y$$

can be constructed with a similar test statistic:

$$D = \text{maximum in } x \text{ of } |F_n^X(x) - F_m^Y(x)|.$$

In this case, the `ks.test()` can be used as

```
ks.test(x,y)
```

where x and y store the data.

We illustrate with some simulated data.

```
> x = rnorm(100,mean=5, sd=2)
> ks.test(x,"pnorm",mean=0,sd=2) # "wrong" parameters

        One-sample Kolmogorov-Smirnov test
data:  x
D = 0.7578, p-value = < 2.2e-16
alternative hypothesis: two.sided

> ks.test(x,"pnorm",mean=5,sd=2) # correct population parameters
...
D = 0.1102, p-value = 0.1759
...
> x = runif(100, min=0, max=5)
> ks.test(x,"punif",min=0,max=6) # "wrong" parameters
...
D = 0.1669, p-value = 0.007588
...
> ks.test(x,"punif",min=0,max=5) # correct population parameters
...
D = 0.0745, p-value = 0.6363
...
```

The p-values are significant only when the parameters do not match the known population ones.

■ **Example 9.3: Difference in SAT scores** The data set stud.recs (UsingR) contains math and verbal SAT scores for some students (sat.m and sat.v). Assume naively that the two samples are independent, are the samples from the same population of scores?

First, we make a q-q plot, a side-by-side boxplot, and a plot of the e.c.d.f.'s for the data, to see whether there is any merit to the question.

```
> data(stud.recs,package="UsingR") # or library(UsingR)
> attach(stud.recs)
> boxplot(list(math=sat.m,verbal=sat.v), main="SAT scores")
> qqplot(sat.m,sat.v, main="Math and verbal SAT scores")
> plot(ecdf(sat.m), main="Math and verbal SAT scores")
> lines(ecdf(sat.v), lty=2)
```

The graphics are in Figure 9.4. The q-q plot shows similarly shaped distributions, but boxplots show that the centers appear to be different. Consequently, the cumulative distribution functions do not look that similar. The Kolmogorov-Smirnov test detects this and returns a small p-value.

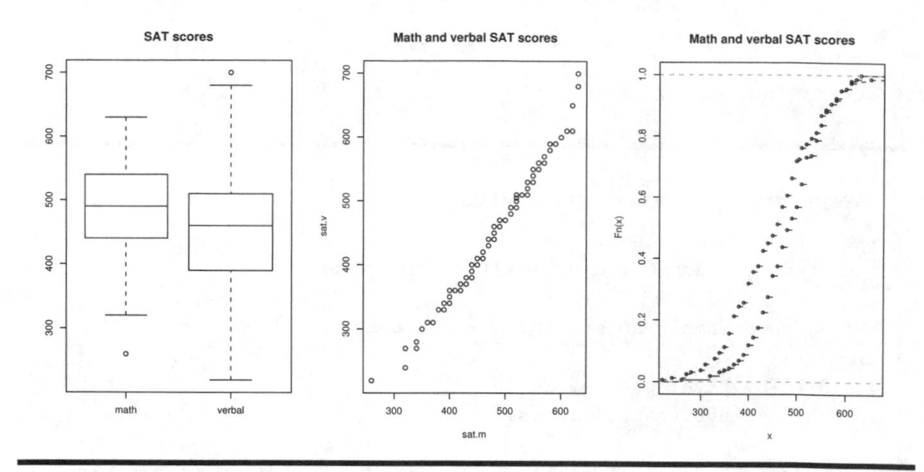

Figure 9.4 Three plots comparing the distribution of math and verbal SAT scores in the `stud.recs (UsingR)` data set.

```
> ks.test(sat.m,sat.v)

        Two-sample Kolmogorov-Smirnov test
data:  sat.m and sat.v
D = 0.2125, p-value = 0.001456
alternative hypothesis: two.sided
...
```

■

9.3.2 The Shapiro-Wilk test for normality

The Kolmogorov-Smirnov test for a univariate data set works when the distribution in the null hypothesis is fully specified prior to our looking at the data. In particular, any assumptions on the values for the parameters should not depend on the data, as this can change the sampling distribution. Figure 9.5 shows the sampling distribution of the Kolmogorov-Smirnov statistic for $\text{Normal}(0,1)$ data and the sampling distribution of the Kolmogorov-Smirnov statistic for the same data when the sample values of $\bar{x}$ and s are used for the parameters of the normal distribution (instead of 0 and 1). The figure was generated with this simulation:

```
> res.1 = c(); res.2 = c()
> for(i in 1:500) {
+ x = rnorm(25)
+ res.1[i] = ks.test(x,pnorm)$statistic
+ res.2[i] = ks.test(x,pnorm,mean(x),sd(x))$statistic
+ }
> plot(density(res.1),main="K-S sampling distribution")
> lines(density(res.2),lty=2)
```

(To retrieve just the value of the test statistic from the output of `ks.test()` we take advantage of the fact that its return value is a list with one component named `statistic` containing the desired value. This is why the syntax `ks.test(...)$statistic` is used.)

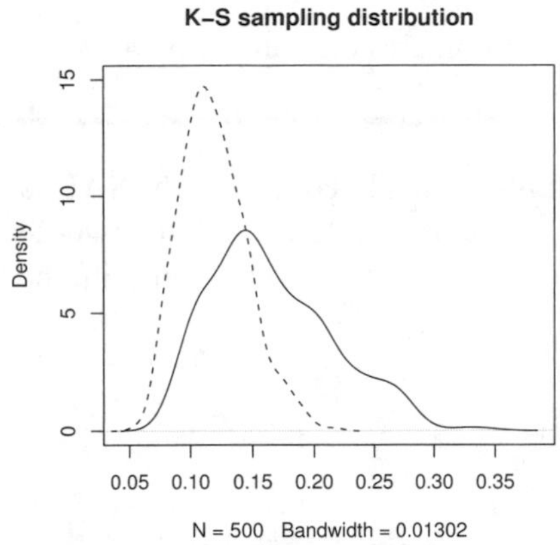

Figure 9.5 The sampling distribution for the Kolmogorov-Smirnov statistic when the parameters are estimated (dashed line) and when not

A consequence is that we can't use the Kolmogorov-Smirnov test to test for normality of a data set unless we know the parameters of the underlying distribution.† The Shapiro-Wilk test allows us to do this. This test statistic is based on the ideas behind the quantile-quantile plot, which we've used to gauge normality. Its definition is a bit involved, but its usage in R is straightforward.

The Shapiro-Wilk test for normality

If $X_1, X_2, \ldots, X_n$ is an *i.i.d.* sample from a continuous distribution, a significance test of

$$H_0 : \text{parent distribution is normal,}$$

$$H_A : \text{the parent distribution is not normal}$$

† The Lilliefors test, implemented by `lillie.test()` in the contributed package `nortest`, will make the necessary adjustments to use this test statistic. As well, the `nortest` package implements other tests of normality. In many installations of R, `nortest` may be installed from a menubar or with the command `install.packages("nortest")`. See Chapter 1 for further information about package installation.

can be carried out with the Shapiro-Wilk test statistic.

In R, the function `shapiro.test()` will perform the test. The usage is simply

```
shapiro.test(x)
```

where the data vector x contains the sample data.

■ **Example 9.4: Normality of SAT scores** For the SAT data in the `stud.recs` (`UsingR`) data set, we saw in Example 9.3 that the two distributions are different. Are they normally distributed? We can answer with the Shapiro-Wilk test:

```
> attach(stud.recs)
> shapiro.test(sat.m)

        Shapiro-Wilk normality test

data:  sat.m
W = 0.9898, p-value = 0.3056

> shapiro.test(sat.v)
...
W = 0.994, p-value = 0.752
> detach(stud.recs)
```

In each case, the p-value is not statistically significant. There is no evidence that the data is not normally distributed. ■

■ **Example 9.5: Is on-base percentage normally distributed?** In Example 2.8 the distribution of the on-base percentage from the 2002 major league baseball season was shown. It appears bell shaped except for one outlier. Does the data come from a normally distributed population?

Using the Shapiro-Wilk test gives us

```
> shapiro.test(OBP)

        Shapiro-Wilk normality test

data:  OBP
W = 0.9709, p-value = 1.206e-07
```

So it is unlikely. Perhaps this is due to the one outlier. We eliminate this and try again

```
> shapiro.test(OBP[OBP<.5])

        Shapiro-Wilk normality test

data:  OBP[OBP < 0.5]
W = 0.9905, p-value = 0.006404
```

The conclusion is the same: the data is not normally distributed. Also, note the dramatic difference in the p-value that just one outlier makes. The statistic is not very resistant. ■

In defining the t-test, it was assumed that the data is sampled from a normal population. This is because the sampling distribution of the t-statistic is known under this assumption. However, this would not preclude us from using the t-test to perform statistical inference on data that has failed a formal test for normality. For small samples the t-test may apply, as the distribution of the t-statistic is robust to small changes in the assumptions on the parent distribution. If the parent distribution is not normal but also not too skewed, then a t test can be appropriate. For large samples, the central limit theorem may apply, making a t-test valid.

9.3.3 Finding parameter values using `fitdistr()`

If we know a data set comes from a known distribution and would like to estimate the parameter values, we can use the convenient `fitdistr()` function from the `MASS` library. This function estimates the parameters for a wide family of distributions. The function is called with these arguments:

```
fitdistr(x, densfun=family.name, start=list(...))
```

We specify the data as a data vector, `x`; the family is specified by its full name, unlike that used in `ks.test`); and, for many of the distributions, reasonable starting values are specified using a named list. The `fitdistr()` function fits the parameters by a method called maximum-likelihood. Often this coincides with using the sample mean or standard deviation to estimate the parameters, but in general it allows for a uniform approach to this estimation problem and associated inferential problems.

■ **Example 9.6: Exploring** `fitdistr()` The data set `babyboom (UsingR)` contains data on the births of 44 children in a one-day period at a hospital in Brisbane, Australia. The variable `wt` records the weights of each newborn. A histogram suggests that the data comes from a normally distributed population. We can use `fitdistr()` to find estimates for the parameters μ and σ, which for the normal distribution are the population mean and standard deviation.

```
> data(babyboom, package="UsingR") # or library(UsingR)
> fitdistr(babyboom$wt,"normal")
   mean        sd
  3275.95    522.00
 (  78.69) (  55.65)
```

These estimates include standard errors in parentheses, using a normal approximation. These can be used to give confidence intervals for the estimates.

This estimate for the mean and standard deviation could also be done directly, as it coincides with the sample mean and sample standard deviation. However,

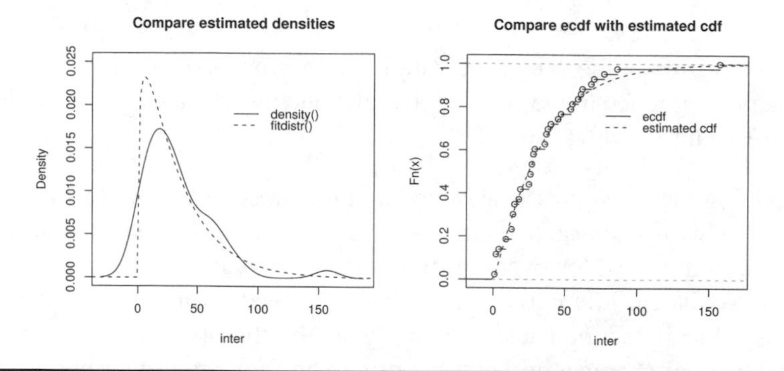

Figure 9.6 Both figures illustrate the inter-arrival times of the `babyboom` data set. Figure on left shows empirical density and the fit of the gamma distribution given by `fitdistr()`. Figure on right shows same relationship using cumulative distribution functions.

the standard errors are new. To give a different usage, we look at the variable `running.time`, which records the time of day of each birth. The time differences between successive births are called the inter-arrival times. To make a densityplot (Figure 9.6), we first find the inter-arrival times using `diff()`:

```
> inter = diff(babyboom$running.time)
> plot(density(inter), ylim=c(0,0.025), # adjust ylim for next plot
+ main="Compare estimated densities", xlab="inter")
```

We fit the gamma distribution to the data. The gamma distribution generalizes the exponential distribution. It has two parameters, a shape and a rate. A value of 1 for the shape coincides with the exponential distribution. The `fitdistr()` function does not need starting values for the gamma distribution.

```
> fitdistr(inter,"gamma")
     shape          rate
  1.208593    0.036350
 (0.233040) (0.008625)
Warning messages:
1: NaNs produced in: dgamma(x, shape, scale, log)
2: NaNs produced in: dgamma(x, shape, scale, log)
> curve(dgamma(x,shape=1.208593, rate=0.036350), add=T, lty=2)
> legend(100,.020,legend=c("density()","fitdistr()"),lty=1:2)
```

The warning message informs us that the fitting encounted some difficulties.

Finally, we compare the cumulative distribution functions with the following commands (the graphic on the right in Figure 9.6):

```
> plot(ecdf(inter),
+ main="Compare ecdf with estimated cdf", xlab="inter")
> curve(pgamma(x,shape=1.208593, rate=0.036350), add=T)
> legend(70,.8,legend=c("ecdf","estimated cdf"),lty=1:2)
```

■

9.3.4 Problems

9.16 In two examples in Chapter 7, data on CEOs is compared. The data is repeated in Table 9.12. Are the parent distributions the same? Answer this using a test of significance.

Table 9.12 CEO pay data for 2000 and 2002

Year	Compensation in $10,000s										
2001	110	12	2.5	98	1017	540	54	4.3	150	432	
2002	312	316	175	200	92	201	428	51	289	1126	822

9.17 Carry out a Shapiro-Wilk test for the mother's height, `ht`, and weight, `wt`, in the `babies (UsingR)` data set. Remember to exclude the cases when `ht==99` and `wt==999`. Are the data sets normally distributed?

9.18 The `brightness (UsingR)` data set contains brightness measurements for 966 stars from the Hipparcos catalog. Is the data normal? Compare the result with a significance test to the graphical investigation done by

```
> hist(brightness, prob=TRUE)
> lines(density(brightness))
> curve(dnorm(x, mean(brightness), sd(brightness)), add=TRUE)
```

9.19 The variable `temperature` in the data set `normtemp (UsingR)` contains normal body temperature measurements for 130 healthy, randomly selected individuals. Is normal body temperature normally distributed?

9.20 The `rivers` data set contains the length of 141 major rivers in North America. Fit this distribution using the gamma distribution and `fitdistr()`. How well does the gamma distribution fit the data?

9.21 Find parameter estimates for μ and σ for the variables `sat.m` and `sat.v` in the `stud.recs (UsingR)` data set. Assume the respective populations are normally distributed.

9.22 How good is the Kolmogorov-Smirnov test at rejecting the null when it is false? The following command will do 100 simulations of the test when the data is not normal, but long-tailed and symmetric.

```
> res = sapply(1:100,
+ function(x) ks.test(rt(25,df=3),"pnorm")$p.value)
```

(The syntax above is using the fact that `ks.test()` returns a list of values with one component named `p.value`.) What percentage of the trials have a p-value less than 0.05?

Try this with the exponential distribution (that is, replace `rt(25,df=3)` with `rexp(25)-1`). Is it better when the data is skewed?

9.23 A key to understanding why the Kolmogorov-Smirnov statistic has a sampling distribution that does not depend on the underlying parent population (as long as it is continuous) is the fact that if $F(x)$ is the c.d.f. for a random variable X, then $F(X)$ is uniformly distributed on $[0,1]$.

This can be proved algebraically using inverse functions, but instead we see how to simulate the problem to gain insight. The following line will illustrate this for the normal distribution:

```
> qqplot(pnorm(rnorm(100)),runif(100))
```

The `qqplot()` should be nearly straight if the distribution is uniform. Change the distribution to some others and see that you get a nearly straight line in each case. For example, the t-distribution with 5 degrees of freedom would be done with

```
> qqplot(pt(rt(100,df=5),df=5),runif(100))
```

Try the uniform distribution, the exponential distribution, and the lognormal distribution (`lnorm`).

9.24 Is the Shapiro-Wilk test resistant to outliers? Run the following commands and decide whether the presence of a single large outlier changes the ability of the test to detect normality.

```
> shapiro.test(c(rnorm(100),5))
> shapiro.test(c(rnorm(1000),5))
> shapiro.test(c(rnorm(4000),5))
```

Chapter 10

Linear regression

In Chapter 3 we looked at the simple linear regression model,

$$y_i = \beta_0 + \beta_1 x_i + \varepsilon_i,$$

as a way to summarize a linear relationship between pairs of data (x_i, y_i). In this chapter we return to this model. We begin with a review and then further the discussion using the tools of statistical inference. Additionally, we will see that the methods developed for this model extend readily to the multiple linear regression model where there is more than one predictor.

10.1 The simple linear regression model

Many times we assume that an increase in the predictor variable will correspond to an increase (or decrease) in the response variable. A basic model for this is a simple linear regression model:

$$Y_i = \beta_0 + \beta_1 x_i + \varepsilon_i.$$

The Y variable is called the response variable and the x variable the predictor variable, covariate, or regressor.

As a statistical model, this says that the value of Y_i depends on three things: that of x_i, the function $\beta_0 + \beta_1 x$, and the value of the random variable ε_i. The model says that for a given value of x, the corresponding value of Y is found by first using the function on x and then adding the random error term ε_i.

To be able to make statistical inference, we assume that the error terms, ε_i, are *i.i.d.* and have a $\mathsf{Normal}(0, \sigma)$ distribution. This assumption can be rephrased as an assumption on the randomness of the response variable. If the x values are fixed, then the distribution of Y_i is normal with mean $\mu_{y|x} = \beta_0 + \beta_1 x_i$ and variance σ^2. This can be expressed as Y_i has a $\mathsf{Normal}(\beta_0 + \beta_1 x_i, \sigma)$ distribution. If the x values are random, the model assumes that, conditionally on knowing these random values, the same is true about the distribution of the Y_i.

10.1.1 Model formulas for linear models

Before using `R` to find estimates, we need to learn how `R` represents statistical models. Linear models are fit using `R`'s model formulas, of which we have already seen a few examples.

The basic format for a formula is

```
response ~ predictor
```

The `~` (tilde) is read "is modeled by" and is used to separate the response from the predictor(s). The response variable can have regular mathematical expressions applied to it, but for the predictor variables the regular notations `+`, `-`, `*`, `/`, and `^` have different meanings. A `+` means to add another term to the model, `-` means to drop a term, more or less coinciding with the symbols' common usage. But `*`, `/`, and `^` are used differently. If we want to use regular mathematical notation for the `predictor` we must insulate the symbols' usage with the `I()` function, as in `I(x^2)`.

10.1.2 Examples of the linear model

At first, the simple linear regression model appears to be solely about a straight-line relationship between pairs of data. We'll see that this isn't so, by looking at how the model accommodates many of the ideas previously mentioned.

Simple linear regression If (x_i, y_i) are related by the linear model

$$Y_i = \beta_0 + \beta_1 x_i + \varepsilon_i$$

as above, then the model is represented in `R` by the formula `y ~ x`. The intercept term, β_0, is implicitly defined.

If for some reason the intercept term is not desired, it can be dropped from the model by including the term `-1`, as in `y ~ x - 1`.

The mean of an *i.i.d.* sample In finding confidence intervals or performing a significance test for the mean of an *i.i.d.* sample, $Y_1, Y_2, \ldots, Y_n$, we often assumed normality of the population. In terms of a statistical model this could be viewed as

$$Y_i = \mu + \varepsilon_i,$$

where the ε_i are $\mathsf{Normal}(0, \sigma)$.

The model for this in R is `y ~ 1`. As there is no predictor variable, the intercept term is explicitly present.

The paired t-test In Chapter 8, we considered the paired t-test. This test applies when two samples are somehow related and the differences between the two samples is random. That is, $Y_i - X_i$, is the quantity of interest. This corresponds to the statistical model

$$Y_i = X_i + \varepsilon_i.$$

If we assume ε_i has mean 0, then we can model the mean difference between Y and X by μ, and our model becomes

$$Y_i = \mu + X_i + \varepsilon_i.$$

Our significance test with $H_0 : \mu_1 = \mu_2$ turns into a test of $\mu = 0$.

The model formula to fit this in R uses an offset, which we won't discuss again, but for reference it would look like `y ~ offset(x)`.

In Chapter 11 we will see that this model can be used for a two-sample t-test. Later in this chapter we will extend the model to describe relationships that are not straight lines and relationships involving multiple predictors.

10.1.3 Estimating the parameters in simple linear regression

One goal when modeling is to "fit" the model by estimating the parameters based on the sample. For the regression model the method of least squares is used. With an eye toward a more general usage, suppose we have several predictors, $x_1, x_2, \ldots, x_k$; several parameters, $\beta_0, \beta_1, \ldots, \beta_p$; and some function, f, which gives the mean for the variables Y_i. That is, the statistical model

$$Y_i = f(x_{1i}, x_{2i}, \ldots, x_{ki} \mid \beta_1, \beta_2, \ldots, \beta_p) + \varepsilon_i.$$

The method of least squares finds values for the β's that minimize the squared difference between the actual values, y_i, and those predicted by the function f. That is, the following sum is minimized:

$$\sum_i [y_i - f(x_{1i}, x_{2i}, \ldots, x_{ki} \mid \beta_0, \beta_1, \ldots, \beta_p)]^2.$$

For the simple linear regression model, the formulas are not difficult to write (they are given below). For the more general model, even if explicit formulas are known, we don't present them.

The simple linear regression model for Y_i has three parameters, β_0, β_1, and σ^2. The least-squares estimators for these are

$$\widehat{\beta}_1 = \frac{\sum(x_i - \bar{x})(y_i - \bar{y})}{\sum(x_i - \bar{x})^2}, \tag{10.1}$$

$$\widehat{\beta}_0 = \bar{y} - \widehat{\beta}_1 \bar{x}, \qquad \text{and} \tag{10.2}$$

$$\widehat{\sigma}^2 = \frac{1}{n-2} \sum [y_i - (\widehat{\beta}_0 + \widehat{\beta}_1 x_i)]^2. \tag{10.3}$$

We call $\widehat{y} = \widehat{\beta}_0 + \widehat{\beta}_1 x$ the prediction line; a value $\widehat{y}_i = \widehat{\beta}_0 + \widehat{\beta}_1 x_i$ the predicted value for x_i; and the difference between the actual and predicted values, $e_i = y_i - \widehat{y}_i$, the **residual**. The **residual sum of squares** is denoted RSS and is equal to $\sum_i e_i^2$. See Figure 3.10 for a picture.

Quickly put, the regression line is chosen to minimize the RSS; it has slope $\widehat{\beta}_1$, intercept $\widehat{\beta}_0$, and goes through the point $(\bar{x}, \bar{y})$. Furthermore, the estimate for σ^2 is $\widehat{\sigma}^2 = \text{RSS}/(n-2)$.

Figure 10.1 shows a data set simulated from the equation $Y_i = 1 + 2x_i + \varepsilon_i$, where $\beta_0 = 1, \beta_1 = 2$, and $\sigma^2 = 3$. Both the line $y = 1 + 2x$ and the regression line $\widehat{y} = 0.329 + 2.158 \cdot x$, predicted by the data, are drawn. They are different, of course, as one of them depends on the random sample. Keep in mind that the data is related by the true model, but if all we have is the data, the estimated model is given by the regression line. Our task of inference is to decide how much the regression line can tell us about the underlying true model.

10.1.4 Using `lm()` to find the estimates

In Chapter 3 we learned how to fit the simple linear regression model using `lm()`. The basic usage is of the form

```
lm(formula, data=..., subset=...)
```

As is usual with functions using model formulas, the `data=` argument allows the variable names to reference those in the specified data frame, and the `subset=` argument can be used to restrict the indices of the variables used by the modeling function.

By default, the `lm()` function will print out the estimates for the coefficients. Much more is returned, but needs to be explicitly asked for. Usually, we store the results of the model in a variable, so that it can subsequently be queried for more

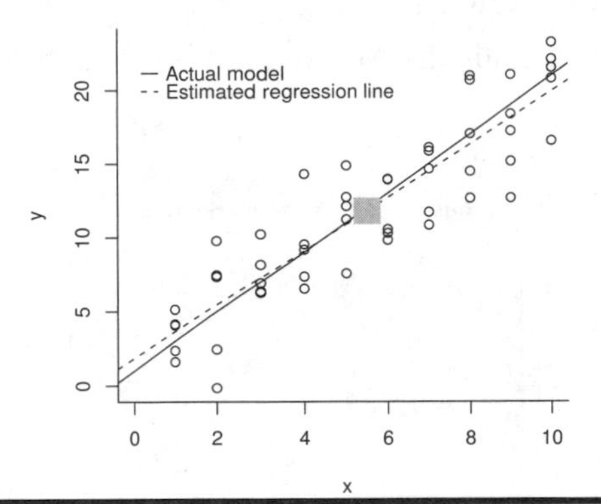

Figure 10.1 Simulation of model $Y_i = 1 + 2x_i + \varepsilon_i$. The regression line based on the data is drawn with dashes. The big square marks the value $(\bar{x}, \bar{y})$.

information.

■ **Example 10.1: Maximum heart rate** Many people use heart-rate monitors when exercising in order to achieve target heart rates for optimal training. The maximum safe heart rate is generally thought to be 220 minus one's age in years. A 25-year-old would have a maximum heart rate of 195 beats per minute.

This formula is said to have been devised in the 1970s by Sam Fox and William Haskell while en route to a conference. Fox and Haskell had plotted a graph of some data, and had drawn by hand a line through the data, guessing the slope and intercept.* Their formula is easy to compute and comprehend and has found widespread acceptance.

It may be wrong, though. In 2001, Tanaka, Monahan, and Seals found that 209 - 0.7 times one's age is a better fit.

The following data is simulated to illustrate:

```
> age = rep(seq(20,60,by=5), 3)
> mhr = 209 - 0.7*age + rnorm(length(age),sd=4)
> plot(mhr ~ age, main="Age versus maximum heart rate")
```

The scatterplot (Figure 10.2) shows that the data lends itself to the linear model. The regression coefficients are found using the `lm()` function.

```
> res.mhr = lm(mhr ~ age)
> res.mhr
Call:
lm(formula = mhr ~ age)

Coefficients:
```

* source `http://www.drmirkin.com`.

```
(Intercept)          age
     208.36        -0.76
```

The `lm()` function, by default, displays the formula and the estimates for the $\widehat{\beta_i}$.

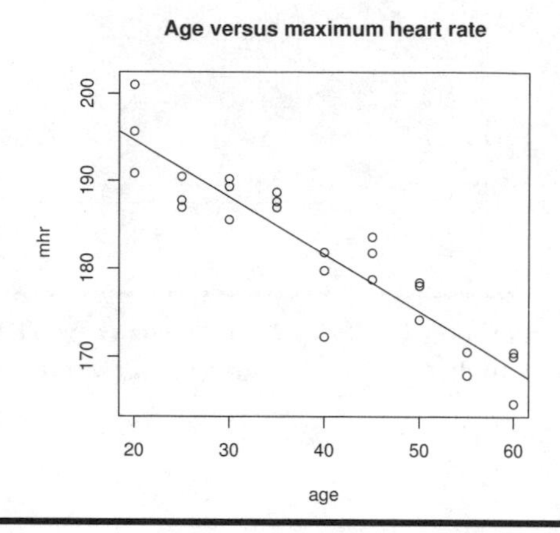

Figure 10.2 Age versus maximum heart rate

These estimates can be used with the `abline()` function to add the regression line, as in Figure 10.2.

```
> abline(res.mhr)                    # add regression line
```

A predicted value can be made directly using the estimates. For example, the predicted maximum heart rate for a 39-year-old would be

```
> 208.36 - 0.76 * 39
[1] 178.7
```

■

Extractor functions for `lm()`

The `lm()` function is reticent, but we can coax out more information as needed. This is done using extractor functions. Useful ones are summarized in Table 10.1.

These functions are called with the result of a modeling function, such as `lm()`. There are other types of modeling functions in R; these so-called "generic functions" may be used with them to return similar information.

To illustrate, the estimate for σ^2 can be found using the `resid()` function to retrieve the residuals from the model fitting:

```
> sum( resid(res.mhr)^2 ) / (length(age) - 2) # RSS/(n-2)
[1] 14.15
```

Or, the RSS part can be found directly with `deviance()`:

Table 10.1 Extractor functions for the result of `lm()`

`summary()`	returns summary information about the regression
`plot()`	makes diagnostic plots
`coef()`	returns the coefficients
`residuals()`	returns the residuals (can be abbreviated `resid()`)
`fitted()`	returns fitted values, $\widehat{y}_i$
`deviance()`	returns RSS
`predict()`	performs predictions
`anova()`	finds various sums of squares
`AIC()`	is used for model selection

```
> deviance(res.mhr)/ (length(age) - 2)
[1] 14.15
```

10.1.5 Problems

10.1 For the `Cars93` (`MASS`) data set, answer the following:

1. For `MPG.highway` modeled by `Horsepower`, find the simple regression coefficients. What is the predicted mileage for a car with 225 horsepower?
2. Fit the linear model with `MPG.highway` modeled by `Weight`. Find the predicted highway mileage of a 6,400 pound HUMMER H2 and a 2,524 pound MINI Cooper.
3. Fit the linear model `Max.Price` modeled by `Min.Price`. Why might you expect the slope to be around 1?

Can you think of any other linear relationships among the variables?

10.2 For the data set `MLBattend (UsingR)` concerning major league baseball attendance, fit a linear model of `attendance` modeled by `wins`. What is the predicted increase in attendance if a team that won 80 games last year wins 90 this year?

10.3 People often predict children's future height by using their 2-year-old height. A common rule is to double the height. Table 10.2 contains data for eight people's heights as 2-year-olds and as adults. Using the data, what is the predicted adult height for a 2-year-old who is 33 inches tall?

Table 10.2 Height as two-year old and as an adult

Age 2 (in.)	39	30	32	34	35	36	36	30
Adult (in.)	71	63	63	67	68	68	70	64

10.4 The `galton (UsingR)` data set contains data collected by Francis Galton in 1885 concerning the influence a parent's height has on a child's height. Fit a linear model for a child's height modeled by his parent's height. Make a scatterplot with a regression line. (Is this dataset a good candidate for using `jitter()`?) What is the value of $\widehat{\beta}_1$, and why is this of interest?

10.5 Formulas (10.1), (10.2), and the prediction line equation can be rewritten in terms of the correlation coefficient, r, as

$$\frac{\widehat{y}_i - \bar{y}}{s_y} = r\frac{x_i - \bar{x}}{s_x}.$$

Thus the five summary numbers: the two means, the standard deviations, and the correlation coefficient are fundamental for regression analysis.

This is interpreted as follows. Scaled differences of $\widehat{y}_i$ from the mean $\bar{y}$ are less than the scaled differences of x_i from $\bar{x}$, as $|r| \leq 1$. That is, "regression" toward the mean, as unusually large differences from the mean are lessened in their prediction for y.

For the data set `galton (UsingR)` use `scale()` on the variables `parent` and `child`, and then model the height of the child by the height of the parent. What are the estimates for r and β_1?

10.2 Statistical inference for simple linear regression

If we are convinced that the simple regression model is appropriate for our data, then statistical inferences can be made about the unknown parameters. To assess whether the simple regression model is appropriate for the data we use a graphical approach.

10.2.1 Testing the model assumptions

The simple linear regression model places assumptions on the data set that we should verify before proceeding with any statistical inference. In particular, the linear model should be appropriate for the mean value of the y_i, and the error distribution should be normally distributed and independent.

Just as we looked at graphical evidence when investigating assumptions about normally distributed populations when performing a t-test, we will consider graphical evidence to assess the appropriateness of a regression model for the data. Four of the graphs we consider are produced by using the `plot()` function as an extractor function for `lm()` function. Others we can produce as desired.

The biggest key to the aptness of the model is found in the residuals. The residuals are not an *i.i.d.* sample, as they sum to 0 and they do not have the same variance. The **standardized residuals** rescale the residuals to have unit variance. These appear in some of the diagnostic plots provided by `plot()`.

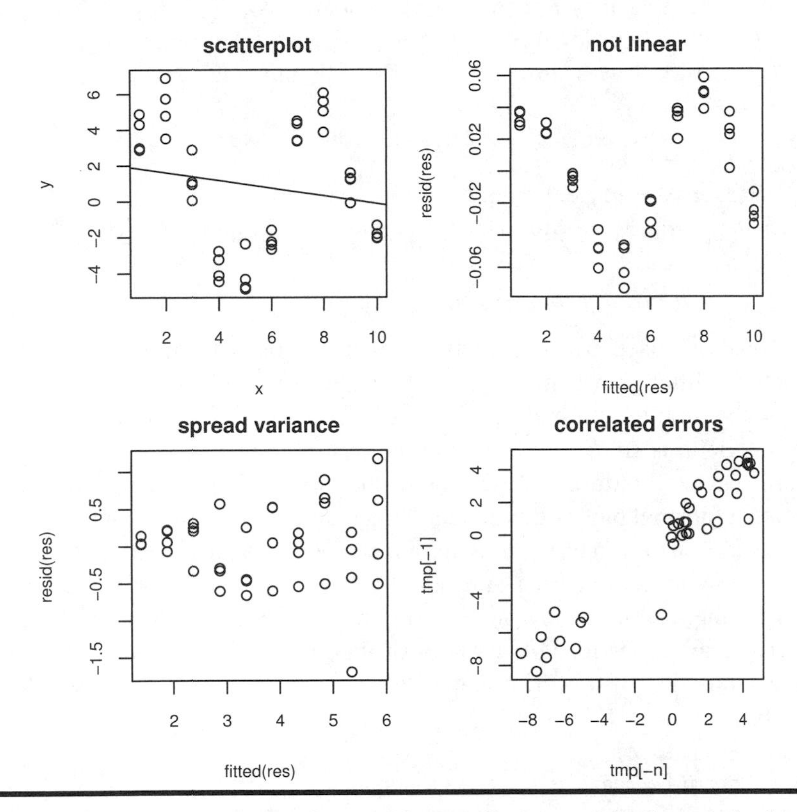

Figure 10.3 Four graphs showing problematic linear models. Scatterplot in upper left shows linear model is incorrect. Fitted versus residual plot in upper right shows a nonlinear trend. Fitted versus residual plot in lower left shows nonconstant variance. Lag plot in lower right shows correlations in error terms.

Assessing the linear model for the mean

A scatterplot of the data with the regression line can show quickly whether the linear model seems appropriate for the data. If the general trend is not linear, either a transformation or a different model is called for. An example of a cyclical trend (which calls for a transformation of the data) is the upper-left plot in Figure 10.3 and is made with these commands:

```
x = rep(1:10,4)
y = rnorm(40, mean=5*sin(x), sd=1)
plot(y ~ x); abline(lm(y~x))
```

When there is more than one predictor variable, a scatterplot will not be as useful.

A residual plot can also show whether the linear model is appropriate and can be made with more than one predictor. As well, it can detect small deviations from the model that may not show up in a scatterplot. The upper-right plot in Figure 10.3 shows a residual plot that finds a sinusoidal trend that will not show up in a scatterplot. It was simulated with these commands:

```
> x = rep(1:10,4)
> y = rnorm(40,mean = x + .05*sin(x),sd=.01) # small trend
> res = lm(y~x)
> plot(fitted(res),resid(res))
```

The residual plot is one of the four diagnostic plots produced by `plot()`.

Assessing normality of the residuals

The residuals are used to assess whether the error terms in the model are normally distributed. Although a histogram can be used to investigate normality, we've seen that the quantile-normal plot is better at visualizing differences from normality. Deviations from a straight line indicate nonnormality. Quantile-normal plots are made with `qqnorm()`. One of the diagnostic plots produced by `plot()` is a quantile-normal plot of the standardized residuals.

In addition to normality, an assumption of the model is also that the error terms have a common variance. A residual plot can show whether this is the case. When it is, the residuals show scatter about a horizontal line. In many data sets, the variance increases for larger values of the predictor. The commands below create a simulation of this. The graph showing the effect is in the lower-left of Figure 10.3.

```
> x = rep(1:10,4)
> y = rnorm(40, mean = 1 + 1/2*x, sd = x/10)
> res = lm(y ~ x)
> plot(fitted(res),resid(res))
```

The scale-location plot is one of the four diagnostic plots produced by `plot()`. It also shows the residuals, but in terms of the square root of the absolute value of the standardized residuals. The graph should show points scattered along the y-axis, as we scan across the x-axis, but the spread of the scattered points should not get larger or smaller.

In some data sets, there is a lack of independence in the residuals. For example, the errors may accumulate. A lag plot may be able to show this. For an independent sequence, the lag plot should be scattered, whereas many dependent sequences will show some pattern. This is illustrated in the lower-right plot in Figure 10.3, which was made as follows:

```
> x = rep(1:10,4)
> epsilon = rnorm(40,mean=0,sd=1)
> y = 1 + 2*x + cumsum(epsilon) # cumsum() correlates errors
> res = lm(y ~ x)
> tmp = resid(res)
```

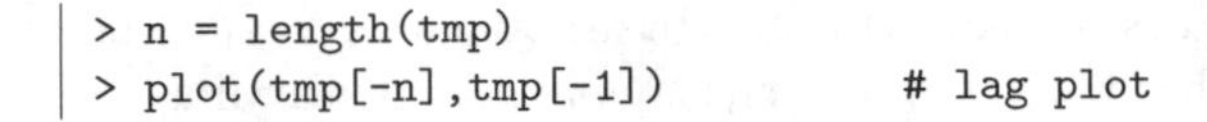

```
> n = length(tmp)
> plot(tmp[-n],tmp[-1])          # lag plot
```

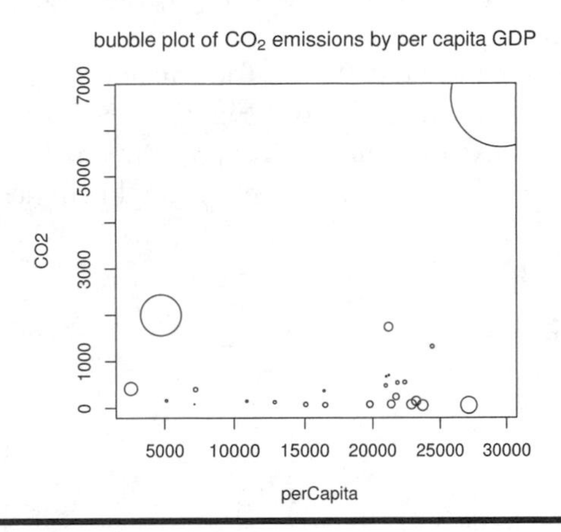

Figure 10.4 Bubble plot of CO_2 emissions by per capita GDP with area of points proportional to Cook's distance

Influential points

As we observed in Chapter 3, the regression line can be greatly influenced by a single observation that is far from the trend set by the data. The difference in slopes between the regression line with all the data and the regression line with the ith point missing will mostly be small, except for influential points. The Cook's distance is based on the difference of the predicted values of y_i for a given x_i when the point (x_i, y_i) is and isn't included in the calculation of the regression coefficients. The predicted amounts are used for comparison, as comparing slopes isn't applicable for multivariate models. The Cook's distance is computed by the extractor function `cooks.distance()`.

One of the diagnostic plots produced by `plot()` will show the Cook's distance for the data points plotted using spikes. Another way to display this information graphically is to make the size of the points in the scatterplot depend on this distance using the `cex=` argument. This type of plot is referred to as a bubble plot and is illustrated using the `emissions (UsingR)` data set in Figure 10.4. The graphic is made with the following commands:

```
> res = lm(CO2 ~ perCapita, emissions)
> plot(CO2 ~ perCapita, emissions,
+       cex = 10*sqrt(cooks.distance(res)),
+       main = expression(          # make subscript on CO2 in title
+          paste("bubble plot of ",CO[2],
+                " emissions by per capita GDP")
+       ))
```

The square root of the distances is used, so the area of the points is proportional to the Cook's distance rather than to the radius. (The argument to `main=` illustrates how to use mathematical notation in the title of a graphic. See the help page `?plotmath` for details.)

For the maximum-heart-rate data, the four diagnostic plots produced by R with the command `plot(res.mhr)` are in Figure 10.5. The `par(mfrow=c(2,2))` command was used to make the four graphs appear in one figure. This command sets the number of rows and columns in a multi-graphic figure.

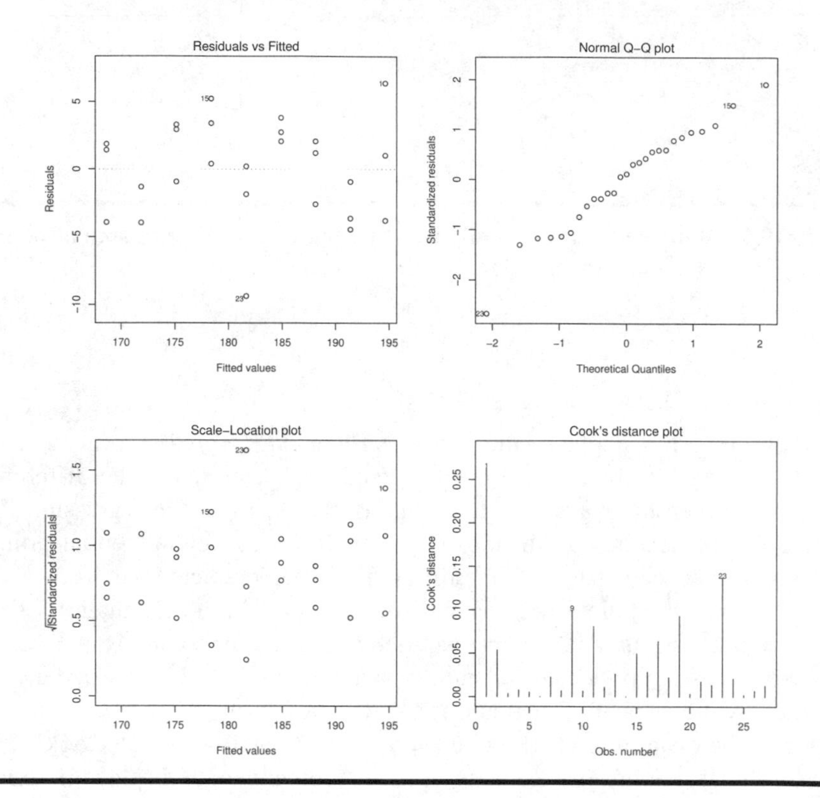

Figure 10.5 Four diagnostic plots for the maximum-heart-rate data produced by the extractor function `plot()`

10.2.2 Statistical inferences

If the linear model seems appropriate for the data, statistical inference is possible. What is needed is an understanding of the sampling distribution of the estimators.

To investigate these sampling distributions, we performed simulations of the model $Y_i = x_i + \varepsilon_i$, using `x = rep(1:10,10)` and `y = rnorm(100,x,5)`. Figure 10.6 shows the resulting regression lines for the different simulations. For reference, a single result of the simulation is plotted using a scatterplot. There is wide variation among the regression lines. In addition, histograms of the simulated values of $\widehat{\beta}_0$ and $\widehat{\beta}_1$ are shown.

We see from the figure that the estimators are random but not arbitrary. Both $\widehat{\beta}_0$ and $\widehat{\beta}_1$ are normally distributed, with respective means β_0 and β_1. Furthermore, $(n-2)\widehat{\sigma}^2/\sigma^2$ has a χ^2-distribution with $n-2$ degrees of freedom.

We will use the fact that the following statistics have a t-distribution with $n-2$ degrees of freedom:

$$\frac{\widehat{\beta}_0 - \beta_0}{\mathsf{SE}(\widehat{\beta}_0)}, \quad \frac{\widehat{\beta}_1 - \beta_1}{\mathsf{SE}(\widehat{\beta}_1)}. \tag{10.4}$$

The standard errors are found from the known formulas for the variances of the $\widehat{\beta}_i$:

$$\mathsf{SE}(\widehat{\beta}_0) = \widehat{\sigma}\left(\sum \frac{x_i^2}{\sum(x_i - \bar{x})^2}\right)^{1/2}, \quad \mathsf{SE}(\widehat{\beta}_1) = \frac{\widehat{\sigma}}{\sqrt{\sum(x_i - \bar{x})^2}}. \tag{10.5}$$

(Recall that, $\widehat{\sigma}^2 = \mathsf{RSS}/(n-2)$.)

Marginal t-tests

We can find confidence intervals and construct significance tests from the statistics in (10.4) and (10.5). For example, a significance test for

$$H_0: \beta_1 = b, \quad H_A: \beta_1 \neq b$$

is carried out with the test statistic

$$T = \frac{\widehat{\beta}_1 - b}{\mathsf{SE}(\widehat{\beta}_1)}.$$

Under H_0, T has the t-distribution with $n-2$ degrees of freedom.

A similar test for β_0 would use the test statistic $(\widehat{\beta}_0 - b)/\mathsf{SE}(\widehat{\beta}_0)$.

When the null hypothesis is $\beta_1 = 0$ or $\beta_0 = 0$ we call these **marginal t-tests**, as they test whether the parameter is necessary for the model.

The F-test

An alternate test for the null hypothesis $\beta_1 = 0$ can be done using a different but related approach that generalizes to the multiple-regression problem.

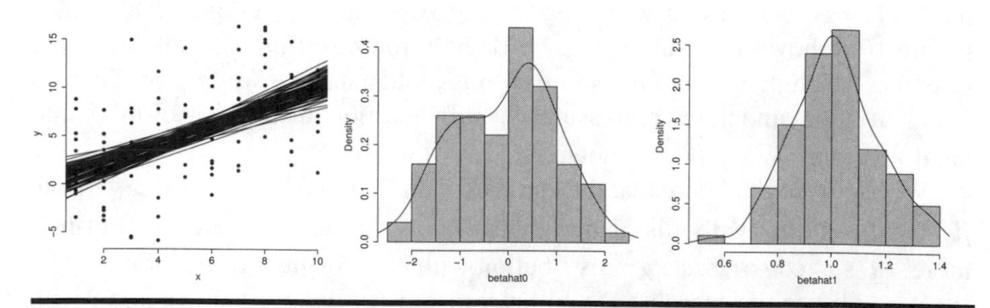

Figure 10.6 The plot on the left shows regression lines for 100 simulations from the model $Y_i = x_i + \varepsilon_i$. The plotted points show a single realization of the paired data during the simulation. The center and right plots are histograms of $\widehat{\beta}_0$ and $\widehat{\beta}_1$.

The total variation in the y values about the mean is

$$\mathsf{SST} = \text{total sum of squares} = \sum (y_i - \bar{y})^2.$$

Algebraically, this can be shown to be the sum of two easily interpreted terms:

$$\sum (y_i - \bar{y})^2 = \sum (y_i - \widehat{y}_i)^2 + \sum (\widehat{y}_i - \bar{y})^2. \tag{10.6}$$

The first term is the residual sum of squares, or RSS. The second is the total variation for the fitted model about the mean and is called the regression sum of squares, SSReg. Equation 10.6 becomes

$$\mathsf{SST} = \mathsf{RSS} + \mathsf{SSReg}.$$

For each term, a number—called the degrees of freedom—is assigned that depends on the sample size and the number of estimated values in the term. For the SST there are n data points and one estimated value, $\bar{y}$, leaving $n-1$ degrees of freedom. For RSS there are again n data points but two estimated values, $\widehat{\beta}_0$ and $\widehat{\beta}_1$, so $n-2$ degrees of freedom. This leaves 1 degree of freedom for the SSReg, as the degrees of freedom are additive in this case. When a sum of squares is divided by its degrees of freedom it is referred to as a **mean sum of squares**.

We rewrite the form of the prediction line:

$$\widehat{y}_i = \bar{y} + \widehat{\beta}_1 (x_i - \bar{x}).$$

If $\widehat{\beta}_1$ is close to 0, $\widehat{y}_i$ and $\bar{y}$ are similar in size, so we would have $\mathsf{SST} \approx \mathsf{RSS}$. In this case SSReg would be small. Whereas, if $\widehat{\beta}_1$ is not close to 0, then SSReg is not small. So, SSReg would be a reasonable test statistic for the hypothesis

$H_0 : \beta_1 = 0$. What do small and big mean? As usual, we need to scale the value by the appropriate factor. The F statistic is the ratio of the mean regression sum of squares divided by the mean residual sum of squares.

$$F = \frac{\mathsf{SSReg}/1}{\mathsf{RSS}/(n-2)} = \frac{\mathsf{SSReg}}{\widehat{\sigma}^2}. \tag{10.7}$$

Under the null hypothesis $H_0 : \beta_1 = 0$, the sampling distribution of F is known to be the F-distribution with 1 and $n-2$ degrees of freedom.

This allows us to make the following significance test.

F-test for $\beta_1 = 0$

A significance test for the hypotheses

$$H_0 : \beta_1 = 0, \qquad H_A : \beta_1 \neq 0$$

can be made with the the test statistic

$$F = \frac{\mathsf{SSReg}}{\widehat{\sigma}^2}.$$

Under the null hypothesis, F has F-distribution with 1 and $n-2$ degrees of freedom. Larger values of F are more extreme, so the p-value is given by $\mathsf{P}(F \geq \text{ observed value} \mid H_0)$.

The F-statistic can be rewritten as

$$F = \left(\frac{\widehat{\beta}_1}{\mathsf{SE}(\widehat{\beta}_1)} \right)^2.$$

Under the assumption $\beta_1 = 0$, this is the square of one of the t-distributed random variables of Equation 10.4. For simple linear regression the two tests of $H_0 : \beta_1 = 0$, the marginal t-test and the F-test, are equivalent. However, we will see that with more predictors, the two tests are different.

R^2 — *the coefficient of determination*

The decomposition of the total sum of squares into the residual sum of squares and the regression sum of squares in Equation 10.6 allows us to interpret how well the regression line fits the data. If the regression line fits the data well, then the residual sum of squares, $\sum(y_i - \widehat{y}_i)^2$, will be small. If there is a lot of scatter about the regression line, then RSS will be big. To quantify this, we can divide by the total sum of squares, leading to the definition of the **coefficient of determination**:

$$R^2 = 1 - \frac{\sum(y_i - \widehat{y}_i)^2}{\sum(y_i - \bar{y}_i)^2} = \frac{\sum(\widehat{y}_i - \bar{y})^2}{\sum(y_i - \bar{y})^2}. \tag{10.8}$$

This is close to 1 when the linear regression fit is good and close to 0 when it is not.

When the simple linear regression model is appropriate this value is interpreted as the proportion of the total response variation explained by the regression. That is, $R^2 \cdot 100\%$ of the variation is explained by the regression line. When R^2 is close to 1, most of the variation is explained by the regression line, and when R^2 is close to 0, not much is.

This interpretation is similar to that given for the Pearson correlation coefficient, r, in Chapter 3. This is no coincidence: for the simple linear regression model $r^2 = R^2$.

The **adjusted** R^2 divides the sums of squares by their degrees of freedom. For the simple regression model, these are $n-2$ for RSS and $n-1$ for SST. This is done to penalize models that get better values of R^2 by using more predictors. This is of interest when multiple predictors are used.

10.2.3 Using `lm()` to find values for a regression model

R can be used in different ways to do the above calculations.

Confidence intervals

We can find a 95% confidence interval for β_0 with

$$\widehat{\beta}_0 \pm t^* \mathsf{SE}(\widehat{\beta}_0).$$

In our example, this could be found with

```
> n = length(age)
> betahat0 = coef(res)[1]          # first coefficient
> sigmahat = sqrt( sum(resid(res)^2) / (n -2))
> SE = sigmahat * sqrt(sum(age^2) / (n* sum( (age - mean(age))^2)))
> tstar = qt(1 - 0.05/2,df= n - 2)
> c(betahat0 - tstar*SE, betahat0 + tstar*SE)
(Intercept) (Intercept)
      203.5       213.2
```

Standard error

The standard error above

```
> SE
[1] 2.357
```

is given as part of the `summary()` function applied to the output of `lm()`. Find it in the `Coefficients:` part of the output under the column labeled `Std. Error`.

```
> summary(res)

Call:
lm(formula = mhr ~ age)

Residuals:
   Min      1Q Median      3Q     Max
 -9.21   -2.47   1.13    2.65    7.79

Coefficients:
            Estimate Std. Error t value Pr(>|t|)
(Intercept) 208.3613     2.3571    88.4  < 2e-16 ***
age          -0.7595     0.0561   -13.5  5.2e-13 ***
---
Signif. codes:  0 '***' 0.001 '**' 0.01 '*' 0.05 '.' 0.1 ' ' 1

Residual standard error: 3.76 on 25 degrees of freedom
Multiple R-Squared: 0.88,       Adjusted R-squared: 0.875
F-statistic:  183 on 1 and 25 DF,  p-value: 5.15e-13
```

By reading the standard error from this output, a 95% confidence interval for β_1 may be more easily found than the one for β_0 above:

```
> betahat1 =  -0.7595             # read from summary
> SE =  0.0561                    # read from summary
> tstar = qt(1 - 0.05/2,df= n - 2)
> c(betahat1 - tstar*SE, betahat1 + tstar*SE)
[1] -0.875 -0.644
```

Significance tests

The `summary()` function returns more than the standard errors. For each coefficient a marginal t-test is performed. This is a two-sided hypothesis test of the null hypothesis that $\beta_i = 0$ against the alternative that $\beta_i \neq 0$. We see in this case that both are rejected with very low p-values. These small p-values are flagged in the output of `summary()` with significance stars.

Other t-tests are possible. For example, we can test the null hypothesis that the slope is -1 with the commands

```
> T.obs = (betahat1 - (-1))/SE
> T.obs
[1] 4.287
> 2*pt(-4.287,df = n - 2)         # or use lower.tail=F with 4.287
[1] 0.0002364
```

This is a small p-value, indicating that the model with slope -1 is unlikely to have produced this data or anything more extreme than it.

Finding $\widehat{\sigma}^2$, R^2

The estimate for $\widehat{\sigma}$ is marked `Residual standard error` and is labeled with $25 = 27 - 2$ degrees of freedom. The value of $R^2 =$ `cor(age,mhr)^2` is given along with an adjusted value.

F*-test for* $\beta_1 = 0$.

Finally, the F-statistic is calculated. As this is given by $(\widehat{\beta}_1/\mathsf{SE}(\widehat{\beta}_1))^2$ it can be found directly with

```
> (-0.7595 / 0.0561)^2
[1] 183.3
```

The significance test $H_0 : \beta_1 = 0$ with two-sided alternative is performed and again returns a tiny p-value.

The sum of squares to compute F are also given as the output of the `anova()` extractor function.

```
> anova(res)
Analysis of Variance Table

Response: mhr
          Df Sum Sq Mean Sq F value  Pr(>F)
age        1   2596    2596     183 5.2e-13 ***
Residuals 25    354      14
---
Signif. codes:  0 ‘***’ 0.001 ‘**’ 0.01 ‘*’ 0.05 ‘.’ 0.1 ‘ ’ 1
```

These values in the column headed `Sum Sq` are SSReg and RSS. The total sum of squares, SST, would be the sum of the two. Although the ratio of the mean sums of squares, $2596/14$, is not exactly 183, 183 is the correct value, as numbers have been rounded to integers.

Predicting the response with `predict()`

The function `predict()` is used to make different types of predictions.

A template for our usage is

```
predict(res, newdata=..., interval=..., level = ...)
```

The value of `res` is the output of a modeling function, such as `lm()`. We call this `res` below, but we can use any valid name. Any changes to the values of the predictor are given to the argument `newdata=` in the form of a data frame with names that match those used in the model formula. The arguments `interval=` and `level=` are set when prediction or confidence intervals are desired.

The simplest usage, `predict(res)`, returns the predicted values (the $\widehat{y}_i$'s) for the data. Predictions for other values of the predictor are specified using a data frame, as this example illustrates:

```
> predict(res, newdata=data.frame(age=42))
[1] 176.5
```

This finds the predicted maximum heart rate for a 42-year-old. The `age=` part of the data frame call is important. Variable names in the data frame supplied to the `newdata=` argument must exactly match the variable names used when the model object was produced.

Prediction intervals

The value of $\widehat{y}$ can be used to predict two different things: the value of a single estimate of y for a given x or the average value of many values of y for a given x. If we think of a model with replication (repeated y's for a given x, such as in Figure 10.6), then the difference is clear: one is a prediction for a given point, the other a prediction for the average of the points.

Statistical inference about the predicted value of y based on the sample is done with a **prediction interval**. As y is not a parameter, we don't call this a confidence interval. The form of the prediction interval is similar to that of a confidence interval:

$$\widehat{y} \pm t^* \mathsf{SE}.$$

For the prediction interval, the standard error is

$$\mathsf{SE} = \widehat{\sigma}\sqrt{1 + \frac{1}{n} + \frac{(x-\bar{x})^2}{s_{xx}}}. \tag{10.9}$$

The value of t^* comes from the t-distribution with $n-2$ degrees of freedom.

The prediction interval holds for all x simultaneously. It is often plotted using two lines on the scatterplot to show the upper and lower limits.

The `predict()` function will return the lower and upper endpoints for each value of the predictor. We specify `interval="prediction"` (which can be shortened) and a confidence level with `level=`. (The default is 0.95.)

For the heart-rate example we have:

```
> pred.res = predict(res, int = "pred")
> pred.res
fit   lwr   upr
1  193.2 185.0 201.4
2  189.4 181.3 197.5
...
```

A matrix is returned with columns giving the data we want. We cannot access these with the data frame notation `pred.res$lwr`, as the return value is not a data frame. Rather we can access the columns by name, like `pred.res[,'lwr']`, or by column number, as in

```
> pred.res[,2]                        # the 'lwr' column
    1     2     3     4     5     6     7     8     9    10
185.0 181.3 177.6 173.9 170.1 166.3 162.4 158.5 154.6 185.0
...
```

We want to plot both the lower and upper limits. In our example, we have the predicted values for the given values of `age`. As the `age` variable is not sorted, simply plotting will make a real mess. To remedy this, we specify the values of the `age` variable for which we make a prediction. We use the values `sort(unique(age))`, which gives just the x values in increasing order.

```
> age.sort = sort(unique(age))
> pred.res = predict(res.mhr, newdata = data.frame(age = age.sort),
+   int="pred")
> pred.res[,2]
    1     2     3     4     5     6     7     8     9
185.0 181.3 177.6 173.9 170.1 166.3 162.4 158.5 154.6
```

Now we can add the prediction intervals to the scatterplot with the `lines()` function (`matlines()` offers a one-step alternative). The result is Figure 10.7.

```
> plot(mhr ~ age); abline(res)
> lines(age.sort,pred.res[,2], lty=2) # lower curve
> lines(age.sort,pred.res[,3], lty=2) # upper curve
```

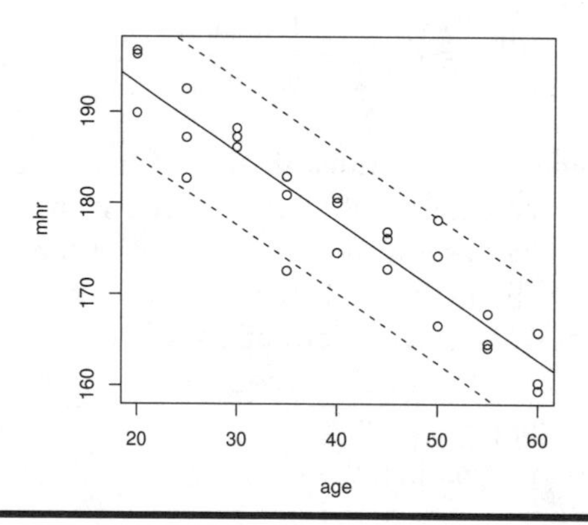

Figure 10.7 Regression line with 95% prediction intervals drawn for age versus maximum heart rate

There is a slight curve in the lines drawn, which is hinted at in equation 10.9. This implies that estimates near the value $(\bar{x}, \bar{y})$ have a smaller variance. This is expected: there is generally more data near this value, so the variances should be smaller.

Confidence intervals for $\mu_{y|x}$

A confidence interval for the mean value of y for a given x is given by

$$\widehat{Y} \pm t^* \mathsf{SE}(\widehat{Y}).$$

Again, t^* is from the t-distribution with $n-2$ degrees of freedom. The standard error used is now

$$\mathsf{SE}(\widehat{Y}) = \widehat{\sigma}\sqrt{\frac{1}{n} + \frac{(x-\bar{x})^2}{s_{xx}}}.$$

The standard error for the prediction interval differs by an extra term of plus 1 inside the square root. This may appear minor, but is not. If we had so much data (large n) that the estimates for the β's have small variance, we would not have much uncertainty in predicting the mean amount, but we would still have uncertainty in predicting a single deviation from the mean due to the error term in the model.

The values for this confidence interval are also returned by `predict()`. In this case, we use the argument `interval="confidence"`.

10.2.4 Problems

10.6 The cost of a home is related to the number of bedrooms it has. Suppose Table 10.3 contains data recorded for homes in a given town. Make a scatterplot, and fit the data with a regression line. On the same graph, test the hypothesis that an extra bedroom is worth $60,000 versus the alternative that it is worth more.

Table 10.3 Number of bedrooms and sale price of a home in thousands

price	$300	$250	$400	$550	$317	$389	$425	$289	$389
bedrooms	3	3	4	5	4	3	6	3	4

10.7 The more beer you drink, the more your blood alcohol level (BAL) rises. Table 10.4 contains a data set on beer consumption. Make a scatterplot with a regression line and 95% prediction intervals drawn. Test the hypothesis that one beer raises your BAL by 0.02% against the alternative that it raises it less.

10.8 For the same blood-alcohol data, do a significance test that the intercept is 0 with a two-sided alternative.

Table 10.4 Beer consumption and blood alcohol level

beers	5	2	9	8	3	7	3	5	3	5
BAL	0.10	0.03	0.19	0.12	0.04	0.095	0.07	0.06	0.02	0.05

10.9 The lapse rate is the rate at which temperature drops as you increase elevation. Some hardy students were interested in checking empirically whether the lapse rate of 9.8 °C/km was accurate. To investigate, they grabbed their thermometers and their Suunto wrist altimeters and recorded the data in Table 10.5 on their hike. Draw a scatterplot with regression line and investigate whether the lapse rate is 9.8 °C/km/km. (It helps to convert to the rate of change °F per feet, which is 5.34 degrees per 1,000 feet.) Test the hypothesis that the lapse rate is 5.34 degrees per 1,000 feet against a two-sided alternative.

Table 10.5 Elevation and temperature measurements

elevation (ft)	600	1000	1250	1600	1800	2100	2500	2900
temperature (°F)	56	54	56	50	47	49	47	45

10.10 A seal population is counted over a ten-year period. The counts are reported in Table 10.6. Make a scatterplot and find the regression line. What is the predicted value for 1963? Would you use this to predict the population in 2004? Why or why not?

Table 10.6 Seal population from 1952 to 1962

year	pop.	year	pop.	year	pop	year	pop
1952	724	1955	1,392	1958	1,212	1961	1,980
1953	176	1956	1,392	1959	1,672	1962	2,116
1954	920	1957	1,448	1960	2,068		

10.11 For the `homedata (UsingR)` data set, find the regression equation to predict the year-2000 value of a home from its year-1970 value. Make a prediction for an $80,000 home in 1970. Comment on the appropriateness of the regression model by investigating the residuals.

10.12 The `deflection (UsingR)` data set contains deflection measurements for various loads. Fit a linear model to `Deflection` as a function of `load`. Plot the data and the regression line. How well does the line fit? Investigate with a residual plot.

10.13 The `alaska.pipeline (UsingR)` data set contains measurements of defects on the Alaska pipeline that are taken first in the field and then in the laboratory. The measurements are done in six batches. Fit a linear model for the lab-defect size as modeled by the field-defect size. Find the coefficients. Discuss the appropriateness of the model.

10.14 In athletic events in which people of various ages participate, performance is sometimes related to age. Multiplying factors are used to compare the performance of a person of a given age to another person of a different age. The data set `best.times (UsingR)` features world records by age and distance in track and field.

We split the records by distance, allowing us to compare the factors for several distances.

```
> attach(best.times)
> by.dist = split(best.times,as.factor(Dist))
> detach(best.times)
```

This returns a list of data frames, one for each distance. We can plot the times in the 800-meter run:

```
> plot(Time ~ age, by.dist[['800']])
```

It is actually better to apply `scale()` first, so that we can compare times.

Through age 70, a linear regression model seems to fit. It can be found with

```
> lm(scale(Time) ~ age, by.dist[['800']], subset = age < 70)

Call:
lm(formula = scale(Time) ~ age, data = by.dist[["800"]],
  subset = age < 70)

Coefficients:
(Intercept)          age
    -1.2933       0.0136
```

Using the above technique, compare the data for the 100-meter dash, the 400-meter dash, and the 10,000-meter run. Are the slopes similar?

10.15 The `galton (UsingR)` data set contains data collected by Francis Galton in 1885 concerning the influence a parent's height has on a child's height. Fit a linear model modeling a child's height by his parents'. Do a test of significance to see whether β_1 equals 1 against a two-sided alternative.

10.16 Find and plot both the prediction and the confidence intervals for the heart-rate example. Simulate your own data.

10.17 The `alaska.pipeline (UsingR)` data set appears appropriate for a linear model, but the assumption of equal variances does not seem appropriate. A log-transformation of each variable does seem to have equal variances. Fit the model

$$\log(\texttt{lab.defect}) = \beta_0 + \beta_1 \cdot \log(\texttt{field.defect}) + \varepsilon.$$

Investigate the residuals and determine whether the assumption of equal variance seems appropriate.

10.18 The following commands will simulate the regression model $Y_i = 1 + 2x_i + \varepsilon_i$:

```
> res = matrix(0,nrow=200,ncol=2)
> for(i in 1:200) {
+ x = rep(1:10,4); y = rnorm(40,1 + 2*x,3)
+ res[i,] = coef(lm(y ~ x))
+ }
> plot(res[,1],res[,2])
```

(We first create `res` as a matrix to store the two values for the coefficients. Alternately, you can create two different vectors for this.)

Run the simulation and comment on the shape of the scatterplot. What does it say about the correlation between $\widehat{\beta_0}$ and $\widehat{\beta_1}$?

10.19 In a simple linear regression, confidence intervals for β_0 and β_1 are given separately in terms of the t-distribution as $\widehat{\beta_i} \pm t^* \mathsf{SE}(\widehat{\beta_i})$. They can also be found *jointly*, giving a **confidence ellipse** for the parameters as a pair. This can be found easily in R with the `ellipse` package.[†] If `res` is the result of the `lm()` function, then `plot(ellipse(res),type="l")` will draw the confidence ellipse.

For the `deflection (UsingR)` data set, find the confidence ellipse for `Deflection` modeled by `Load`.

10.3 Multiple linear regression

Multiple linear regression allows for more than one regressor to predict the value of Y. Lots of possibilities exist. These regressors may be separate variables, products of separate variables, powers of the same variable, or functions of the same variable. In the next chapter, we will consider regressors that are not numeric but categorical. They all fit together in the same model, but there are additional details. We see, though, that much of the background for the simple linear regression model carries over to the multiple regression model.

[†] The `ellipse` package is not part of the standard R installation, but it is on CRAN. You can install it with the command `install.packages("ellipse")`. See Appendix A for details.

10.3.1 Types of models

Let Y be a response variable and let $x_1, x_2, \ldots, x_p$ be p variables that we will use for predictors. For each variable we have n values recorded. The multiple regression model we discuss here is

$$Y_i = \beta_0 + \beta_1 x_{1i} + \cdots + \beta_p x_{pi} + \varepsilon_i.$$

There are $p+1$ parameters in the model labeled $\beta_0, \beta_1, \ldots, \beta_p$. They appear in a linear manner, just like a slope or intercept in the equation of a line. The x_i's are predictor variables, or covariates. They may be random; they may be related, such as powers of each other; or they may be correlated. As before, it is assumed that the ε_i values are an *i.i.d.* sample from a normal distribution with mean 0 and unknown variance σ^2. In terms of the Y variable, the values Y_i are an independent sample from a normal distribution with mean $\beta_0 + \beta_1 x_{1i} + \cdots + \beta_p x_{pi}$ and common variance σ^2. If the x variables are random, this is true after conditioning on their values.

■ **Example 10.2: What influences a baby's birth weight?** A child's birth weight depends on many things; among them the parents' genetic makeup, gestation period, and mother's activities during pregnancy. The `babies (UsingR)` data set lets us investigate some of these relationships.

This data set contains many variables to consider. We first look at the quantitative variables as predictors. These are gestation period; mother's age, height, and weight; and father's age, height, and weight.

A first linear model might incorporate all of these at once:

$$\text{wt} = \beta_0 + \beta_1 \cdot \text{gestation} + \beta_2 \cdot \text{mother's age} + \cdots + \beta_7 \cdot \text{father's weight} + \varepsilon_i.$$

Why should this have a linear model? It seems intuitive that birth weight would vary monotonically with the variables, so a linear model might be a fairly good approximation. We'll want to look at some plots to make sure our model seems appropriate. ■

■ **Example 10.3: Polynomial regression** In 1609, Galileo proved mathematically that the horizontal distance traveled by an object with an initial horizontal velocity is a parabola. He based his insight on an experimental setup consisting of a ball placed at a certain height on a ramp and then released. The distance traveled was then measured. This experiment was chosen to reduce the effects of friction. (This example appears in Ramsey and Schafer's *The Statistical Sleuth*, Duxbury 1997, where a schematic of the experimental apparatus is drawn.) The data consists of two variables. Let's call them y for distance traveled and x for initial height. Galileo may have considered any of these polynomial models:

$$y_i = \beta_0 + \beta_1 x_i + \varepsilon_i,$$

$$y_i = \beta_0 + \beta_1 x_i + \beta_2 x_i^2 + \varepsilon_i, \text{ or}$$

$$y_i = \beta_0 + \beta_1 x_i + \beta_2 x_i^2 + \beta_3 x_i^3 + \varepsilon_i.$$

The ε_i would cover error terms that are presumably independent and normally distributed. The quadratic model (the second model) is correct under perfect conditions, as Galileo demonstrated, but the data may suggest a different model if the conditions are not perfect. ■

■ **Example 10.4: Predicting classroom performance** College admissions offices are faced with the problem of predicting future performance based on a collection of measures, such as grade-point average and standardized test scores. These values may be correlated. There may also be other variables that describe why a student does well, such as type of high school attended or student's work ethic.

Initial student placement is also a big issue. If a student does not place into the right class, he may become bored and leave the school. Successful placement is key to retention. For New York City high school graduates, available at time of placement are SAT scores and Regents Exam scores. High school grade-point average may be unreliable or unavailable.

The data set `stud.recs (UsingR)` contains test scores and initial grades in a math class for several randomly selected students. What can we predict about the initial grade based on the standardized scores?

An initial model might be to fit a linear model for grade with all the other terms included. Other restricted models might be appropriate. For example, are the verbal SAT scores useful in predicting grade performance in a future math class? ■

10.3.2 Fitting the multiple regression model using `lm()`

As seen previously, the method of least squares is used to estimate the parameters in the multiple regression model. We don't give formulas for computing the $\widehat{\beta}$'s but note that, since there are $p+1$ estimated parameters, the estimate for the variance changes to

$$\widehat{\sigma}^2 = \frac{\mathsf{RSS}}{n-(p+1)}.$$

To find these estimates in R, again the `lm()` function is used. The syntax for the model formula varies depending on the type of terms in the model. For these problems, we use + to add terms to a model, - to drop terms, and `I()` to insulate terms so that the usual math notations apply.

For example, if x, y, and z are variables, then the following statistical models have the given R counterparts:

$z_i = \beta_0 + \beta_1 x_i + \beta_2 y_i + \varepsilon_i$ is expressed as `z ~ x + y`

$z_i = \beta_0 + \beta_1 x_i + \beta_2 x_i^2 + \varepsilon_i$ is expressed as `z ~ x + I(x^2)`

Once the model is given, the `lm()` function follows the same format as before:

```
lm(formula, data=..., subset=...)
```

To illustrate with an artificial example, we simulate the relationship $z_i = \beta_0 + \beta_1 x_i + \beta_2 y_i + \varepsilon_i$ and then find the estimated coefficients:

```
> x = 1:10; y = rchisq(10,3); z = 1 + x + y + rnorm(10)
> lm(z ~ x + y)

Call:
lm(formula = z ~ x + y)

Coefficients:
(Intercept)            x            y
      1.684        0.881        1.076
```

The output of `lm()` stores much more than is seen initially (which is just the formula and the estimates for the coefficients). It is recommended that the return value be stored. Afterward, the different extractor functions can be used to view the results.

■ **Example 10.5: Finding the regression estimates for baby's birth weight**

Fitting the birth-weight model is straightforward. The basic model formula is

```
wt ~ gestation + age + ht + wt1 + dage + dht + dwt
```

We've seen with this data set that the variables have some missing values that are coded not with `NA` but with very large values that are obvious when plotted, but not when we blindly use the functions. In particular, gestation should be less than 350 days, mother's age and height less than 99, and weight less than 999, etc. We can avoid these cases by using the `subset=` argument as illustrated. Recall that we combine logical expressions with & for "and" and | for "or."

```
> res.lm = lm(wt ~ gestation + age + ht + wt1 + dage + dht + dwt ,
+ data = babies,
+ subset= gestation < 350 & age < 99 & ht < 99 & wt1 < 999 &
```

```
+ dage < 99 & dht < 99 & dwt < 999)
> res.lm
Call:
...
Coefficients:
(Intercept)    gestation          age           ht          wt1
  -105.4576       0.4625       0.1384       1.2161       0.0289
       dage          dht          dwt
     0.0590      -0.0663       0.0782
```

A residual plot (not shown) shows nothing too unusual:

```
> plot(fitted(res.lm), resid(res.lm))
```

The diagnostic plots found with `plot(res.lm)` indicate that observation 261 might be a problem. Looking at `babies[261,]`, it appears that this case is an outlier, as it has a very short gestation period. It could be handled separately. ■

The `subset=` argument is very useful, though repeated uses may make us wish that we could use it just once prior to modeling. In this case the `subset()` function is available.

Using `update()` with model formulas

When comparing models, we may be interested in adding or subtracting a term and refitting. Rather than typing in the entire model formula again, R provides a way to add or drop terms from a model and have the new model fit. This process is called updating and is done with the `update()` function. The usage is

```
update(model.object, formula = . ~ . + new.terms)
```

The `model.object` is the output of some modeling command, such as `lm()`. The `formula=` argument uses a `.` to represent the previous value. In the template above, the `.` to the left of the ~ indicates that the previous left side of the model formula should be reused. The right-hand-side `.` refers to the previous right-hand side. In the template, the `+ new.terms` means to add terms. Use `- old.terms` to drop terms.

■ **Example 10.6: Discovery of the parabolic trajectory** The data set `galileo (UsingR)` contains two variables measured by Galileo (described previously). One is the initial height and one the horizontal distance traveled.

A plot of the data illustrates why Galileo may have thought to prove that the correct shape is described by a parabola. Clearly a straight line does not fit the data well. However, with modern computers, we can investigate whether a cubic term is warranted for this data.

To do so we fit three polynomial models. The `update()` function is used to add terms to the previous model to give the next model. To avoid a different interpretation of `^`, the powers are insulated with `I()`.

```
> init.h = c(600,700,800,950,1100,1300,1500)
> h.d = c(253, 337, 395, 451, 495, 534, 573)
```

```
> res.lm = lm(h.d ~ init.h)
> res.lm2 = update(res.lm,  . ~ . + I(init.h^2))
> res.lm3 = update(res.lm2, . ~ . + I(init.h^3))
```

To plot these, we will use `curve()`, but first we define a simple function to help us plot polynomials when we know their coefficients. The result is in Figure 10.8. The linear model is a poor fit, but both the quadratic and cubic fits seem good.

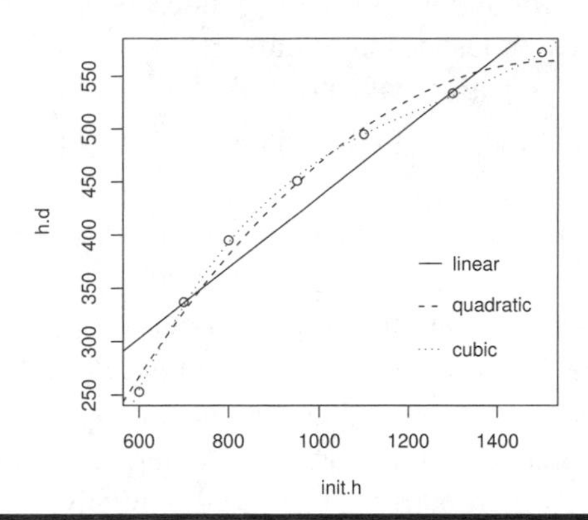

Figure 10.8 Three polynomial models fit to the Galileo data

```
> polynomial = function(x,coefs) {
+ tot = 0
+ for(i in 1:length(coefs)) tot = tot + coefs[i]*x^{i-1}
+ tot
+ }
> plot(h.d ~ init.h)
> curve(polynomial(x,coef(res.lm )), add=TRUE, lty=1)
> curve(polynomial(x,coef(res.lm2)), add=TRUE, lty=2)
> curve(polynomial(x,coef(res.lm3)), add=TRUE, lty=3)
> legend(1200,400,legend=c("linear","quadratic","cubic"),lty=1:3)
```

■

10.3.3 Interpreting the regression parameters

In many cases, interpretation in simple regression is straightforward. Changes in the predictor variable correspond to changes in the response variable in a linear manner: a unit change in the predictor corresponds to a $\widehat{\beta}_1$-unit change in the response.

However, in multiple regression this picture may not be applicable, as we may not be able to change just a single variable. As well, when more variables are added to a model, if the variables are correlated then the sign of the coefficients can change, leading to a different interpretation.

The language often used is that we "control" the other variables while seeking a primary predictor variable.

■ **Example 10.7: Does taller mean higher paid?** A University of Florida press release from October 16, 2003, reads:

> "Height matters for career success," said Timothy Judge, a UF management professor. . . .
>
> Judge's study, which controlled for gender, weight, and age, found that mere inches cost thousands of dollars. Each inch in height amounted to about $789 more a year in pay, the study found.

The mathematical model mentioned would be

$$\text{pay} = \beta_0 + \beta_1 \text{ height} + \beta_2 \text{ gender} + \beta_3 \text{ weight} + \beta_4 \text{ age} + \varepsilon.$$

(In the next chapter we see how to interpret the term involving the categorical variable `gender`.) The data gives rise to the estimate $\widehat{\beta}_1 = 789$. The authors interpret this to mean that each extra inch of height corresponds to a $789 increase in expected pay. So someone who is 4 inches taller, say 6 feet versus 5 feet 8 inches, would be expected to earn $3,156 more annually. ($\widehat{Y}$ is used to predict expected values.) The word "controlled" means that we included these variables in the model.

Unlike in a science experiment, where we may be able to specify the value of a variable, a person cannot simply grow an inch to see if his salary goes up. This is an observational study, so causal interpretations are not necessarily valid. ■

10.3.4 Statistical inferences

As in the simple linear regression case, if the model is correct, statistical inference can be made about the coefficients. In general, the estimators for a linear model are unbiased and normally distributed; from this, t-tests and confidence intervals can be constructed for the estimators, once we learn the standard errors. As before, these are output by the `summary()` function.

■ **Example 10.8: Galileo, continued** For the Galileo data example, the `summary()` of the quadratic fit contains

```
> summary(res.lm2)
...
Coefficients:
             Estimate Std. Error t value Pr(>|t|)
(Intercept) -2.40e+02   6.90e+01   -3.48   0.0253 *
init.h       1.05e+00   1.41e-01    7.48   0.0017 **
I(init.h^2) -3.44e-04   6.68e-05   -5.15   0.0068 **
---
```

```
Signif. codes:  0 ‘***’ 0.001 ‘**’ 0.01 ‘*’ 0.05 ‘.’ 0.1 ‘ ’ 1
...
```

For each $\widehat{\beta}$, the standard errors are given, as is the marginal t-test, which tests for the null hypothesis that the $\widehat{\beta}$ is 0. All three have small p-values and are flagged as such with significance stars.

Finding a confidence interval for the parameters is straightforward, as the values $(\widehat{\beta}_i - \beta_i)/\mathsf{SE}(\widehat{\beta}_i)$ have a t-distribution with $n-(p+1)$ degrees of freedom if the linear model applies.

For example, a 95% confidence interval for β_1 would be

```
> alpha = 0.05
> tstar = qt(1 - alpha/2, df = 4) # n = 7; p=2; df = n-(p+1)
> c(1.05 - tstar*0.141, 1.05 + tstar*0.141)
[1] 0.6585 1.4415
```

■

10.3.5 Model selection

If there is more than one possible model for a relationship, how do we know which to prefer? There are many criteria for selecting a model. We mention two here that are easily used within R.

Partial F-test

Consider these two **nested models** for Y_i:

$$Y_i = \beta_0 + \beta_1 x_{1i} + \cdots + \beta_k x_{ki} + \varepsilon_i \tag{10.10}$$

$$Y_i = \beta_0 + \beta_1 x_{1i} + \cdots + \beta_k x_{ki} + \beta_{k+1} x_{(k+1)i} + \cdots + \beta_p x_{pi} + \varepsilon_i.$$

The first model has $k+1$ parameters, and the second has $p+1$ with $p > k$ (not counting σ). Recall that the residual sum of squares, RSS, measures the variation between the data and the model. For the model with p predictors, $\mathsf{RSS}(p)$ can only be less than $\mathsf{RSS}(k)$ for the model with k predictors. Call the difference the **extra sum of squares**.

If the new parameters are not really important, then there should be little difference between the sums of squares when computed with or without the new parameters. If they are important, then there should be a big difference. To measure big or small, we can divide by the residual sum of squares for the full model. That is,

$$\frac{\mathsf{RSS}(k) - \mathsf{RSS}(p)}{\mathsf{RSS}(p)}$$

should measure the influence of the extra parameters. If we divide the extra sum of squares by $p-k$ and the residual sum of squares by $n-(p+1)$ (the respective degrees of freedom), then the statistic becomes

$$F = \frac{(\mathsf{RSS}(k) - \mathsf{RSS}(p))/(p-k)}{\mathsf{RSS}(p)/(n-(p+1)))} = \frac{(\mathsf{RSS}(k) - \mathsf{RSS}(p))/(p-k)}{\widehat{\sigma}^2}. \tag{10.11}$$

This statistic is actually a more general example of that in equation 10.7 and has a similar sampling distribution. Under the null hypothesis that the extra β's are 0 ($\beta_{k+1} = \cdots = \beta_p = 0$), and the ε_i are *i.i.d.* with a $\mathsf{Normal}(0, \sigma^2)$ distribution, F will have the F-distribution with $p-k$ and $n-(p+1)$ degrees of freedom.

This leads to the following significance test.

Partial F-test for null hypothesis of no effect

For the nested models of Equation 10.10, a significance test for the hypotheses

$$H_0: \beta_{k+1} = \beta_{k+2} = \cdots = \beta_p = 0 \quad \text{and} \quad H_A: \text{ at least one } \beta_j \neq 0 \text{ for } j > k$$

can be performed with the test statistic (10.11):

$$F = \frac{\text{extra sum of squares}/(p-k)}{\widehat{\sigma}^2}.$$

Under H_0, F has the F-distribution with $p-k$ and $n-(p+1)$ degrees of freedom. Large values of F are in the direction of the alternative. This test is called the **partial F-test**.

The `anova()` function will perform the partial F-test. If `res.lm1` and `res.lm2` are the return values of two nested models, then

```
anova(res.lm1, res.lm2)
```

will perform the test and produce an analysis of variance table.

■ **Example 10.9: Discovery of the parabolic trajectory revisited** In Example 10.6 we fitted the data with three polynomials and graphed them. Referring to Figure 10.8, we see that the parabola and cubic clearly fit better than the linear. But which of those two fits better? We use the partial F-test to determine whether the extra cubic term is significant.

To do this, we use the `anova()` function on the two results `res.lm2` and `res.lm3`. This yields

```
> anova(res.lm2,res.lm3)
Analysis of Variance Table

Model 1: h.d ~ init.h + I(init.h^2)
Model 2: h.d ~ init.h + I(init.h^2) + I(init.h^3)
  Res.Df RSS Df Sum of Sq    F Pr(>F)
1      4 744
2      3  48  1       696 43.3 0.0072 **
---
Signif. codes:  0 '***' 0.001 '**' 0.01 '*' 0.05 '.' 0.1 ' ' 1
```

The F-test is significant ($p = 0.0072$), indicating that the null hypothesis ($\beta_3 = 0$) does not describe the data well. This suggests that the underlying relationship from Galileo's data is cubic and not quadratic. Perhaps the apparatus introduced drag. ■

The Akaike information criterion

In the partial F-test, the trade-off between adding more parameters to improve the model fit and making a more complex model appears in the $n-(p+1)$ divisor. Another common criterion with this trade-off is Akaike's information criterion (AIC). The AIC is computed in R with the `AIC()` extractor function. The details of the statistic involve the likelihood function, a more advanced concept, but the usage is straightforward: models with lower AICs are preferred. An advantage to the AIC is that it can be used to compare models that are not nested. This is a restriction of the partial F-test.

The extractor function `AIC()` will compute the value for a given model, but the convenient `stepAIC()` function from the `MASS` library will step through the submodels and do the comparisons for us.

■ **Example 10.10: Predicting grades based on standardized tests**

The data set `stud.recs (UsingR)` contains five standardized test scores and a numeric value for the initial grade in a subsequent math course. The goal is to use the test-score data to predict the grade that a student will get. If the grade is predicted to be low, perhaps an easier class should be recommended.

First, we view the data using paired scatterplots

```
> pairs(stud.recs)
```

The figure (not shown) indicates strong correlations among the variables.

We begin by fitting the entire model. In this case, the convenient `.` syntax on the right-hand side is used to indicate all the remaining variables.

```
> res.lm = lm(num.grade ~ ., data = stud.recs)
> res.lm

Call:
lm(formula = num.grade ~ ., data = stud.recs)
```

```
Coefficients:
(Intercept)        seq.1        seq.2        seq.3        sat.v
   -0.73953     -0.00394     -0.00272      0.01565     -0.00125
      sat.m
    0.00590
```

Some terms are negative, which seems odd. Looking at the summary of the regression model we have

```
> summary(res.lm)
...
Coefficients:
            Estimate Std. Error t value Pr(>|t|)
(Intercept) -0.73953    1.21128   -0.61    0.543
seq.1       -0.00394    0.01457   -0.27    0.787
seq.2       -0.00272    0.01503   -0.18    0.857
seq.3        0.01565    0.00941    1.66    0.099 .
sat.v       -0.00125    0.00163   -0.77    0.443
sat.m        0.00590    0.00267    2.21    0.029 *
...
```

The marginal t-tests for whether the given parameter is 0 or not are "rejected" only for the `seq.3` (sequential 3 is the last high school test taken) and `sat.m` (the math SAT score). It is important to remember that these are tests concerning whether the value is 0 given the other predictors. They can change if predictors are removed.

The `stepAIC()` function can step through the various submodels and rank them by AIC. This gives

```
> library(MASS)                # load in MASS package for stepAIC
> stepAIC(res.lm)
Start:  AIC= 101.2
... lots skipped ...
Coefficients:
(Intercept)        seq.3        sat.m
   -1.14078      0.01371      0.00479
```

The submodel with just two predictors is selected. As expected, the verbal scores on the SAT are not a good indicator of performance. ■

10.3.6 Problems

10.20 Do Example 10.5 and fit the full model to the data. For which variables is the t-test for $\beta_i = 0$ flagged? What model is selected by AIC?

10.21 Following Example 10.9, fit a fourth-degree polynomial to the `galileo` (`UsingR`) data and compare to the cubic polynomial using a partial F-test. Is the new coefficient significant?

10.22 For the data set `trees`, model the `Volume` by the `Girth` and `Height` variables. Does the model fit the data well?

10.23 The data set `MLBattend (UsingR)` contains attendance data for major league baseball for the years 1969 to 2000. Fit a linear model of `attendance` modeled by `year`, `runs.scored`, `wins`, and `games.behind`. Which variables are flagged as significant? Look at the diagnostic plots and comment on the validity of the model.

10.24 For the `deflection (UsingR)` data set, fit the quadratic model

$$\texttt{Deflection} = \beta_0 + \beta_1 \texttt{Load} + \beta_2 \texttt{Load}^2 + \varepsilon.$$

How well does this model fit the data? Compare to the linear model.

10.25 The data set `kid.weights` contains age, weight, and height measurements for several children. Fit the linear model

$$\texttt{weight} = \beta_0 + \beta_1 \texttt{age} + \beta_2 \texttt{height} + \beta_3 \texttt{height}^2 + \beta_4 \texttt{height}^3 + \beta_5 \texttt{height}^4$$

Use the partial F-test to select between this model and the nested models found by using only first-, second-, and third-degree polynomials for `height`.

10.26 The data set `fat (UsingR)` contains several body measurements that can be done using a scale and a tape measure. These can be used to predict the body-fat percentage (`body.fat`). Measuring body fat requires a special apparatus; if our resulting model fits well, we have a low-cost alternative.

Fit the variable `body.fat` using each of the variables `age`, `weight`, `height`, `BMI`, `neck`, `chest`, `abdomen`, `hip`, `thigh`, `knee`, `ankle`, `bicep`, `forearm`, and `wrist`. Use the `stepAIC()` function to select a submodel. For this submodel, what is the adjusted R^2?

10.27 The data set `Cars93 (MASS)` contains data on cars sold in the United States in the year 1993. Fit a regression model with `MPG.city` modeled by the numeric variables `EngineSize`, `Weight`, `Passengers`, and `price`. Which variables are marked as statistically significant by the marginal t-tests? Which model is selected by the AIC?

10.28 We can simulate the data to see how often the partial F-test or AIC works. For example, a single simulation can be done with the commands

```
> x = 1:10;y = rnorm(10,1+2*x+3*x^2,4)
> stepAIC(lm(y~x+I(x^2)))          # needs library(MASS) at first
```

Do a few simulations to see how often the correct model is selected.

10.29 The data set `baycheck (UsingR)` contains estimated populations for a variety of Bay checkerspot butterflies near California. A common model for population dynamics is the Ricker model, for which t is time in years:

$$N_{t+1} = aN_t e^{bN_t} W_t,$$

where a and b are parameters and W_t is a lognormal multiplicative error. This can be turned into a regression model by dividing by N_t and then taking logs of both sides to give

$$\log(\frac{N_{t+1}}{N_t}) = \log(a) + bN_t + \varepsilon_t.$$

Let y_t be the left-hand side. This may be written as

$$y_t = r(1 - \frac{N_t}{K}) + \varepsilon_t,$$

because r can be interpreted as an unconstrained growth rate and K as a carrying capacity.

Fit the model to the `baycheck (UsingR)` data set and find values for r and K. To find y_t you can do the following:

```
> attach(baycheck)
> n = length(year)
> yt = log(Nt[-1]/Nt[-n])
> nt = Nt[-n]
```

Recall that a negative index means all but that index.

Chapter 11

Analysis of variance

Analysis of variance, **ANOVA**, is a method of comparing means based on variations from the mean. We begin by doing ANOVA the traditional way, but we will see that it is a special form of the linear model discussed in the previous chapter. As such, it can be approached in a unified way, with much of the previous work being applicable.

11.1 One-way ANOVA

A one-way analysis of variance is a generalization of the t-test for two independent samples, allowing us to compare means for several independent samples. Suppose we have k populations of interest. From each we take a random sample. These samples are independent if the knowledge of one sample does not effect the distribution of another. Notationally, for the ith sample, let $X_{i1}, X_{i2}, \ldots, X_{in_i}$ designate the sample values.

The one-way analysis of variance applies to normally distributed populations. Suppose the mean of the ith population is μ_i and its standard deviation is σ_i. We use a σ if these are all equivalent. A statistical model for the data with common standard deviation is

$$X_{ij} = \mu_i + \varepsilon_{ij},$$

where the error terms, ε_{ij}, are independent with $\mathsf{Normal}(0, \sigma)$ distribution.

■ **Example 11.1: Number of calories consumed by month** Consider 15 subjects split at random into three groups. Each group is assigned a month. For each group we record the number of calories consumed on a randomly chosen day. Figure 11.1 shows the data. We assume that the amounts consumed are normally distributed with common variance but perhaps different means. From the figure, we see that there appears to be more clustering around the means for each month than around the grand mean or mean for all the data. This would indicate that the means may be different. Perhaps more calories are consumed in the winter?

The goal of one-way analysis of variance is to decide whether the difference in the sample means is indicative of a difference in the population means of each sample or is attributable to sampling variation. ■

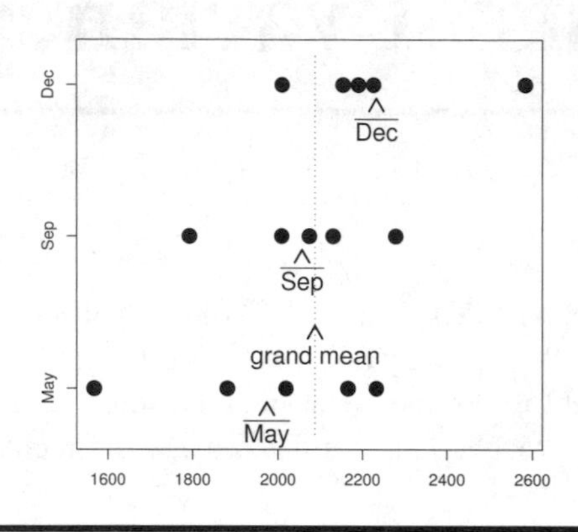

Figure 11.1 Amount of calories consumed by subjects for different months. Sample means are marked, as is the grand mean.

This problem is approached as a significance test. Let the hypotheses be

$$H_0 : \mu_1 = \mu_2 = \cdots = \mu_k, \quad H_A : \mu_i \neq \mu_j \text{ for at least one pair } i \text{ and } j.$$

A test statistic is formulated that compares the variations within a single group to those among the groups.

Let $\bar{x}$ be the grand mean, or mean of all the data, and $\bar{x}_i$ the mean for the ith sample. Then the total sum of squares is given by

$$\mathsf{SST} = \sum_i \sum_j (x_{ij} - \bar{x})^2.$$

This measures the amount of variation from the center of all the data.

An analysis of variance breaks this up into two sums:

$$\sum_i \sum_j (x_{ij} - \bar{x})^2 = \sum_i \sum_j (x_{ij} - \bar{x}_i)^2 + \sum_i n_i (\bar{x}_i - \bar{x})^2. \tag{11.1}$$

The first sum is called the **error sum of squares**, or SSE. The interior sum, $\sum_j (x_{ij} - \bar{x}_i)^2$, measures the variation within the ith group. The SSE is then a measure of the within-group variability. The second term in (11.1) is called the **treatment sum of squares** (SSTr). The word treatment comes from medical experiments where the population mean models the effect of some treatment. The SSTr compares the means for each group, $\bar{x}_i$, with the grand mean, $\bar{x}$. It measures the variability among the means of the samples. We can reexpress Equation 11.1 as

$$\mathsf{SST} = \mathsf{SSE} + \mathsf{SSTr}.$$

From looking at the data in Figure 11.1 we expect that the SSE is smaller than the SSTr, as there appears to be more variation among groups than within groups. If the data came from a common mean, then we would expect SSE and SSTr to be roughly the same. If SSE and SSTr are much different, it would be evidence against the null hypothesis. How can we tell whether the differences are due to the null hypothesis being false or merely to sampling variation? As usual, we tell by finding a test statistic that can discriminate.

Based on our observation, a natural test statistic to test whether $\mu_1 = \mu_2 = \cdots = \mu_k$ would be to compare the two values SSTr and SSE. The F statistic,

$$F = \frac{\mathsf{SSTr}/(k-1)}{\mathsf{SSE}/(n-k)}, \tag{11.2}$$

does so by a ratio. Large values would be consistent with a difference in the means. To get the proper scale, each term is divided by its respective degrees of freedom, yielding the mean sum of squares. The degrees of freedom for the total sum of squares is $n-1$, as only the grand mean is estimated. For the SSE the degrees of freedom are $n-k$, so the degrees of freedom for SSTr is $k-1$.

Under the assumption that the data is normally distributed with common mean and variance, this statistic will have a known distribution: the F-distribution with $k-1$ and $n-k$ degrees of freedom. This is a consequence of the partial F-test discussed in Chapter 10.*

The one-way analysis-of-variance significance test

Suppose we have k independent, *i.i.d.* samples from populations with $\mathsf{Normal}(\mu_i, \sigma)$ distributions, $i = 1, \ldots k$. A significance test of

$$H_0: \mu_1 = \mu_2 = \cdots = \mu_k, \qquad H_A: \mu_i \neq \mu_j \text{ for at least one pair } i \text{ and } j,$$

---

* This can be shown by identifying $\mathsf{RSS}(k)$ with the total sum of squares and $\mathsf{RSS}(p)$ with SSE in (10.11) and simplifying.

can be performed with test statistic

$$F = \frac{\mathsf{SSTr}/(k-1)}{\mathsf{SSE}/(n-k)}.$$

Under H_0, F has the F-distribution with $k-1$ and $n-k$ degrees of freedom. The p-value is calculated from $\mathsf{P}(F \geq \text{observed value} \mid H_0)$.

The R function `oneway.test()` will perform this significance test.

■ **Example 11.2: Number of calories consumed by month, continued** The one-way test can be applied to the example on caloric intake. The two sums can be calculated directly as follows:

```
> may = c(2166, 1568, 2233, 1882, 2019)
> sep = c(2279, 2075, 2131, 2009, 1793)
> dec = c(2226, 2154, 2583, 2010, 2190)
> xbar = mean(c(may,sep,dec))
> SST = 5*((mean(may)-xbar)^2 + (mean(sep)-xbar)^2 + (mean(dec)-xbar)^2)
> SST
[1] 174664
> SSE = (5-1)*var(may) + (5-1)*var(sep) + (5-1)*var(dec)
> SSE
[1] 586720
> F.obs = (SST/(3-1)) / (SSE/(15-3))
> pf(F.obs,3-1,15-3,lower.tail=FALSE)
[1] 0.2094
```

We get a p-value that is not significant. Despite the graphical evidence, the differences can be explained by sampling variation. ■

11.1.1 Using R's model formulas to specify ANOVA models

The calculations for analysis of variance need not be so complicated, as R has functions to compute the values desired. These functions use model formulas. If `x` stores all the data and `f` is a *factor* indicating which group the data value belongs to, then

```
x ~ f
```

represents the statistical model

$$X_{ij} = \mu_i + \varepsilon_{ij}.$$

In Chapter 4 we remarked that the default behavior for `plot()` of the model formula `x ~ f` was to make a boxplot. This is because this graphic easily allows for comparison of centers for multiple samples. The strip chart in Figure 11.1 is good for a small data set, but the boxplot is preferred when there are larger data sets.

11.1.2 Using `oneway.test()` to perform ANOVA

The function `oneway.test()` is used as

```
oneway.test(x ~ f, data=..., var.equal=FALSE)
```

As with the `t.test` function, the argument `var.equal=` is set to `TRUE` if appropriate. By default it is `FALSE`.

Before using `oneway.test()` with our example of caloric intake, we put the data into the appropriate form: a data vector containing the values and a factor indicating the sample the corresponding value is from. This can be done using `stack()`.

```
> d = stack(list(may=may,sep=sep,dec=dec)) # need names for list
> names(d)                         # stack returns two variables
[1] "values" "ind"
> oneway.test(values ~ ind, data=d, var.equal=TRUE)

        One-way analysis of means

data:  values and ind
F = 1.786, num df = 2, denom df = 12, p-value = 0.2094
```

We get the same p-value as in our previous calculation, but with much less effort.

11.1.3 Using `aov()` for ANOVA

The alternative `aov()` function will also perform an analysis of variance. It returns a model object similar to `lm()` but has different-looking outputs for the `print()` and `summary()` extractor functions. These are analysis-of-variance tables that are typical of other computer software and statistics books.

Again, it is called with a model formula, but with no specification of equal variances:

```
> res = aov(values ~ ind, data = d)
> res                              # uses print()
Call:
   aov(formula = values ~ ind, data = d)

Terms:
                   ind Residuals
Sum of Squares  174664    586720
Deg. of Freedom      2        12

Residual standard error: 221.1
Estimated effects may be unbalanced
```

It returns the two sums of squares calculated in Example 11.2 with their degrees of freedom. The `Residual standard error`, $\widehat{\sigma}$, is found by the square root of $\mathsf{RSS}/(n-k)$, which in this example is

```
> sqrt(586720/12)
[1] 221.1
```

The result of `aov()` has more information than shown, just as the result of `lm()` does. For example, the `summary()` function returns

```
> summary(res)
            Df Sum Sq Mean Sq F value Pr(>F)
ind          2 174664   87332    1.79   0.21
Residuals   12 586720   48893
```

These are the values needed to perform the one-way test. This tabular layout is typical of an analysis of variance.

■ **Example 11.3: Effect of grip on cross-country skiing** Researchers at Montana State University performed a study on how various ski-pole grips affect cross-country skiing performance. There are three basic grip types: classic, modern, and integrated. For each of the grip types, a skier has upper-body power output measured three times. The data is summarized in Table 11.1.

Table 11.1 Upper-body power output (watts) by ski-pole grip type

Grip type	classic	integrated	modern
	168.2	166.7	160.1
	161.4	173.0	161.2
	163.2	173.3	166.8

[a] simulated from study values

Does there appear to be a difference in power output due to grip type?

We can investigate the null hypothesis that the three grips will produce equal means with an analysis of variance. We assume that the errors are all independent and that the data is sampled from normally distributed populations with common variance but perhaps different means.

First we enter in the data. Instead of using `stack()`, we enter in all the data at once and create a factor using `rep()` to indicate grip type.

```
> UBP = c(168.2,161.4,163.2,166.7,173.0,173.3,160.1,161.2,166.8)
> grip.type = rep(c("classic","integrated","modern"),c(3,3,3))
> grip.type = factor(grip.type)
> boxplot(UBP ~ grip.type, ylab="Power (watts)",
+ main="Effect of cross country grip")
```

(We use `rep()` repeatedly. In particular, if `u` and `v` are data vectors of the same length, then `rep(u,v)` repeats `u[i]`—the ith value of `u`—`v[i]` times.)

The boxplot in Figure 11.2 indicates that the integrated grip has a significant advantage. But is this due to sampling error? We use `aov()` to carry out the analysis of variance.

```
> res = aov(UBP ~ grip.type)
> summary(res)
            Df Sum Sq Mean Sq F value Pr(>F)
grip.type    2  116.7    58.3    4.46  0.065 .
Residuals    6   78.4    13.1
---
Signif. codes:  0 '***' 0.001 '**' 0.01 '*' 0.05 '.' 0.1 ' ' 1
```

We see that there is a small p-value that is significant at the 10% level. (Although, in most cases, samples with only three observations will fail to pick up on actual differences.) ■

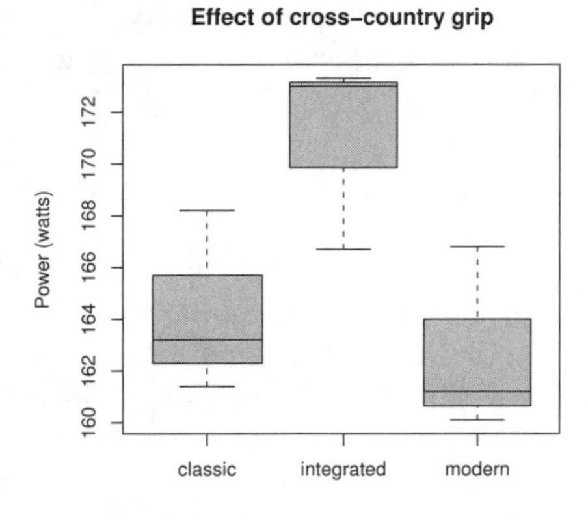

Figure 11.2 Effect of cross-country ski pole grip on measured power output

11.1.4 The nonparametric Kruskal-Wallis test

The Wilcoxon rank-sum test was discussed as a nonparametric alternative to the two-sample t-test for independent samples. Although the populations had no parametric assumption, they were assumed to have densities with a common shape but perhaps different centers.

The Kruskal-Wallis test, a nonparametric test, is analogous to the rank-sum test for comparing the population means of k independent samples.

In particular, if $f(x)$ is a density of a continuous random variable with mean 0, the assumption on the data is that X_{ij} is drawn independently of the others from a population with density $f(x-\mu_i)$. The hypotheses tested are

$$H_0: \mu_1 = \mu_2 = \cdots = \mu_k, \qquad H_A: \mu_i \neq \mu_j \text{ for at least one pair } i \text{ and } j.$$

The test statistic involves the ranks of all the data. Let r_{ij} be the respective rank of a data point when all the data is ranked from smallest to largest, $\bar{r}_i$ be the mean of the ranks for each group, and $\bar{r}$ the grand mean. The test statistic is:

$$T = \frac{12}{n(n+1)} \sum_i n_i(\bar{r}_i - \bar{r})^2. \tag{11.3}$$

Statistical inference is based on the fact that T has an asymptotic χ^2-distribution with $k-1$ degrees of freedom.

Kruskal-Wallis test for equivalence of means

Assume k populations, the ith one with density $f(x-\mu_i)$. Let $X_{ij}, i = 1,\dots,k, j = 1,\dots,n_i$ denote k independent, *i.i.d.* random samples from these populations. A significance test of

$$H_0: \mu_1 = \mu_2 = \cdots = \mu_k, \quad H_A: \mu_i \neq \mu_j \text{ for at least one pair } i \text{ and } j,$$

can be performed with the test statistic T given by (11.3). The asymptotic distribution of T under H_0 is the χ^2-distribution with $k-1$ degrees. This is used as the approximate distribution for T when there are at least five observations in each category. Large values of T support the alternative hypothesis.

The `kruskal.test()` function will perform the test. The syntax is

```
kruskal.test(x ~ f, data=..., subset=...)
```

■ **Example 11.4: Multiple tests** An instructor wishing to cut down on cheating makes three different exams and distributes them randomly to her students. After collecting the exams, she grades them. The instructor would like to know whether the three exams are equally difficult. She will decide this by investigating whether the scores have equal population means. That is, if she could give each exam to the entire class, would the means be similar? The test scores are in Table 11.2. Is there a difference in the means?

We enter the data and then use `stack()` to put it in the proper format:

```
> x = c(63, 64, 95, 64, 60, 85)
> y = c(58, 56, 51, 84, 77)
```

Table 11.2 Test scores for three separate exams

test 1	63	64	95	64	60	85	
test 2	58	56	51	84	77		
test 3	85	79	59	89	80	71	43

```
> z = c(85, 79, 59, 89, 80, 71, 43)
> d = stack(list("test 1"=x,"test 2"=y,"test 3"=z))
> plot(values ~ ind, data=d, xlab = "test", ylab="grade")
```

The boxplots in Figure 11.3 show that the assumption of independent samples from a common population, which perhaps is shifted, is appropriate.

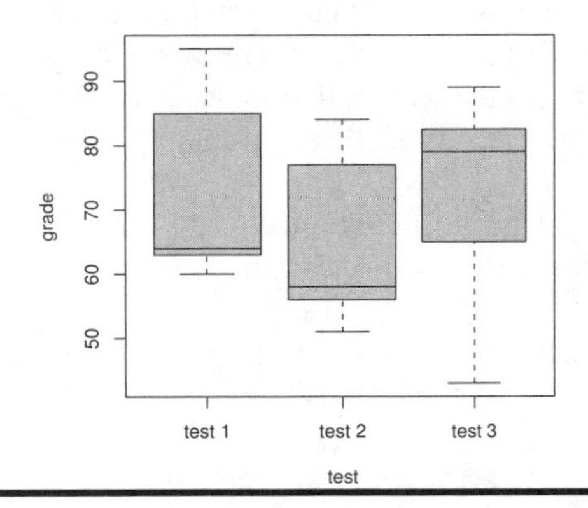

Figure 11.3 Boxplots comparing grades for three separate exams

The Kruskal-Wallis test returns

```
> kruskal.test(values ~ ind, data=d)

        Kruskal-Wallis rank sum test

data:  values by ind
Kruskal-Wallis chi-squared = 1.775, df = 2, p-value =
0.4116
```

This large p-value indicates no reason to doubt the null hypothesis of equally difficult exams. ■

11.1.5 Problems

11.1 The `morley` data set contains speed-of-light measurements by Michaelson and Morley. There were five experiments, each consisting of multiple runs. Perform a one-way analysis of variance to see if each of the five experiments has the same population mean.

11.2 For the data set `Cars93 (MASS)` perform a one-way analysis of variance of `MPG.highway` for each level of `DriveTrain`. Does the data support the null hypothesis of equal population means?

11.3 The data set `female.inc (UsingR)` contains income data for females age 15 or over in the United States for the year 2001, broken down by `race`. Perform a one-way analysis of variance of `income` by `race`. Is there a difference in the mean amount earned? What is the p-value? What test did you use and why?

11.4 The data set `carsafety (UsingR)` contains car-crash data. For several makes of car the number of drivers killed per million is recorded in `Drivers.deaths`. The number of drivers of other cars killed in accidents with these cars, per million, is recorded in `Other.deaths`. The variable `type` is a factor indicating the type of car.

Perform a one-way analysis of variance of the model `Drivers.deaths ~ type`. Is there a difference in population means? Did you assume equal variances? Normally distributed populations?

Repeat with an analysis of variance of the model `Other.deaths ~ type`. Is there a difference in population means?

11.5 The data set `hall.fame (UsingR)` contains statistics for several major league baseball players. Perform a one-way test to see whether the mean batting average, `BA`, is the same for Hall of Fame members (`Hall.Fame.Membership`) as for other players.

Table 11.3 Production of a chemical

Lab 1	4.13	4.07	4.04	4.07	4.05
Lab 2	3.86	3.85	4.08	4.11	4.08
Lab 3	4.00	4.02	4.01	4.01	4.04
Lab 4	3.88	3.89	3.91	3.96	3.92

11.6 A manufacturer needs to outsource the production of a chemical. Before deciding on a laboratory, the manufacturer asks four laboratories to manufacture five batches each. A numeric measurement is assigned to each batch. The data is given in Table 11.3. Perform a one-way analysis of variance to see if there is a difference in the population means. Is the data appropriate for `oneway.test()`? `kruskal.test()`?

11.7 A manufacturer of point-of-sale merchandise tests three types of ENTER-button markings. They wish to minimize wear, as customers get annoyed when the markings on this button wear off. They construct a test of the three types, and conduct several trials for each. The results, in unspecified units, are recorded in Table 11.4. Is there a difference in wear time among the three types? Answer this using a one-way ANOVA.

Table 11.4 Wear times for point-of-sale test

Type 1	303	293	296	299	298	
Type 2	322	326	315	318	320	320
Type 3	309	327	317	315		

11.8 Perform a Kruskal-Wallis test on the data in the data set `PlantGrowth`, where `weight` is modeled by the factor `group`. Is there a significant difference in the means?

11.9 Perform a one-way analysis of variance on the data in Example 11.4. Is there a different conclusion from the example?

11.2 Using `lm()` for ANOVA

The mathematics behind analysis of variance is the same as that behind linear regression. Namely, it uses least-squares estimates based on a linear model. As such, it makes sense to unify the approaches. To do so requires a new idea in the linear model.

To illustrate, we begin with an example comprising just two samples, to see how t-tests are handled with the `lm()` function.

■ **Example 11.5: ANOVA for two independent samples**

Suppose we have two independent samples from normally distributed populations. Let $X_{11}, X_{12}, \ldots, X_{1n}$ record the first and $X_{21}, X_{22}, \ldots, X_{2n}$ the second. Assume the population means are μ_1 and μ_2 and the two samples have a common variance. We may perform a two-sided significance test of $\mu_1 = \mu_2$ with a t-test.

We illustrate with simulated data:

```
> mu1 = 0; mu2 = 1
> x = rnorm(15,mu1); y = rnorm(15,mu2)
> t.test(x,y, var.equal=TRUE)
        Two Sample t-test
data:  x and y
t = -2.858, df = 28, p-value = 0.007961
alternative hypothesis: true difference in means is not equal to 0
95 percent confidence interval:
 -2.0520 -0.3386
sample estimates:
mean of x mean of y
   0.0157    1.211
```

We see that the p-value is small, as expected.

We can approach this test differently, in a manner that generalizes to the case when there are more than two independent samples. Combine the data into a single data vector, Y, and a factor keeping track of which sample, 1 or 2, the data is from. This presumes some ordering on the data after it is stored in Y. For example, we can let the first n_1 values be from the first sample and the second n_2 from the last. This is what `stack()` does. Using this order, let $1_1(i)$ be an indicator function that is 1 if the level of the factor for the ith data value is 1. Similarly, define $1_2(i)$. Then we can rewrite our model as

$$Y_i = \mu_1 1_1(i) + \mu_2 1_2(i) + \varepsilon_i.$$

When the data for the first sample is considered, $1_2(i) = 0$, and this model is simply $Y_i = \mu_1 + \varepsilon_i$. When the second sample is considered, the other dummy variable is 0, and the model considered is $Y_i = \mu_2 + \varepsilon_i$.

We can rewrite the model to use just the second indicator variable. We use different names for the coefficients:

$$Y_i = \beta_1 + \beta_2 1_2(i) + \varepsilon_i.$$

Now when the data for the first sample is considered the model is $Y_i = \beta_1 + \varepsilon_i$, so β_1 is still μ_1. However, when the second sample is considered, we have $Y_i = \beta_1 + \beta_2 + \varepsilon_i$, so $\mu_2 = \beta_1 + \beta_2$. That is, $\beta_2 = \mu_2 - \mu_1$. We say that level 1 is a reference level, as the mean of the second level is represented in reference to the first.

It turns out that statistical inference is a little more natural when we pick one of the means to serve as a reference. The resulting model looks just like a linear-regression model where x_i is $1_2(i)$. We can fit it that way and interpret the coefficients accordingly. The model is specified the same way, as with `oneway.test()`, `y ~ f`, where `y` holds the data and `f` is a factor indicating which group the data is for.

To model, first we stack, then we fit with `lm()`.

```
> d = stack(list(x=x,y=y))       # need named list.
> d
   values ind
1  -0.5263   x
2  -0.9709   x
...
> res = lm(values ~ ind, data = d)
> summary(res)
...
Coefficients:
            Estimate Std. Error t value Pr(>|t|)
(Intercept)    0.157      0.261    0.60    0.553
indy           1.054      0.369    2.86    0.008 **
---
Signif. codes:  0 '***' 0.001 '**' 0.01 '*' 0.05 '.' 0.1 ' ' 1

Residual standard error: 1.01 on 28 degrees of freedom
Multiple R-Squared: 0.226,      Adjusted R-squared: 0.198
F-statistic: 8.17 on 1 and 28 DF,  p-value: 0.00796
```

Look at the variable `indy`, which means the `y` part of `ind`. The marginal t-test tests the null hypothesis that $\beta_2 = 0$, which is equivalent to the test that $\mu_1 = \mu_2$. This is why the t-value of 2.86 coincides (up to a sign and rounding) with $t = -2.858$ from the output of `t.test(x,y)`.

The F-statistic also tests the hypothesis that $\beta_2 = 0$. In this example, it is identical to the marginal t-test, as there are only two samples.

Alternatively, we can try to fit the model using two indicator functions, $Y_i = \mu_1 1_1(i) + \mu_2 1_2(i) + \varepsilon_i$.

This model is specified in R by dropping the implicit intercept term with a - 1 in the model formula.

```
> res = lm(values ~ ind - 1, data = d)
> summary(res)
...
Coefficients:
     Estimate Std. Error t value Pr(>|t|)
indx    0.157      0.261    0.60     0.55
indy    1.211      0.261    4.64  7.4e-05 ***
---
Signif. codes:  0 '***' 0.001 '**' 0.01 '*' 0.05 '.' 0.1 ' ' 1

Residual standard error: 1.01 on 28 degrees of freedom
Multiple R-Squared: 0.439,      Adjusted R-squared: 0.399
F-statistic:   11 on 2 and 28 DF,  p-value: 0.000306
```

Now the estimates have a clear interpretation in terms of the means, but the marginal t-tests are less useful, as they are testing simply whether the respective means are 0, rather than whether their difference is 0. The F-statistic in this case is testing whether both β's are 0. ■

11.2.1 Treatment coding for analysis of variance

The point of the above example is to use indicator variables to represent different levels of a factor in the linear model. When there are k levels, $k-1$ indicator variables are used. For example, if the model is

$$X_{ij} = \mu_i + \varepsilon_{ij}, \quad i = 1, \ldots, k, \tag{11.4}$$

then this can be fit using

$$Y_i = \beta_1 + \beta_2 1_2(i) + \cdots + \beta_k 1_k(i) + \varepsilon_i. \tag{11.5}$$

The mean of the reference level, μ_1, is coded by β_1, and the other β's are differences from that. That is, $\beta_i = \mu_i - \mu_1$ for $i = 2, \ldots, k$.

This method of coding is called **treatment coding** and is used by default in R with unordered factors. It is not the only type of coding, but it is the only one we will discuss.[†]

Treatment coding uses a reference level to make comparisons. This is chosen to be the first level of the factor coding the group. To change the reference level we can use the `relevel()` function in the following manner:

```
f = relevel(f, ref=...)
```

The argument `ref=` specifies the level we wish to be the reference level.

■ **Example 11.6: Child's birth weight and mother's smoking history**

The `babies (UsingR)` data set contains information on birth weight of a child and whether the mother smoked. The birth weight, `wt`, is coded in ounces, and `smoke` is a numeric value: 0 for never, 1 for smokes now, 2 for smoked until current pregnancy, 3 for smoked previously but not now, and 9 if unknown.

To do an analysis of variance on this data set, we use `subset()` to grab just the desired data and then work as before, only we use `factor()` to ensure that smoking is treated as a factor. First, we see whether there appears to be a difference in the means with a boxplot (Figure 11.4).

```
> library(UsingR)
> df = subset(babies,select=c("wt","smoke"))
> plot(wt ~ factor(smoke), data=df,     # notice factor() for boxplot
+ main="Birthweight by smoking level")
```

[†] For more detail see `?contrasts` and the section on contrasts in the manual *An Introduction to R* that accompanies R.

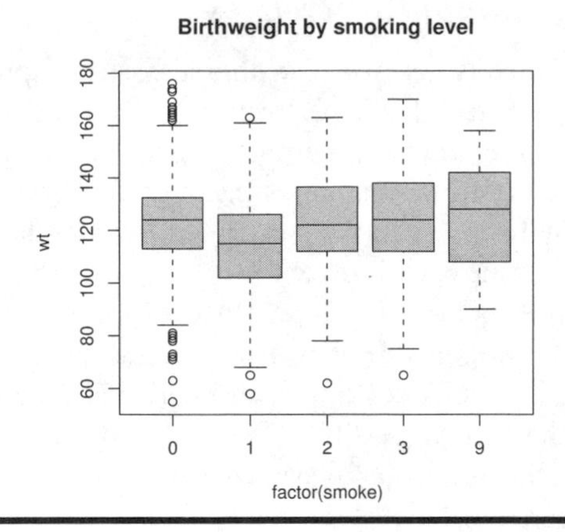

Figure 11.4 Birth weight by smoking history

Perhaps the assumption of normality isn't correct, but we ignore that. If the test is valid, it looks like level 1 (smokes now) has a smaller mean. Is this due to sampling? We fit the model as follows:

```
> res = lm(wt ~ factor(smoke), data=df)
> summary(res)
...
Coefficients:
                Estimate Std. Error t value Pr(>|t|)
(Intercept)      122.778      0.760  161.60  < 2e-16 ***
factor(smoke)1    -8.668      1.107   -7.83  1.1e-14 ***
factor(smoke)2     0.307      1.970    0.16     0.88
factor(smoke)3     1.659      1.904    0.87     0.38
factor(smoke)9     3.922      5.655    0.69     0.49
---
Signif. codes:  0 '***' 0.001 '**' 0.01 '*' 0.05 '.' 0.1 ' ' 1

Residual standard error: 17.7 on 1231 degrees of freedom
Multiple R-Squared: 0.0588,     Adjusted R-squared: 0.0557
F-statistic: 19.2 on 4 and 1231 DF,  p-value: 2.36e-15
```

The marginal t-tests indicate that the level 1 of the `smoke` factor is important, whereas the others may not contribute. That is, this is strong evidence that a mother's smoking during pregnancy decreases a baby's birth weight. The treatment coding quantifies this in terms of differences from the reference level of never smoked. The estimate, -8.668, says that the birth weight of a baby whose mother smoked during her pregnancy is predicted to be 8.688 grams less than that of a baby whose mother never smoked. ■

11.2.2 Comparing multiple differences

When analysis of variance is performed with `lm()`, the output contains numerous statistical tests. The F-test that is performed uses for the null hypothesis that $\beta_2 = \beta_3 = \cdots = \beta_k = 0$ against an alternative that one or more differ from 0. That is, that one or more of the treatments has an effect compared to the reference level. The marginal t-tests that are performed are two-sided tests with a null hypothesis that $\beta_i = \beta_1$. One each is done for $i = 2, \ldots, k$. These test whether any of the additional treatments have a different effect from the reference one when controlled by the other variables. However, we may wish to ask other questions about the various parameters. For example, comparisons not covered by the standard output are "Do the β_2 and β_3 differ?" and "Are β_1 and β_2 half of β_3?" We show next how to handle simultaneous pairwise comparisons of the parameters, such as the first comparison.

If we know ahead of time that we are looking for a pairwise difference, then a simple t-test is appropriate (as in the case where we are considering just two independent samples). However, if we look at the data and then decide to test whether the second and third parameters differ, then our t-test is shaky. Why? Remember that any test is correct only with some probability—even if the models are correct. This means that sometimes they fail, and the more tests we perform, the more likely one or more will fail. When we look at the data, we are essentially performing lots of tests, so there is more chance of failing.

In this case, to be certain that our t-test has the correct significance level, we adjust it to include all the tests we can possibly consider. This adjustment can be done by hand with the simple, yet often overly conservative Bonferroni adjustment. This method uses a simple probability bound to ensure the proper significance level.

However, with `R` it is straightforward to perform Tukey's generally more useful and powerful "honest significant difference" test. This test covers all pairwise comparisons at one time by simultaneously constructing confidence intervals of the type

$$(\bar{y}_i - \bar{y}_j) \pm q^* \sqrt{\frac{1}{2} s^2 \left(\frac{1}{n_i} + \frac{1}{n_j} \right)}. \tag{11.6}$$

The values $\bar{y}_i$ are the sample means for the i-th level and q^* is the quantile for a distribution known as the studentized range distribution. This choice of q^* means that all these confidence intervals hold simultaneously with probability $1 - \alpha$.

This procedure is implemented in the `TukeyHSD()` function as illustrated in the next example.

■ **Example 11.7: Difference in takeoff times at the airport**

We investigate the takeoff times for various airlines at Newark Liberty airport. As with other busy airports, Newark's is characterized by long delays on the runway due to requirements that plane departures be staggered. Does this affect all the airlines equally? Without suspecting that any one airline is favored, we can perform a simultaneous pair-wise comparison to investigate.

First, we massage the data in `ewr (UsingR)` so that we have two variables: one to keep track of the time and the other a factor indicating the airline.

```
> ewr.out = subset(ewr, subset=inorout=="out", select=3:10)
> out = stack(ewr.out)
> names(out) = c("time","airline")
> levels(out$airline)
[1] "AA" "CO" "DL" "HP" "NW" "TW" "UA" "US"
```

In modeling, the reference level comes from the first level reported by the `levels()` function. This is `AA`, or American Airlines.

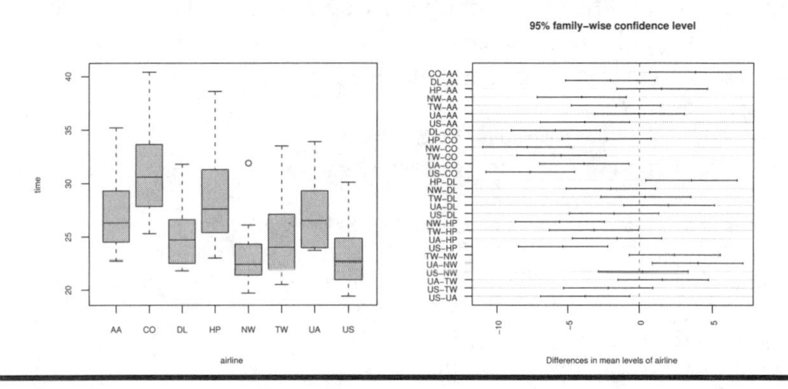

Figure 11.5 Boxplots and plots of confidence intervals given by the Tukey procedure for time it takes to takeoff at Newark Liberty airport by airline

Now plot (the boxplots in Figure 11.5) and fit the linear model as follows:

```
> plot(time ~ airline, data=out)
> res = lm(time ~ airline, data=out)
> summary(res)

Call:
lm(formula = time ~ airline, data = out)
...
Coefficients:
            Estimate Std. Error t value Pr(>|t|)
(Intercept)  27.0565     0.7204   37.56  < 2e-16 ***
airlineCO     3.8348     1.0188    3.76  0.00023 ***
airlineDL    -2.0522     1.0188   -2.01  0.04550 *
airlineHP     1.5261     1.0188    1.50  0.13595
airlineNW    -4.0609     1.0188   -3.99  9.8e-05 ***
airlineTW    -1.6522     1.0188   -1.62  0.10667
```

```
airlineUA    -0.0391    1.0188  -0.04 0.96941
airlineUS    -3.8304    1.0188  -3.76 0.00023 ***
---
Signif. codes:  0 '***' 0.001 '**' 0.01 '*' 0.05 '.' 0.1 ' ' 1

Residual standard error: 3.45 on 176 degrees of freedom
Multiple R-Squared: 0.355,     Adjusted R-squared: 0.329
F-statistic: 13.8 on 7 and 176 DF,  p-value: 3.27e-14
```

The boxplots show many differences. Are they statistically significant? We assume for now that the data is actually a collection of independent samples (rather than monthly averages of varying sizes) and proceed using the TukeyHSD() function.

```
> TukeyHSD(res)
Error in TukeyHSD(res) : no applicable method for "TukeyHSD"
```

Oops, the TukeyHSD() function wants aov() to fit the linear model, not lm(). The commands are the same.

```
> res.aov = aov(time ~ airline, data=out)
> TukeyHSD(res.aov)
  Tukey multiple comparisons of means
    95% family-wise confidence level

Fit: aov(formula = time ~ airline, data = out)

$airline
          diff      lwr      upr
CO-AA  3.83478   0.7093  6.96025
DL-AA -2.05217  -5.1776  1.07330
...
US-TW -2.17826  -5.3037  0.94721
US-UA -3.79130  -6.9168 -0.66583

> plot(TukeyHSD(res.aov), las=2)
```

The output of TukeyHSD() is best viewed with the plot of the confidence intervals (Figure 11.5). This is created by calling plot() on the output. The argument las=2 turns the tick-mark labels perpendicular to the axes.

Recall the duality between confidence intervals and tests of hypothesis discussed in Chapter 8. For a given confidence level and sample, if the confidence interval excludes a population parameter, then the two-sided significance test of the same parameter will be rejected. Applying this to the Newark airport example, we see several statistically significant differences at the $\alpha = .05$ level, the first few being CO-AA and NW-AA (just visible on the graph shown). ■

11.2.3 Problems

11.10 The data set MLBAttend (UsingR) contains attendance data for major league baseball between the years 1969 and 2000. Use `lm()` to perform a t-test on `attendance` for the two levels of `league`. Is the difference in mean attendance significant? Compare your results to those provided by `t.test()`.

11.11 The `Traffic (MASS)` data set contains data on road deaths in Sweden during 1961 and 1962. An investigation into the effect of an enforced speed limit on the number of traffic fatalities was conducted. The `y` variable contains the number of deaths for a given day, the `year` variable is the year of the data, and `limit` is a factor indicating when the speed limit was enforced.

Use `lm()` to perform a t-test to investigate whether the `year` has an effect on the number of deaths. Repeat to test whether the variable `limit` has an effect.

11.12 For the data in Table 11.4, perform the one-way ANOVA using `lm()`. Compare to the results of `oneway.test()`.

11.13 For the `mtcars` data set, perform a one-way analysis of variance of the response variable `mpg` modeled by `cyl`, the number of cylinders. Use `factor()`, as `cyl` is stored as a numeric variable.

11.14 The data set `npdb (UsingR)` contains malpractice award information. The variable `amount` contains the amount of a settlement, and the variable `year` contains the year of the award. We wish to investigate whether the dollar amount awarded was steady during the years 2000, 2001, and 2002.

1. Make boxplots of `amount` broken up by `year`. Why is the data not suitable for a one-way analysis of variance?
2. Make boxplots of `log(amount)` broken up by `year`. Is this data suitable for a one-way analysis of variance?
3. Perform an analysis of variance of `log(amount)` by `factor(year)` for the years 2000, 2001, and 2002. Is the null hypothesis of no difference in mean award amount reasonable given this data?

11.15 For the `mtcars` data set, perform a one-way analysis of variance of the response variable `mpg` modeled by `am`, which is 0 for automatic and 1 for manual. Use `factor()`, as `am` is stored as a numeric variable.

11.16 Perform the Tukey procedure on the data set `morley` after modeling `Speed` by `expt`. Which differences are significant? Do they include all the ones flagged by the marginal t-tests returned by `lm()` on the same model?

11.17 The `carsafety (UsingR)` data set shows a difference in means through an analysis of variance when the variable `Other.deaths` is modeled by `type`. Perform the Tukey HSD method to see what pairwise differences are flagged at a 95% confidence level. What do you conclude?

11.18 The `InsectSprays` data set contains a variable `count`, which counts the number of insects and a factor `spray`, which indicates the treatment given.

First perform an analysis of variance to see whether the treatments make a difference. If so, perform the Tukey HSD procedure to see which pairwise treatments differ.

11.3 ANCOVA

An analysis of covariance (ANCOVA) is the term given to models where both categorical and numeric variables are used as predictors. Performing an ANCOVA in R is also done using `lm()`.

■ **Example 11.8: Birth weight by mother's weight and smoking history**

In Example 11.6 we performed an analysis of variance of a baby's birth weight modeled by whether the mother smoked. In this example, we also regress on the numeric measurement of the mother's weight. First we make a plot, marking the points with different characters depending on the value of `smoke`. As `smoke` is stored as a numeric variable, the different plot symbols for those numbers are used.

```
> plot(wt ~ wt1, data = babies, pch=smoke, subset = wt1 < 800)
```

The graph in Figure 11.6 indicates a possible linear relationship. The analysis of covariance model, fit next, is essentially the model

$$\text{birth weight} = \beta_1 + \beta_2 \text{mom's weight} + \beta_3 1_{\text{mom smokes now}}$$

This model is a parallel-lines model. For those mothers who don't smoke, the intercept is given by β_1; for those who do, the intercept is $\beta_1 + \beta_3$. The slope is given by β_2. The actual model we fit is different, as there are four levels to the `smoke` variable, so there would be three indicator variables, each indicating a difference in the intercept.

In R, we fit the model as follows, using `factor()` to coerce `smoke` to be a factor:

```
> res = lm(wt ~ wt1 + factor(smoke), data = babies,
+ subset = wt1 < 800)
> summary(res)

Call:
lm(formula = wt ~ wt1 + factor(smoke), data = babies, subset = wt1 <
    800)

Residuals:
    Min       1Q  Median       3Q      Max
-68.928 -10.901   0.437  11.014  52.685
```

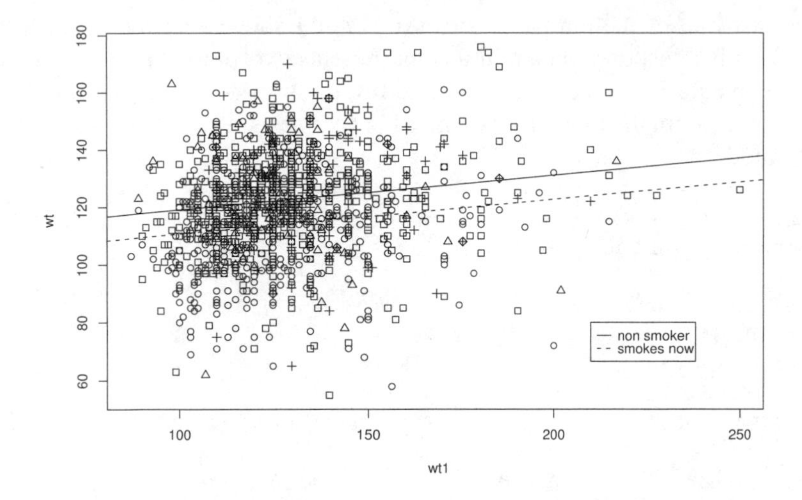

Figure 11.6 Parallel-lines model showing reference slope and slope for smokers (dashed line)

```
Coefficients:
                 Estimate Std. Error t value Pr(>|t|)
(Intercept)      107.0674     3.2642   32.80  < 2e-16 ***
wt1                0.1204     0.0245    4.93  9.6e-07 ***
factor(smoke)1  -8.3971     1.1246   -7.47  1.6e-13 ***
factor(smoke)2   0.7944     1.9974    0.40     0.69
factor(smoke)3   1.2550     1.9112    0.66     0.51
factor(smoke)9   2.8683     5.6452    0.51     0.61
---
Signif. codes:  0 ‘***’ 0.001 ‘**’ 0.01 ‘*’ 0.05 ‘.’ 0.1 ‘ ’ 1

Residual standard error: 17.7 on 1194 degrees of freedom
Multiple R-Squared: 0.0775,     Adjusted R-squared: 0.0736
F-statistic: 20.1 on 5 and 1194 DF,  p-value: <2e-16
```

We read this output the same way we read the output of any linear regression. For each coefficient, the marginal t-test of $\beta_i = 0$ against a two-sided alternative is performed. Three variables are flagged as highly significant. The third one for the variable `factor(smoke)1` says that the value of this coefficient, -8.3971, is statistically different from 0. This value is an estimate of the difference between the intercept for the data of nonsmoking mothers (level 0) and the data of mothers who answered “smokes now” (level 1).

We plot the data with two different but parallel regression lines in Figure 11.6.

```
> plot(wt ~ wt1, pch=smoke, data=babies, subset = wt1 < 800)
> abline(107.0674, 0.1204)
```

```
> abline(107.0674 - 8.3971, 0.1204, lty=2)
```

The last line of the output of `summary(res)` shows that the F-test is rejected. This is a test of whether all the coefficients except the intercept are 0. A better test would be to see whether the additional `smoke` variable is significant once we control for the mother's weight. This is done using `anova()` to compare the two models.

```
> res.1 = lm(wt ~ wt1, data = babies, subset = wt1 < 800)
> anova(res.1,res)
Analysis of Variance Table

Model 1: wt ~ wt1
Model 2: wt ~ wt1 + factor(smoke)
  Res.Df    RSS   Df Sum of Sq     F Pr(>F)
1   1198 394572
2   1194 372847    4     21725 17.4  7e-14 ***
---
Signif. codes:  0 '***' 0.001 '**' 0.01 '*' 0.05 '.' 0.1 ' ' 1
```

The small p-value indicates that the additional term is warranted. ■

11.3.1 Problems

11.19 The `nym.2002 (UsingR)` data set contains data on the finishers of the 2002 New York City Marathon. Do an ANCOVA of `time` on the numeric variable `age` and the factor `gender`. How much difference is there between the genders?

11.20 For the `mtcars` data set, perform an ANCOVA of `mpg` on the weight, `wt`, and the transmission type, `am`. You should use `factor(am)` in your model to ensure that this variable is treated as a factor. Is the transmission type significant?

11.21 Perform an ANCOVA for the `babies (UsingR)` data set modeling birth weight (`wt`) by gestation (`gestation`), mother's weight (`wt1`), mother's height (`ht`), and mother's smoking status (`smoke`).

11.22 From the `kid.weights (UsingR)` data set, the body mass index (BMI) can be computed by dividing the weight by the height squared in metric units. The following will add a `BMI` variable:

```
> kid.weights$BMI = (kid.weights$weight/2.54)/
+   (kid.weights$height*2.54/100)^2
```

Model the BMI by the `age` and `gender` variables. This is a parallel-lines model. Which variables are significant? Use the partial F-test to find the preferred model. Does this agree with the output of `stepAIC()`?

11.23 The `cfb (UsingR)` data set contains information on consumer expenses. In particular, `INCOME` contains income figures, `EDUC` is the number of years of education, and `AGE` is the age of the participant. Perform an ANCOVA modeling `log(INCOME + 1)` by `AGE` and `EDUC`. You need to force `EDUC` to be a factor. Are both variables significant?

11.24 The data set `normtemp (UsingR)` contains body `temperature` and heart rate (`hr`) for 65 randomly chosen males and 65 randomly chosen females (marked by `gender` with 1 for males and 2 for females). Perform an ANCOVA modeling `temperature` by heart rate with gender treated as a factor.

11.4 Two-way ANOVA

"Two-way analysis of variance" is the term given when a numeric response variable is modeled by two categorical predictors. After we fit the model into the regression framework, the t-tests and partial F test will be available for analysis.

Let Y be the response variable and x_1 and x_2 be two categorical predictors, with n_1 and n_2 levels respectively. The simplest generalization of the one-way ANOVA model (11.4) is the two-way additive model:

$$Y_{ijk} = \mu + \alpha_i + \delta_j + \varepsilon_{ijk}. \tag{11.7}$$

The grand mean is μ, α_i the mean for the ith level of x_1, δ_j is the mean for the ith level of x_2, and the error terms, ε_{ijk}, are an *i.i.d.* sequence with a $\mathsf{Normal}(0, \sigma)$ distribution.

Two common significance tests investigate whether the different levels of x_1 and x_2 have an effect on the mean of Y. For the first variable, x_1, the hypotheses are

$$H_0: \alpha_1 = \alpha_2 = \cdots = \alpha_{n_1} = \alpha \qquad H_A: \alpha_i \neq \alpha_j \text{ for at least one pair } i \text{ and } j$$

The equivalent one for x_2 replaces the α's above with δ's.

■ **Example 11.9: Driver differences in evaluating gas mileage** An automotive web site wishes to test the miles-per-gallon rating of a car. It has three drivers and two cars of the same type. Each driver is asked to drive each car three times and record the miles per gallon. Table 11.5 records the data. Ideally, there should be little variation. But is this the case with the data? ■

Table 11.5 Does the driver or car make a difference in mileage?

	Driver					Driver	
Car	*a*	*b*	*c*	Car	*a*	*b*	*c*
A	33.3	34.5	37.4	*B*	32.6	33.4	36.6
	33.4	34.8	36.8		32.5	33.7	37.0
	32.9	33.8	37.6		33.0	33.9	36.7

11.4.1 Treatment coding for additive two-way ANOVA

Before analyzing this model, we incorporate it into our linear-model picture using dummy variables. We follow the same coding (treatment coding) in terms of indicators as the one-way case. Relabel the observations 1 through 18. Let $1_b^{\text{driver}}(i)$ be the indicator that the observation is for driver b, (similarly $1_c^{\text{driver}}(i)$) and $1_B^{\text{car}}(i)$ the indicator that the car is B. Then the additive model becomes

$$Y_i = \beta_1 + \beta_2 1_b^{\text{driver}} + \beta_3 1_c^{\text{driver}} + \beta_4 1_B^{\text{car}} + \varepsilon_i.$$

Again, the ε_i are *i.i.d.* $\mathsf{Normal}(0, \sigma)$.

Recall that with treatment coding we interpret the parameters in terms of differences. For this model, $\beta_1 = \mu + \alpha_A + \delta_a$, or the sum of the grand mean, the mean of the first level of the first variable, and the mean of the first level of the second variable. As $\beta_1 + \beta_2$ is the mean for car A, driver b, this would be $\mu + \alpha_A + \delta_b$ or $\beta_2 = \delta_b - \delta_a$. Similarly, the β_3 and β_4 can be interpreted in terms of differences, as $\beta_3 = \delta_c - \delta_a$ and $\beta_4 = \alpha_B - \alpha_A$.

11.4.2 Testing for row or column effects

To perform the significance test that the row variable has constant mean we can use the partial F-test. In our example, this is the same as saying $\beta_4 = 0$. The partial F-test fits the model with and without β_4 and uses the ratio of the residual sum of squares to make a test statistic. The details are implemented in the `anova()` function.

First we enter the data:

```
> x = c(33.3, 33.4, 32.9, 32.6, 32.5, 33.0, 34.5, 34.8, 33.8,
+ 33.4, 33.7, 33.9, 37.4, 36.9, 37.6, 36.6, 37.0, 36.7)
> car = factor(rep(rep(1:2,c(3,3)) , 3))
> levels(car) = c("A","B")
> driver = factor(rep(1:3,c(6,6,6)))
> levels(driver) = letters[1:3] # make letters not numbers
```

The additive model is fit with

```
> res.add = lm(x ~ car + driver)
```

We want to compare this to the model when $\beta_4 = 0$.

```
> res.nocar = lm(x ~ driver)
```

We compare nested models with `anova()`:

```
> anova(res.add,res.nocar)
Analysis of Variance Table

Model 1: x ~ car + driver
Model 2: x ~ driver
  Res.Df   RSS Df Sum of Sq  F Pr(>F)
1     14  1.31
2     15  2.82 -1      -1.50 16 0.0013 **
---
Signif. codes:  0 '***' 0.001 '**' 0.01 '*' 0.05 '.' 0.1 ' ' 1
```

We see that the difference is significant, leading us to rule out the simpler model.

What about the effect of the car? The two cars should have been identical. Is there a difference? The null hypothesis is now $H_0 : \delta_a = \delta_b = \delta_c$, which can be rewritten as $\beta_2 = \beta_3 = 0$. As such, we fit the model without the β_2 and β_3 terms and compare to the full model as above.

```
> res.nodriver = lm(x ~ car)
> anova(res.add,res.nodriver)
Analysis of Variance Table

Model 1: x ~ car + driver
Model 2: x ~ car
  Res.Df   RSS Df Sum of Sq   F  Pr(>F)
1     14   1.3
2     16  55.1 -2     -53.8 287 4.4e-12 ***
---
Signif. codes:  0 '***' 0.001 '**' 0.01 '*' 0.05 '.' 0.1 ' ' 1
```

This too is flagged as significant.

11.4.3 Testing for interactions

The extra factor in two-way ANOVA introduces another possibility: interaction. For example, as there seems to be a difference in the two cars, perhaps one is sportier, which makes one of the drivers drive faster. That is, there is an interaction when the two factors combine. A statistical model for interactions in the two-way analysis of variance model is

$$Y_{ijk} = \mu + \alpha_i + \delta_j + \gamma_{ij} + \varepsilon_{ijk}, \quad 1 \leq i \leq n_1, 1 \leq j \leq n_2. \tag{11.8}$$

The γ_{ij} terms add to the grand mean and group means when both levels are present.

We again rewrite this in terms of dummy variables. We get extra variables corresponding to all possible combinations of the two factors:

$$Y_i = \beta_1 + \beta_2 1_b^{\text{driver}} + \beta_3 1_c^{\text{driver}} + \beta_4 1_B^{\text{car}} + \beta_5 1_b^{\text{driver}} 1_B^{\text{car}} + \beta_6 1_c^{\text{driver}} 1_B^{\text{car}} + \varepsilon_i. \quad (11.9)$$

Although (11.8) has $1 + n_1 + n_2 + n_1 \cdot n_2$ parameters, this is more than can be identified. Instead, (11.9) has only $n_1 \cdot n_2 = 1 + (n_1 - 1) + (n_2 - 1) + (n_1 - 1)(n_2 - 1)$ parameters needed for the modeling.

A significance test to see if the extra terms from the interaction are necessary can be done with the partial F-test. Before doing so, we introduce a diagnostic plot to see if the extra terms are warranted.

Interaction plots

An interaction plot is a plot that checks to see whether there is any indication of interactions. For two-way analysis of variance there are three variables. To squeeze all three onto one graphic, one of the factors is selected as the trace factor. Different lines will be drawn for each level of this factor. Fix a level, for now, of the trace factor. For each level of the main factor, the mean of the data where both levels occur is plotted as a point. These points are then connected with a line segment. Repeat for the other levels of the trace factor. If the line segments for each level of the trace factor are roughly parallel, then no interaction is indicated. If the lines differ dramatically, then an interaction is indicated.

This graphic is made with the function `interaction.plot()`. The template is

```
interaction.plot(f, trace.factor, y, legend=TRUE)
```

The response variable is stored in `y`, the `f` holds the main factor, and the other is in `trace.factor`. By default, a legend will be drawn indicating the levels of the trace factor.

For our example, Figure 11.7 is made with the following commands. The line segments are nearly parallel, indicating that no interaction is present.

```
> interaction.plot(driver,car,x)
```

Significance test for presence of interactions

To test the hypothesis of no interaction formally we can use the partial F-test. The null hypothesis can be expressed as $\gamma_{ij} = 0$ in (11.8) or, for our car-and-driver example, as $\beta_5 = \beta_6 = 0$ from Equation (11.9). For our car-and-driver example, this is done by comparing the models with and without interaction.

Specifying an interaction in a model formula An interaction can be specified in different ways in the model formula. The symbol `:`, used as `f1:f2`, will introduce the interaction terms for the two factors. Whereas `*`, as in `f1*f2`, will

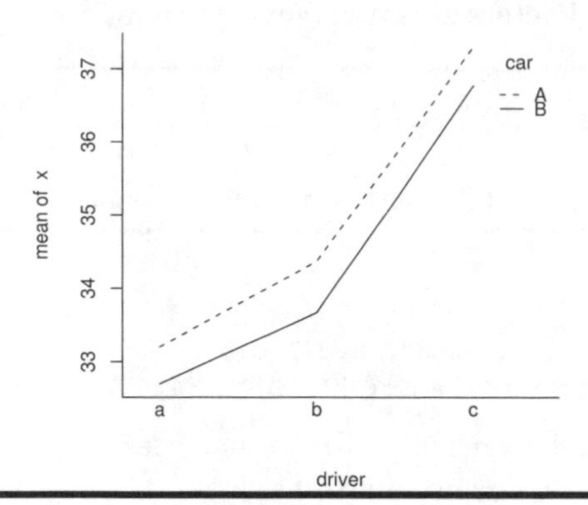

Figure 11.7 Interaction plot for car-and-driver data. The lines are nearly parallel, indicating no interaction.

introduce not only an interaction, but the main effects, `f1 + f2`, as well. Finally, the power notation, `^`, as in `(f1+f2)^2`, will do the main effects and all possible interactions up to order 2. This generalizes with higher powers and more terms. For our example with two factors, all three of these model formulas are equivalent: `f1 + f2 + f1:f2`, `f1*f2`, and `(f1+f2)^2`.

To proceed, we save the model with an interaction and then use `anova()` to compare nested models.

```
> lm.int = lm(x ~ car * driver)
> lm.add = lm(x ~ car + driver)
> anova(lm.add,lm.int)
Analysis of Variance Table

Model 1: x ~ car + driver
Model 2: x ~ car * driver
  Res.Df    RSS Df Sum of Sq    F Pr(>F)
1     14 1.314
2     12 1.280  2     0.034 0.16   0.85
```

The large p-value is consistent with Figure 11.7, indicating no interaction.

■ **Example 11.10: Factors in movie enjoyment** The proprietors of a movie house want to maximize their customers' movie-going experience. In particular, they want to know whether either eating popcorn or sitting in more comfortable seats makes a difference in customer enjoyment. They randomly assign 16 people equally to the four possible combinations and then ask them to rate the same movie on a 0-100 scale. The data is in Table 11.6.

The data is entered in with

```
> x = scan()
1: 92 80 80 78 63 65 65 69 60 59 57 51 60 58 52 65
17:
Read 16 items
```

Table 11.6 Factors affecting movie enjoyment

seat type		good				bad			
popcorn	yes	92	80	80	78	60	59	57	51
	no	63	65	65	69	60	58	52	65

```
> Seat = factor(rep(c("Good","Bad"),c(8,8)))
> Popcorn = factor(rep( rep(c("Y","N"),c(4,4)), 2))
```

We can check our numbers using `xtabs()` and `ftable()`. First we add a variable to keep the data from being summed.‡

```
> replicate = rep(1:4,4)
> ftable(xtabs(x ~ Popcorn + Seat + replicate))
             replicate  1  2  3  4
Popcorn Seat
N       Bad            60 58 52 65
        Good           63 65 65 69
Y       Bad            60 59 57 51
        Good           92 80 80 78
```

It matches up, although we didn't fuss with the order.

Now to see if an interaction term is warranted:

```
> interaction.plot(Seat,Popcorn,x)
```

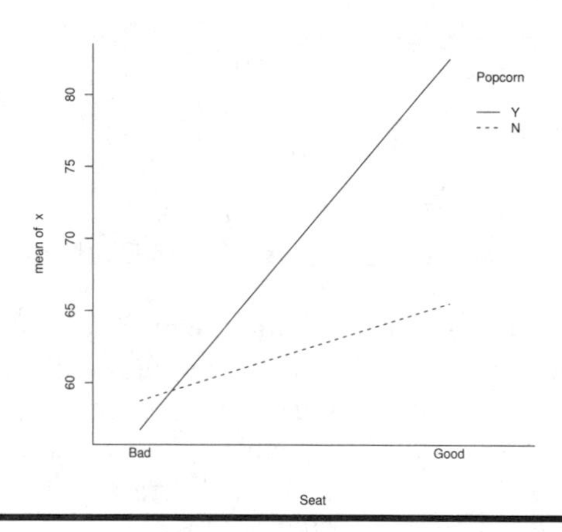

Figure 11.8 Interaction plot indicating presence of an interaction, as lines are not parallel

‡ See `?xtabs` for a similar example.

Figure 11.8 seems to show an interaction, as the slopes are not parallel. We can do a formal test with `anova()`.

```
> res.int = lm(x ~ Seat * Popcorn)
> res.add = lm(x ~ Seat + Popcorn)
> anova(res.int,res.add)
Analysis of Variance Table

Model 1: x ~ Seat * Popcorn
Model 2: x ~ Seat + Popcorn
  Res.Df  RSS Df Sum of Sq    F Pr(>F)
1     12  277
2     13  638 -1      -361 15.6 0.0019 **
---
```

The small p-value casts doubt on the null hypothesis model of no interaction. The `summary()` function gives more detailed information about the interaction model.

```
> summary(res.int)

Call:
lm(formula = x ~ Seat * Popcorn)
...
Coefficients:
                 Estimate Std. Error t value Pr(>|t|)
(Intercept)         58.75       2.40   24.43  1.3e-11 ***
SeatGood             6.75       3.40    1.99   0.0705 .
PopcornY            -2.00       3.40   -0.59   0.5673
SeatGood:PopcornY   19.00       4.81    3.95   0.0019 **
---
Signif. codes:  0 '***' 0.001 '**' 0.01 '*' 0.05 '.' 0.1 ' ' 1

Residual standard error: 4.81 on 12 degrees of freedom
Multiple R-Squared: 0.855,      Adjusted R-squared: 0.819
F-statistic: 23.7 on 3 and 12 DF,  p-value: 2.50e-05
```

It appears that a good seat and popcorn can go a long way toward a moviegoer's satisfaction (at least from this fabricated data). Perhaps new seats and less expensive popcorn will keep the customers coming back. ■

11.4.4 Problems

11.25 A politician's campaign manager is interested in the effects of television and internet advertising. She surveys 18 people and records changes in likability after a small advertising campaign. Additionally, she records the amount of exposure her subjects have to the ad campaigns. The data is in Table 11.7.

Use an analysis of variance to investigate the following questions:

1. Is there any indication that web advertising alone is effective?

2. After controlling for television exposure, is there any indication that web advertising is effective?

Table 11.7 Change in likability of politician

TV ad exposure (viewings)		0			1-2			3+		
Web exposure	N	−1	−4	0	−1	4	1	6	2	7
	Y	1	2	2	7	5	2	3	6	1

11.26 The `grip (UsingR)` data set contains more data than is used in Example 11.3. The data is from four skiers instead of one. You can view the data in a convenient manner with the command

```
> ftable(xtabs(UBP ~ person + replicate + grip.type, data=grip))
```

Perform a two-way analysis of variance on the data. Check first to see whether there are any interactions, then see whether the difference in skier or grip has an effect.

11.27 In the data set `mtcars` the variables `mpg`, `cyl`, and `am` indicate the miles per gallon, the number of cylinders, and the type of transmission respectively. Perform a two-way ANOVA modeling `mpg` by the `cyl` and `am`, each treated as categorical data.

Is there an indication of an interaction? Do both the number of cylinders and the type of transmission make a difference?

11.28 The data set `ToothGrowth` has measurements of tooth growth (`len`) of guinea pigs for different dosages of Vitamin C (`dose`) and two different delivery methods (`supp`).

Perform a two-way analysis of variance of tooth growth modeled by dosage and delivery method. First, fit the full model including interactions and use the F-test to compare this with the additive model.

11.29 The data set `OrchardSprays` contains measurements on the effectiveness of various sprays on repelling honeybees. The variable `decrease` measures effectiveness of the spray, `treatment` records the type of treatment, and `rowpos` records the row in the field the measurement comes from.

Make an interaction plot of the mean of `decrease` with `treatment` as a trace factor. Then fit the additive analysis-of-variance model and the model with interaction. Compare the two models using `anova()`. Is the interaction model suggested by the results of the modeling?

Chapter 12

Two extensions of the linear model

The linear-regression ideas are building blocks for many other statistical models. The R project's archive (CRAN, `http://cran.r-project.org`) warehouses over 300 add-on packages to R, many of which implement extensions to the linear regression model covered in the last two chapters. In this chapter, we look at two extensions: logistic-regression models and nonlinear models. Our goal is to illustrate that most of the techniques used for linear models carry over to these (and other) models.

The logistic-regression model covers the situation where the response variable is a binary variable. Logistic regression, which is a particular case of a generalized linear model, arises in several areas, including, for example, analyzing survey data. The nonlinear models we discuss use a function to describe the mean response that is not linear in the parameters.

12.1 Logistic regression

A binary variable is one that can have only two values, "success" or "failure," often coded as 1 or 0. In the ANOVA model we saw that we can use binary variables as predictors in a linear-regression model by using factors. But what if we want to use a binary variable as a response variable?

■ **Example 12.1: Spam** Junk e-mail, or spam, is a real nuisance, but it must make some business sense, as the internet is flooded with it. Let's look at the situation from the spammer's perspective.

The spammer's problem is that very few people will open spam. How to entice someone to do so? Is it worth the expense of buying an e-mail list that includes names? Does the subject line make a difference? Imagine a test is done in which 5,000 e-mails are sent out in four different ways. The subject heading on some includes a first name, on some an offer, on some both, and on some neither. The number that are opened by the recipient is measured by an embedded image in the e-mail body that can be tracked via a web server.

Table 12.1 Number of spam e-mails opened

		Offer in subject	
		yes	no
First name	yes	20 of 1,250	15 of 1,250
in subject	no	17 of 1,250	8 of 1,250

If Table 12.1 contains data on the number of e-mails opened for each possible combination, what can we say about the importance of including a name or an offer in the subject heading? ■

For simplicity, assume that we have two variables, X and Y, where Y is a binary variable coded as a 0 or 1. For example, 1 could mean a spam message was opened. If we try to model the response with $Y_i = \beta_0 + \varepsilon_i$ or $Y_i = \beta_0 + \beta_1 x_i + \varepsilon_i$, then, as Y_i is either 0 or 1, the ε_i can't be an *i.i.d.* sample from a normal population. Consequently, the linear model won't apply. As having only two answers puts a severe restriction on the error term, instead the probability of success is modeled.

Let $\pi_i = \mathsf{P}(Y_i = 1)$. Then π_i is in the range 0 to 1. We might try to fit the model $\pi_i = \beta_0 + \beta_1 x_i + \varepsilon_i$, but again the range on the left side is limited, whereas that on the right isn't. Even if we restrict our values of the x_i, the variation of the ε_i can lead to probabilities outside of $[0,1]$.

Let's change tack. For a binary random variable, the probability is also an expected value. That is, after conditioning on the value of x_i, we have $\mathsf{E}(Y_i|x_i) = \pi_i$. In the simple linear model we called this $\mu_{y|x}$, and we had the model $Y_i = \mu_{y|x} + \varepsilon_i$. Interpreting this differently will let us continue. We mentioned that the assumption on the error can be viewed two ways. Either assuming the error terms, the ε_i values, are a random sample from a mean 0 normally distributed population, or, equivalently that each data point Y_i is randomly selected from a $\mathsf{Normal}(\mu_{y|x}, \sigma)$ distribution independently of the others. Thus, we have the following ingredients in simple linear regression:

- The predictors enter in a linear manner through $\beta_0 + \beta_1 x_1$
- The distribution of each Y_i is determined by the mean, $\mu_{y|x}$, and some scale parameter σ

- There is a relationship between the mean and the linear predictors ($\mu_{y|x} = \beta_0 + \beta_1 x_1$)

The last point needs to be changed to continue with the binary regression model. Let $\eta = \beta_0 + \beta_1 x_1$. Then the change is to assume that η can be transformed to give the mean by some function $m()$ via $\mu_{y|x} = m(\eta)$, which can be inverted to yield back $\eta = m^{-1}(\mu_{y|x})$. The function $m()$ is called a **link function**, as it links the predictor with the mean.

The logistic function $m(x) = e^x/(1+e^x)$ is often used (see Figure 12.1), and the corresponding model is called **logistic regression**. For this, we have

$$\pi_i = m(\beta_0 + \beta_1 x_i) = \frac{e^{\beta_0+\beta_1 x_i}}{1+e^{\beta_0+\beta_1 x_i}}.$$

The logistic function turns values between $-\infty$ and ∞ into values between 0 and 1, so the numbers specifying the probabilities will be between 0 and 1. When $m()$ is inverted we have

$$\log\left(\frac{\pi_i}{1-\pi_i}\right) = \beta_0 + \beta_1 x_i. \tag{12.1}$$

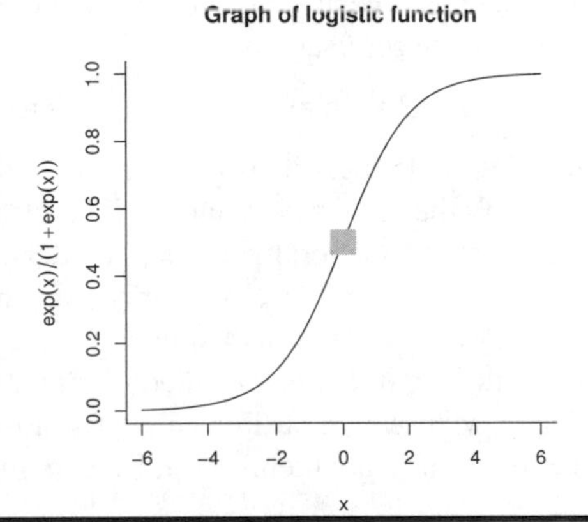

Figure 12.1 Graph of logistic function, $m(x) = e^x/(1+e^x)$. The inflection point is marked with a square.

This log term is called the **log-odds ratio**. The odds associated to some probability are $p/(1-p)$, which is evident if we understand that an event having odds a to b means that in $a+b$ *i.i.d.* trials we expect a wins. Thus the probability of success should be $a/(a+b)$. Reversing, if the probability of success is $a/(a+b)$, then the ratio becomes $(a/(a+b))/(1-a/(a+b))$ or a/b, which is the ratio of the odds.

To finish the model, we need to specify the distribution of Y_i. It is Bernoulli with success probability π_i, so that no extra parameters, such as a standard deviation, are needed.

12.1.1 Generalized linear models

Logistic regression is an example of a **generalized linear model**. The key ingredients are as above: a response variable Y and some predictor variables $x_1, x_2, \ldots, x_p$. The predictors enter into the model via a single linear function:

$$\eta = \beta_0 + \beta_1 x_1 + \cdots + \beta_p x_p.$$

The mean of Y given the x values is related to η by an invertible link function $m()$ as $\mu = m(\eta)$ or $m^{-1}(\mu) = \eta$. Finally, the distribution of Y is given in terms of its mean and, perhaps, a scale parameter such as σ.

Thus, the model is specified by the coefficients β_i, a link function $m()$, and a probability distribution that may have an additional scale parameter.

12.1.2 Fitting the model using `glm()`

Generalized linear models are fit in R using the `glm()` function. Its usage is similar to that of `lm()`, except that we need to specify the probability distribution and the link function. A template for usage is

```
res = glm(formula, family=..., data=...)
```

The formula is specified as though it were a linear model. The argument `family=` allows us to specify the distribution and the link. Details are in the help page `?family` and in the section "Generalized linear models" in the manual *An Introduction to R* accompanying R. We will use only two: the one for logistic regression and one to compare the results with simple linear regression.

For logistic regression the argument is specified by `family=binomial`, as the default link function is what we want. For comparison to simple linear regression, the link function is just an identity, and the family is specified as `family=gaussian`.*

As an illustration, let's compare using `glm()` and `lm()` to analyze a linear model. We will use simulated data so we already "know" the answer.

■ **Example 12.2: Comparing `glm()` and `lm()`** We first simulate data from the model that Y_i has a $\mathsf{Normal}(x_{1i} + 2x_{2i}, \sigma)$ distribution.

```
> x1 = rep(1:10,2)
> x2 = rchisq(20,df=2)
> y = rnorm(20,mean=x1 + 2*x2, sd=2)
```

---

* Gaussian is a mathematical term named for Carl Gauss that describes the normal distribution.

We fit this using `lm()` as follows:

```
> res.lm = lm(y ~ x1 + x2)
> summary(res.lm)
...
Coefficients:
            Estimate Std. Error t value Pr(>|t|)
(Intercept)   -0.574      1.086   -0.53      0.6
x1             1.125      0.143    7.89  4.4e-07 ***
x2             1.971      0.254    7.75  5.6e-07 ***
---
Signif. codes:  0 '***' 0.001 '**' 0.01 '*' 0.05 '.' 0.1 ' ' 1
...
```

Both the coefficients for `x1` and `x2` are flagged as significantly different from 0 in the marginal *t*-tests.

The above can all be done using `glm()`. The only difference is that the modeling involves specifying the `family=` argument. We show all the output below.

```
> res.glm = glm(y ~ x1 + x2, family=gaussian)
> summary(res.glm)

Coefficients:
            Estimate Std. Error t value Pr(>|t|)
(Intercept)   -0.574      1.086   -0.53      0.6
x1             1.125      0.143    7.89  4.4e-07 ***
x2             1.971      0.254    7.75  5.6e-07 ***
---
Signif. codes:  0 '***' 0.001 '**' 0.01 '*' 0.05 '.' 0.1 ' ' 1

(Dispersion parameter for gaussian family taken to be 3.239)

    Null deviance: 387.747  on 19  degrees of freedom
Residual deviance:  55.057  on 17  degrees of freedom
AIC: 85.01

Number of Fisher Scoring iterations: 2
```

The same coefficients are found. This is not surprising, but technically a different method is used. For each coefficient, a two-sided significance test is done with null hypothesis that the value is 0. For this model, the results are identical, as with `lm()`. No information about the F statistic is given, as the theory does not apply here in general. Rather, the AIC is given. Recall that this could be used for model selection. Lower values are preferred. ■

Now we fit a logistic model.

■ **Example 12.3: Premature babies** According to the web site `http://www.keepkidshealthy.com`, risk factors associated with premature births include smoking and maternal malnutrition. Do we find this to be the case with the data in the `babies (UsingR)` data set?

We'll need to manipulate the data first. First we extract just the variables of interest, using the `subset=` argument to eliminate the missing values.

```
> babies.prem = subset(babies,
+ subset= gestation < 999 & wt1 < 999 & ht < 99 & smoke < 9,
+ select=c("gestation","smoke","wt1","ht"))
```

A birth is considered premature if the gestation period is less than 37 full weeks.

```
> babies.prem$preemie = as.numeric(babies.prem$gestation < 7*37)
> table(babies.prem$preemie)
   0    1
1079   96
```

For `glm()` and binomial models the response variable can be numeric, as just defined, or a factor (the first level is "failure," the others are "success").

We will use the body mass index (BMI) as a measure of malnutrition. The BMI is the weight in kilograms divided by the height in meters squared. If there is some dependence, we will investigate further.

```
> babies.prem$BMI = with(babies.prem,(wt1 / 2.2) / (ht*2.54/100)^2)
> hist(babies.prem$BMI)                       # looks okay
```

We can now model the variable `preemie` by the levels of `smoke` and the variable `BMI`. This is similar to an ANCOVA, except that the response variable is binary.

```
> res = glm(preemie ~ factor(smoke) + BMI, family=binomial,
+   data=babies.prem)
> summary(res)
...
Coefficients:
               Estimate Std. Error z value Pr(>|z|)
(Intercept)     -3.4246     0.7113   -4.81  1.5e-06 ***
factor(smoke)1   0.1935     0.2355    0.82     0.41
factor(smoke)2   0.3137     0.3888    0.81     0.42
factor(smoke)3   0.1011     0.4047    0.25     0.80
BMI              0.0401     0.0304    1.32     0.19
---
Signif. codes:  0 '***' 0.001 '**' 0.01 '*' 0.05 '.' 0.1 ' ' 1
...
```

None of the variables are flagged as significant. This indicates that the model with no effects is, perhaps, preferred. (The sampling distribution under the null hypothesis is different from the previous example, so the column gets marked with "`z value`" as opposed to "`t value`.") We check which model is preferred by the AIC using `stepAIC()` from the `MASS` package.

```
> library(MASS)
> stepAIC(res)
Start:  AIC= 672.3
...
Step:  AIC= 666.8
 preemie ~ 1
```

```
Call:
glm(formula = preemie ~ 1, family = binomial, data = babies.prem)

Coefficients:
(Intercept)
      -2.42
...
```

The model of constant mean is chosen by this criteria, indicating that these risk factors do not show up in this data set. ■

■ **Example 12.4: The spam data** Let's apply logistic regression to the data on spam in Table 12.1. Set Y_i to be 1 if the e-mail is opened, and 0 otherwise. Likewise, let x_{1i} be 1 if the e-mail has a name in the subject, and x_{2i} be 1 if the e-mail has an offer in the subject. Then we want to model Y_i by x_{1i} and x_{2i}. To use logistic regression, we first turn the summarized data into 5,000 samples. We use `rep()` repeatedly to do so.

```
> first.name = rep(1:0,c(2500,2500))
> offer = rep(c(1,0,1,0),rep(1250,4))
> opened = c(rep(1:0,c(20,1250-20)), rep(1:0,c(15,1250-15)),
+ rep(1:0,c(17,1250-17)), rep(1:0,c(8,1250-8)))
> xtabs(opened ~ first.name + offer)
          offer
first.name 0  1
         0  8 17
         1 15 20
```

This matches Table 12.1, but the default ordering is different, as 0 or, "no," is first.

We remark that the value of `opened` could have been defined a bit more quickly using a function and `sapply()` to repeat the typing. (See below for further savings in work.)

```
> f = function(x) rep(1:0,c(x,1250-x))
> opened = c(sapply(c(20,15,17,8),f))
```

Now to fit the logistic regression model. We use `factor()` around each predictor; otherwise they are treated as numeric values.

```
> res.glm = glm(opened ~ factor(first.name) + factor(offer),
+ family=binomial)
> summary(res.glm)

Call:
glm(formula = opened ~ factor(first.name) + factor(offer),
    family = binomial)

Deviance Residuals:
   Min       1Q  Median       3Q      Max
```

```
-0.187  -0.158  -0.147  -0.124   3.121

Coefficients:
                   Estimate Std. Error z value Pr(>|z|)
(Intercept)          -4.864      0.259  -18.81   <2e-16 ***
factor(first.name)1   0.341      0.263    1.30    0.195
factor(offer)1        0.481      0.266    1.81    0.071 .
---
Signif. codes:  0 '***' 0.001 '**' 0.01 '*' 0.05 '.' 0.1 ' ' 1

(Dispersion parameter for binomial family taken to be 1)

    Null deviance: 650.02  on 4999  degrees of freedom
Residual deviance: 644.99  on 4997  degrees of freedom
AIC: 651

Number of Fisher Scoring iterations: 6
```

Although only the intercept is flagged as significant at the 0.05 level, suppose the estimates are correct. How can we interpret them? The coding is such that when no first name or offer is included, the log-odds ratio is -4.864. When the first name is included but not the offer, the log-odds ratio is $-4.864+0.341$. When both are included, it's $-4.864+0.341+0.481$. Let o_0 be the odds ratio when neither a name nor an offer is included:

$$o_0 = \text{odds ratio} = \frac{\pi}{1-\pi} = e^{-4.864}.$$

If we include the first name, the odds ratio goes up to $e^{-4.864+0.341} = o_0 \cdot e^{0.341}$, which is an additional factor of $e^{0.341} = 1.406$. So, if the original odds were 2 to 100, they go up to $2(1.406)$ to 100. ■

Avoiding replication In the previous example the data was replicated to produce variables `first.name`, `offer`, and `opened` with 5,000 values, so that all the recorded data was present. The interface for `glm()` conveniently allows for tabulated data when the `binomial` family is used. Not only is tabulated data easier to type in, we can save memory as we don't store large vectors of data.

A two-column matrix is used, with its first column recording the number of successes and its second column the number of failures. In our example, we can construct this matrix using `cbind` as follows:

```
> opened = c(8,15,17,20)
> not.opened = 1250 - opened
> opened.mat = cbind(opened = opened, not.opened=not.opened)
> opened.mat
     opened not.opened
[1,]      8       1242
[2,]     15       1235
```

```
[3,]      17        1233
[4,]      20        1230
```

The predictor variables match the levels for the rows. For example, for the values of 8 and 15 for `opened`, `offer` was 0 and `first.name` was 0 then 1. Continuing gives these values:

```
> offer=c(0,0,1,1)
> first.name = c(0,1,0,1)
```

Finally, the model is fit as before, using `opened.mat` in place of `opened`.

```
> glm(opened.mat ~ first.name + offer, family=binomial)

Call:  glm(formula = opened.mat ~ first.name + offer,
           family = binomial)

Coefficients:
(Intercept)   first.name         offer
     -4.864        0.341         0.481

Degrees of Freedom: 3 Total (i.e. Null);  1 Residual
Null Deviance:      5.77
Residual Deviance: 0.736         AIC: 24.7
```

12.2 Nonlinear models

The linear model is called "linear" because of the way the coefficients β_i enter into the formula for the mean. These coefficients simply multiply some term. A nonlinear model allows for more complicated relationships. For example, an exponential model might have the response modeled as

$$Y_i = \beta_0 e^{-\beta_1 x_i} + \varepsilon_i.$$

Here, $\mu_{y|x} = \beta_0 e^{-\beta_1 x}$ is not linear in the parameters due to the β_1. It does not appear as an additive term like $\beta_1 x_i$.

Variations on the exponential model are

$$Y_i = \beta_0 x_i e^{-\beta_1 x_i} + \varepsilon_i \quad \text{and} \quad Y_i = \beta_0 (e^{-\beta_1 x_i}(1 - \beta_2) + \beta_2) + \varepsilon_i.$$

The exponential model, with $\beta_1 > 0$, may be used when the response variable decays as the predictor increases. The second model has a growth-then-decay phase, and the third a decay, not to 0 but to some threshold amount $\beta_0 \cdot \beta_2$.

In general, a single-covariate, nonlinear model can be written as follows:

$$Y_i = f(x_i | \beta_0, \beta_1, \ldots, \beta_r) + \varepsilon_i.$$

We have $r+1$ parameters and only one predictor with an additive error. More general models could have more predictors and other types of errors, such as multiplicative.

The possibilities seem endless but in fact are constrained by the problem we are modeling. When using nonlinear models we typically have some idea of which types of models are appropriate for the data and then fit just those. If the model has *i.i.d.* errors that are normally distributed, then using the method of least squares allows us to find parameter estimates and use AIC to compare models.

12.2.1 Fitting nonlinear models with `nls()`

Nonlinear models can be fit in R using `nls()`. The `nls()` function computes nonlinear least squares. Its usage is similar to, but different from `lm()`. A basic template is

```
res = nls(formula, data=..., start=c(...), trace=FALSE)
```

The model formula is different for nonlinear models. The formula again looks like `response ~ mean`, but the `mean` is specified using ordinary math notations. For example, the exponential model for the mean could be written `y ~ N * exp(-r*(t-t0))`, where `N`, `r`, and `t0` are parameters. It is often convenient to use a function to return the mean, such as `y ~ f(x,beta0,beta1,...)`. That is, a function that specifies the parameter values by name.

The method of nonlinear least squares uses an algorithm that usually needs to start with parameter values that are close to the actual ones. The argument `start=c(...)` is where we put the initial guesses for the parameters. This can be a vector or list using named values, such as `start=c(beta0=1,beta1=2)`. Finally, the optional argument `trace=TRUE` can be used if we want to see what is happening during the execution of the algorithm. This can be useful information if the algorithm does not converge. By default it is `FALSE`.

The initial parameter guesses are often found by doing some experimental plots. These can be done quickly using the `curve()` function with the argument `add=TRUE`, as illustrated in the examples. When we model with a given function, it helps to have a general understanding of how the parameters change the graph of the function. For example, the parameters in the exponential model, written $f(t|N,r,t_o) = Ne^{-r(t-t_0)}$, may be interpreted by t_0 being the place where we want time to begin counting, N the initial amount at this time, and r the rate of decay. For this model, the mean of the data decays by $1/e$, or roughly $1/3$ in $1/r$ units of time.

Some models have self-starting functions programmed for them. These typically start with `SS`. A list can be found with the command `apropos("SS")`. These functions do not need starting values.

■ **Example 12.5: Yellowfin tuna catch rate** The data set `yellowfin (UsingR)` contains data on the average number of yellowfin tuna caught per 100 hooks in the tropical Indian Ocean for various years. This data comes from a paper by Myers and Worm (see `?yellowfin`) that uses such numbers to estimate the decline of fish stocks (biomass) since the advent of large-scale commercial fishing. The authors fit the exponential decay model with some threshold to the data.

We can repeat the analysis using R. First, we plot (Figure 12.2).

```
> plot(count ~ year, data=yellowfin)
```

A scatterplot is made, as the data frame contains two numeric variables. The `count` variable does seem to decline exponentially to some threshold. We try to fit the model

$$Y = N\left(e^{-r(t-1952)}(1-d)+d\right)+\varepsilon.$$

(Instead of β_i we give the parameters letter names.)

To fit this in R, we define a function for the mean

```
> f = function(t,N,r,d) N*(exp(-r*(t-1952))*(1-d) +d)
```

We need to find some good starting points for `nls()`. The value of $N = 7$ seems about right, as this is the starting value when $t = 1952$. The value r is a decay rate. It can be estimated by how long it takes for the data to decay by roughly $1/3$. We guess about 10, so we start with $r = 1/10$. Finally, d is the percent of decay, which seems to be $.6/6 = .10$.

We plot the function with these values to see how well they fit.

```
> curve(f(x, N=6, r=1/10, d=0.1), add=TRUE)
```

The fit is good (the solid line in Figure 12.2), so we expect `nls()` to converge with these starting values.

```
> res.yf = nls(count ~ f(year,N,r,d), start=c(N=6,r=1/10, d=.1),
+ data=yellowfin)
> res.yf
Nonlinear regression model
  model:  count ~ f(year, N, r, d)
   data:  yellowfin
      N       r       d
6.02019 0.09380 0.05359
 residual sum-of-squares:  15.48
```

The numbers below the coefficients are the estimates. Using these, we add the estimated line using `curve()` again. This time it is drawn with dashes, and it visually seems to fit all the data a little better.

```
> curve(f(x,N=6.02,r=.0939,d=.0539), add=TRUE, lty=2,lwd=2)
> legend(1980,6,legend=c("exploratory","exponential"),lty=1:2)
```

The value for d estimates that only 5.3% of the initial amount remains. ■

Using `predict()` to plot the prediction line The output of `nls()` has many of the same extractor functions as `lm()`. In particular, the `predict()` function can be used to make predictions for the model. You can use this in place of

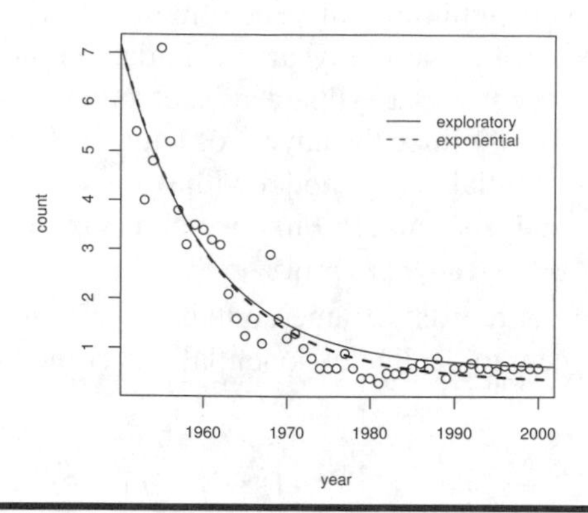

Figure 12.2 Mean catch per 100 hooks of yellowfin tuna in the tropical Indian Ocean. An exponential decay model with threshold is given by the dashed line.

`curve()` to draw the predicted line for the mean response. For example, to draw the line for the yellowfin tuna data, we create a range of values for the `year` variable, and then call `predict()` with a named data frame.

```
> tmp = 1952:2000
> lines(tmp, predict(res.yf, data.frame(year = tmp)))
```

■ **Example 12.6: Sea urchin growth** The `urchin.growth (UsingR)` data set is derived from thesis work by P. Grosjean. It contains growth data of reared sea urchins over time. Typical growth starts at 0 and progresses to some limiting size. Some models for growth include logistic growth

$$g(t|Y_\infty, k, t_0) = Y_\infty \cdot (1 + e^{-k(t-t_0)})^{-1}$$

and a Richards growth model

$$f(t|Y_\infty, k, t_0) = Y_\infty \cdot (1 - e^{-k(t-t_0)})^{m}.$$

The logistic-growth function is identical to that used in logistic regression, although it is written differently. Our goal here is to fit both of these models to the data, assuming *i.i.d.*, additive error terms, and decide between the two based on AIC. As the Richards model has more parameters, its fit should be much better to be considered a superior model for the data.

We follow the same outline as the previous example: define functions, find initial guesses by plotting some candidates, and then use `nlm()` to get the estimates.

We define two functions and plot the jittered scatterplot (Figure 12.3).

```
> g = function(t,Y,k,t0)   Y*(1 + exp(-k*(t-t0)))^(-1)
> f = function(t,Y,k,t0,m) Y*(1 - exp(-k*(t-t0)))^m
> plot(jitter(size) ~ jitter(age,3), data=urchin.growth,
+      xlab="age",ylab="size",main="Urchin growth by age")
```

Next, we try to fit g. The parameters can be interpreted from the scatterplot of the data. The value of Y corresponds to the maximum growth of the urchins, which appears to be around 60. The value of `t0` is where the inflection point of the graph occurs. The inflection point is when the curve stops growing faster. A guess is that it happens around 2 for the data. Finally, k is a growth rate around this point. It should correspond to roughly 1 over the time it takes to grow one-third again after the value at t_0. We guess 1 from the data. With these guesses, we do an exploratory graph with `curve()` (not shown but looks okay).

```
> curve(g(x, Y=60, k=1, t0=2), add=TRUE)
```

We fit the model with `nls()`

```
> res.g = nls(size ~ g(age,Y,k,t0), start=c(Y=60, k=1, t0=2),
+ data = urchin.growth)
> res.g
Nonlinear regression model
  model:  size ~ g(age, Y, k, t0)
   data:  urchin.growth
     Y      k     t0
53.903  1.393  1.958
 residual sum-of-squares:  7299
> curve(g(x, Y=53.903, k=1.393, t0=1.958), add=TRUE)
```

Finally, so we can compare, we find the AIC:

```
> AIC(res.g)
[1] 1559
```

Next, we fit the Richards model. First, we try to use the same values, to see if that will work (not shown).

```
> curve(f(x, Y=53.903, k=1.393, t0=1.958, m = 1), add=TRUE)
> legend(4,20, legend=c("logistic growth","Richards"),lty=1:2)
```

It is not a great fit, but we try these as starting points for the algorithm anyway:

```
> res.f = nls(size ~ f(age,Y,k,t0,m), data = urchin.growth,
+ start=c(Y=53, k=1.393, t0=1.958, m=1))
Error in numericDeriv(form[[3]], names(ind), env) :
        Missing value or an Infinity produced when evaluating the model
```

This is one of the error messages that can occur when the initial guess isn't good or the model doesn't fit well.

Using a little hindsight, we think that the problem might be t_0 and k. For this model, a few exploratory graphs indicate that we should have $t \geq t_0$ for a growth model, as the graphs decay until t_0. So,we should start with $t_0 < 0$. As well, we slow the rate of growth.

```
> res.f = nls(size ~ f(age,Y,k,t0,m),
+ start=c(Y=53, k=.5, t0=0, m=1), data=urchin.growth)
> res.f
```

```
Nonlinear regression model
  model:  size ~ f(age, Y, k, t0, m)
   data:  urchin.growth
      Y        k       t0        m
57.2649  0.7843 -0.8587  6.0636
 residual sum-of-squares:  6922
> curve(f(x, Y=57.26, k=0.78, t0=-0.8587, m = 6.0636), add=TRUE, lty=2)
```

Now we have convergence. The residual sum-of-squares, 6,922, is less than the 7,922 for the logistic model. This is a good thing, but if we add parameters this is often the case.† We compare models here with AIC.

```
> AIC(res.f)
[1] 1548
```

This is a reduction from the other model. As such, we would select the Richards model as a better fit by this criteria. ■

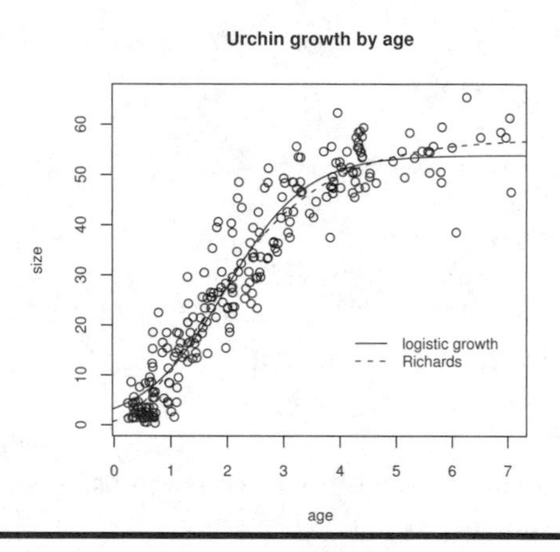

Figure 12.3 Sea urchin growth data, with logistic model fit in solid and Richards model fit in dashed line

Problems

12.1 The data set `tastesgreat (UsingR)` is data from a taste test for *New Goo*, a fictional sports-enhancement product. Perform a logistic regression to investigate whether the two covariates, `age` and `gender`, have a significant effect on the enjoyment variable, `enjoyed`.

† We do not have nested models, for which this would always be the case.

12.2 The data set healthy (UsingR) contains information on whether a person is healthy or not (healthy uses 0 for healthy and 1 for not healthy) and measurements for two unspecified covariates, p and g.

Use stepAIC() to determine which submodel is preferred for the logistic model of healthy, modeled by the two covariates p and g.

12.3 The data set birthwt (MASS) contains data on risk factors associated with low infant birth weight. The variable low is coded as 0 or 1 to indicate whether the birth weight is low (less than 250 grams). Perform a logistic regression modeling low by the variables age, lwt (mother's weight), smoke (smoking status), ht (hypertension), and ui (uterine irritability). Which variables are flagged as significant? Run stepAIC(). Which model is selected?

12.4 The data set hall.fame (UsingR) contains statistics for several major league baseball players over the years. We wish to see which factors contribute to acceptance into the Hall of Fame. To do so, we will look at a logistic regression model for acceptance modeled by the batting average (BA), the number of lifetime home runs (HR), the number of hits (hits), and the number of games played (games).

First, we make binary variable for Hall of Fame membership.

```
> hfm = hall.fame$Hall.Fame.Membership != "not a member"
```

Now, fit a logistic regression model of hfm modeled by the variables above. Which are chosen by stepAIC()?

12.5 The esoph data set contains data from a study on esophageal cancer. The data records the number of patients with cancer in ncases and the number of patients in the control group with ncontrols. The higher the ratio of these two variables the worse the cancer risk. Three factors are recorded: the age of the patient (agegp), alcohol consumption (alcgp), and tobacco consumption (tobgp).

We can fit an age-adjusted model of the effects of alcohol and tobacco consumption with an interaction as follows:

```
> res.full <- glm(cbind(ncases, ncontrols) ~ agegp + tobgp * alcgp,
+                   data = esoph, family = binomial())
```

A model without interaction is fit with

```
> res.add  <- glm(cbind(ncases, ncontrols) ~ agegp + tobgp + alcgp,
+                   data = esoph, family = binomial())
```

Use AIC() to compare the two models to determine whether an interaction term between alcohol and tobacco is hinted at by the data.

12.6 The data set Orange contains circumference measurements for several trees (Tree) based on their age. Use a logistic growth model to fit the data for tree 1. What are the estimates?

12.7 The data set `ChickWeight` contains measurements of `weight` and age (`Time`) for several different chicks (coded with `Chick`). For chick number 1, fit a logistic model for `weight` modeled by `Time`. What are the coefficients?

12.8 The data set `wtloss (MASS)` contains weight measurements of an obese patient recorded during a weight-rehabilitation program. The variable `Weight` records the patient's weight in kilograms, and the variable `Days` records the number of days since the start of the program. A linear model is not a good model for the data, as it becomes increasing harder to lose the same amount of weight each week. A more realistic goal is to lose a certain percentage of weight each week. Fit the nonlinear model

$$\texttt{Weight} = a + b2^{-\texttt{Days}/c}.$$

The estimated value of c would be the time it takes to lose b times half the excess weight.

What is the estimated weight for the patient if he stays on this program for the long run? Suppose the model held for 365 days. How much would the patient be expected to weigh?

12.9 The `reddrum (UsingR)` data set contains length-at-age data for the red drum fish. Try to fit both the models

$$l = b_0(1 - e^{-k(t-t_0)}) \quad \text{and} \quad l = (b_0 + b_1 t)(1 - e^{-k(t-t_0)}).$$

(These are the von Bertalanffy and "linear" von Bertalanffy curves.) Use the AIC to determine which is preferred.

Good starting values for the "linear" curve are 32, $1/4$, $1/2$, and 0.

12.10 The data set `midsize (UsingR)` contains values of three popular mid-size cars for the years 1990 to 2004. The 2004 price is the new-car price, the others values of used cars. For each car, fit the exponential model with decay. Compare the decay rates to see which car depreciates faster. (Use the variable `year=2004-Year` and the model for the mean $\mu_{y|x} = Ne^{-rt}$.)

Appendix A

Getting, installing, and running R

R is being developed for the Unix, Windows (Windows 95, 98, ME, NT4, 2000, or XP), and Mac OS X platforms. For each operating system the installation of R is similar to that of other software programs. This appendix covers the basics of installation. It also includes information about extending the base functionality of R by adding external packages to the system. Once R is installed, more information of this type is available in the *R Administrators Manual* that accompanies the R program. This document can be accessed via the html-based help system started by the function `help.start()`.

A.1 Installing and starting R

R is available through source code, allowing users to compile the program to their liking. However, for most purposes, a convenient binary package is available for installation.

The files for R are available at the Comprehensive R Archive Network, or CRAN, `http://cran.r-project.org`. There is a series of mirror sites to lessen the load on any one server. Choose one close to you. A list is found at `http://cran.r-project.org/mirrors.html`.

What follows are brief instructions, to give you an idea of what is involved in installing R. For each operating system mentioned, more complete instructions are available with the accompanying files.

A.1.1 Binary installation under Windows

R's version for Windows has its own installer. To begin, download R from the /bin/windows/base directory of a CRAN mirror.* The file to download is large, over 20 megabytes. It is titled rwXXXX.exe, where XXXX contains the current version information, such as 2000 for version 2.0.0. This is a self-extracting file, which contains the necessary installation program. After being downloaded to the desktop, R will be installed when you double-click on the icon for the downloaded file. The directory for installation can be adjusted during installation.

Once installed, R can be started with its GUI when you double-click the desktop icon, or from the R submenu under the start menu.

A.1.2 Binary installation under Linux

The Linux operating system is packaged into many different distributions. Familiar ones are Debian, RedHat, and Gentoo. Installation for each follows its usual installation procedure. The /bin/linux directory of a CRAN mirror contains several different binary builds of R.

There is an up-to-date Debian package for R that can be installed with the apt-get command. You just need to add the CRAN directory to a configuration file. (Look under /bin/linux/debian) for details.) The main files are contained in r-base and r-base-core. In addition, several contributed CRAN packages can be installed this way, rather than through the installation methods described later in this appendix. This makes updating R even easier.

The Debian distribution has proved popular for making bootable CD-ROMs that contain the Linux operating system. In particular, the Quantian Scientific Computing Environment (http://dirk.eddelbuettel.com/quantian.html) contains R and many other pre-configured, open-source scientific software packages. To use Quantian, you need to download the ISO image, burn it to a CD-ROM, and then boot your computer from this CD-ROM. This boots to the KDE desktop environment, from which R may be run from a shell or from within an ESS session.

The RedHat Linux distribution has binary files distributed in rpm format. These files can be found on a CRAN mirror under the /bin/linux/redhat directory. Installation can be done from the shell with the following command:

```
rpm -i filename.rpm
```

This also applies to SuSE Linux and other rpm-based distributions. The help files for the rpm mention issues people have with external libraries. If this installation fails, the help files are the first place to look for solutions.

The Gentoo Linux installation is a single command at the shell:

* For example, if the mirror is the main CRAN site, the url is http://cran.r-project.org/bin/windows/base.

```
emerge R
```

Technically, this isn't a binary installation, as R is compiled from source, but it is just as straightforward.

On a UNIX machine, once R is installed you can start it from the shell with the command "R."

A.1.3 Binary installation under Mac OS X

The `/bin/macosx` directory of a CRAN site contains a disk image `R.dmg` that should be downloaded. Once that's done, the Finder application should open to the image. The file `R.pkg` is double-clicked to begin the installation process. This places an R application in your Applications directory. Starting R with its Aqua GUI is done by double-clicking on this icon. R can also be run from within the terminal application or by using ESS to run R within Emacs. The appropriate symbolic link may need to be made prior to this so that the correct file is on the path.

A.1.4 Installing from the source code

R can be installed from the source code. First the source code is downloaded, and then uncompressed. For a UNIX machine, the following commands are issued from the UNIX command line. First unpack the source and change directory (using gnu tar):

```
tar zxvf R-x.y.z.tgz
cd R-x.y.z
```

Then the most basic compilation can be done with the commands

```
./configure
make
```

The `configure` command automatically figures out dependencies for your machine. Many options can be set to override the defaults. They are listed if you type the command `./configure --help`. If the compilation is successful, then the program can be installed by the command `make install`.

A.1.5 Startup files

R's startup process allows you to load in desired commands and previous R sessions.

When R starts it looks for the file `.Rprofile` in the current directory and then in the user's home directory. If this file is found, the R commands in the file are sourced into the new R session. This allows you to make permanent adjustments to the settings for `options()` or `par()`, load frequently used libraries, and define helpful functions that can be used in every session.

After this, R then loads a saved image of the user workspace (if there is one) from the file .RData. If you save your session when quitting, then R will load it back in. This preserves any function definitions and data sets that you may have been working on.

See ?Startup for more information, including site-wide configuration files.

A.2 Extending R using additional packages

R has a number of additional packages that extend its base functionality. Some of these are recommended and are already part of most installations; others need to be installed. Many, but not all, of these packages reside on CRAN.

Installing a package can be done from the main GUIs, from the command line within R, or from the shell that R will run under.

The Windows and Mac OS X GUIs have menu items that query the available packages at CRAN and allow you to install them using your mouse. If you have the proper administrative authority, this method is very straightforward. As many external packages require a compiler, and most Windows installations don't have one, the Windows installation looks for binary versions of a package.

If a GUI option is not available, additional packages can be installed from within R. The key functions for package management are: `install.packages()` to install a package; `update.packages()` to update all your packages (such as when you upgrade R); and `library()` to load a package.

The basic usage of `install.packages()` and `library()` is to call the function with a valid package name. For example, if your computer is connected to the internet and you have the necessary permissions, the following commands will install the Rcmdr package by downloading it from CRAN and then load the package.

```
> install.packages("Rcmdr", dependencies=TRUE)
> library(Rcmdr)                    # load the package
```

In this example, the argument `dependencies=TRUE` is used to specify that packages that the Rcmdr package relies on should also be installed.

If a package is not on CRAN, you may be able to install it in this manner by specifying the extra argument `contriburl=` to `install.packages()`. For example, these commands will install the package that accompanies this book:

```
> where = "http://www.math.csi.cuny.edu/UsingR"
> install.packages("UsingR",contriburl=where)
```

If these methods fail, a package can be downloaded to the local machine and installed from there. Under Windows this last step can be initiated from the menu bar. For UNIX installations, a package can be installed from the command line with a command like:

```
R CMD INSTALL aPackage_0.1.tar.gz
```

The actual package name would replace `aPackage_0.1.tar.gz`.

If you do not have administrative privileges on the machine, you can install packages to a directory where you do have write permissions. When installing packages you can specify this directory with the argument `lib=`. When loading this package with `library()` you specify the directory with the argument `lib.loc=`. This argument is also used with `update.packages()`. For example, these commands will install and load the `ellipse` package into a session, keeping the files in a subdirectory of the user's home directory (the tilde, ~, expands to the home directory in UNIX).

```
> install.packages("ellipse",lib="~/R/")
> library("ellipse",lib.loc="~/R/")
```

A.2.1 Upgrading an existing installation

About every six months the latest and greatest version of R is released. At this point, it is recommended that you upgrade. When you do this, you may need to reinstall your packages or update the existing packages. The `update.packages()` command will allow you to perform the upgrade.

Appendix B

Graphical user interfaces and R

R, unlike many commercial offerings, lacks a common graphical user interface (GUI). The reasons for this are many, but primarily, the multi-platform nature of R make development difficult, as does the fact that most "power users" of R (the likely developers of a GUI) prefer the flexibility and power of typing-centric methods.

However, there are a number of GUI components available. We try only to cover the GUIs for Windows and Mac OS X, and the GUI provided by the add-on package RCmdr. For details about additional GUI components, including the promising SciViews-R and JGR projects, consult the RGui page linked to on the home page of R (http://www.r-project.org).

B.1 The Windows GUI

There are two ways to run R under Windows: from the shell or from within its GUI. The Rgui.exe provides the GUI and is the one that is associated with the R icon on the desktop. When the icon is double-clicked, the Windows GUI starts. It consists of a few basic elements: a window with a menu bar and a container to hold windows for the console (the command line), help pages, and figures. (This describes the default multi-document interface (MDI). An option for a single-document interface (SDI) may be set under the File::Options... menu.)

The initial RGui window looks something like Figure B.1. The window after making a plot and consulting the help pages looks like Figure B.2.

The workings of the GUI are similar to many other Windows applications. We selectively describe a few of them.

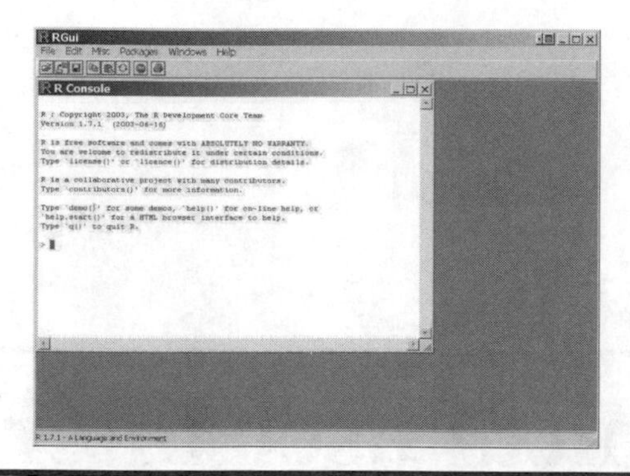

Figure B.1 Initial RGui window for Windows

The menu bar changes depending on which window has the focus. The console has the most extensive menu bar. A few useful commands are:

- `File::source...`: Opens a directory-browsing dialog to find a file to be "`sourced`." This is useful when working with files containing `R` code that may be edited with an external editor.
- `File::Load Workspace...`: Will load an `.Rdata` file, which stores `R` sessions. Storage is done at the end of a session or can be initiated by the menu item `File::Save Workspace...`
- `File::Quit`: Will quit the program after querying to save the session.
- `Edit::copy and paste`: Will paste the currently selected text into the console. This avoids the tediousness of copying and pasting.
- `Edit::data editor`: This is a dialog for the `edit()` command. It is used with existing data sets.
- `Misc::List Objects`: Will list all the objects in the current workspace. Equivalent to `ls()`.
- `Misc::Stop current output`: This will stop the current output. This is useful if something is taking too long.
- `Packages::load package`: This will open a dialog to allow us to load a package by clicking on its name. This replaces commands such as `library(MASS)`.
- `Packages::Install Package(s) from CRAN`: This will show a list of packages on CRAN that will install into Windows and allow us to install one by selecting it (see `?install.packages`). The `Packages::Update packages from CRAN` will update an existing package (see the help page `?update.packages`).
- `Window`: The submenu allows us to manipulate the `R` windows for the console, help pages, and graphics device.
- `Help::Console`: A description of useful console shortcuts.
- `Help::FAQ on R for Windows`: File of frequently asked questions.

The plot window has its own menu. An important item is the `History` fea-

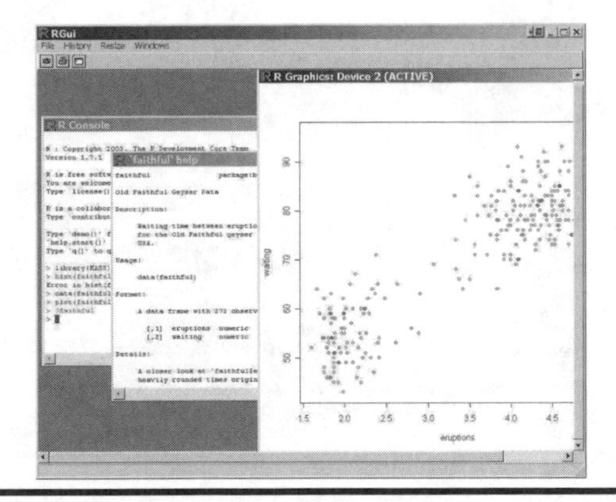

Figure B.2 Multi-document window showing console, plot window, and help page

ture, which allows us to record the graphs that are made and play them back using the PG UP and PG DOWN keys. In addition, the plot window has a binding for left mouse clicks, allowing us to save and print the current graphic.

B.2 The Mac OS X GUI

Users of Mac OS X can use `R` as a native application or as an X11 application. As of `R` version 2.0.0, when running `R` as a native application, a Cocoa GUI is available. This consists of a console window (Figure B.3); a menu bar; a graphical device, `quartz()`; and a workspace browser. The interface is being actively developed, and may be improved upon in the future. The screenshots are from a pre-alpha version of a new GUI design for R 2.0.0.

The console has icons for a few common tasks, such as halting `R` during its execution, sourcing a file, opening the graph window, showing the error logs, showing the history, etc.

The menu bar contains many of the same items as the Windows menu bar. In particular, you can load and install packages (Figure B.4), browse your workspace (Figure B.5, change directories, source files, and access the help system.

The `quartz()` plot device uses anti-aliasing, which gives nice-looking plots and images. The device also uses the native pdf format of Mac OS X graphics. Bitmap copies of a graphic can be produced by copying the graphic into the clipboard. Consult the FAQ for more details. The FAQ can be found under the help menu. The `quartz()` device can be used interactively with `identify()` and `locator()`. Use the `Esc` key to break, as there is no guarantee of a second mouse button. New quartz devices can be opened, and switching between open devices is available using the `Window` menu.

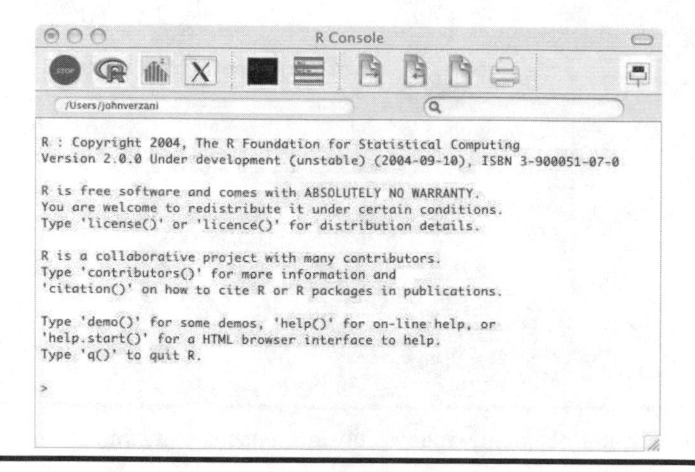

Figure B.3 The R console in the Cocoa GUI for Mac OS X

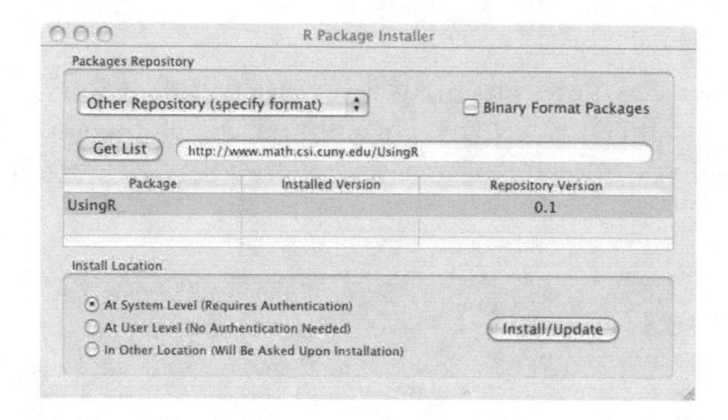

Figure B.4 Package manager for Cocoa GUI in Mac OS X

B.3 Rcdmr

The `tcltk` package can be loaded into R, which allows R to use the GUI elements provided by the tcltk widget collection. This set of widgets is available for the platforms R supports.

The `Rcmdr` package by John Fox uses these bindings to provide an easy to learn and useful interface to many common functions used in data analysis. It is designed for new and casual users of R, like the target audience of this book.

`Rcmdr` is installed easily, as it resides on CRAN, though it requires many other packages for functionality such as that provided by the `car` package. Installing these packages can be done automatically with the command:

```
> install.packages("Rcmdr", dependencies=TRUE)
```

If you forget to install all the dependencies, the first thing `Rcmdr` will do when run is ask if you want them to be installed.

`Rcmdr` is started by loading the package with the command

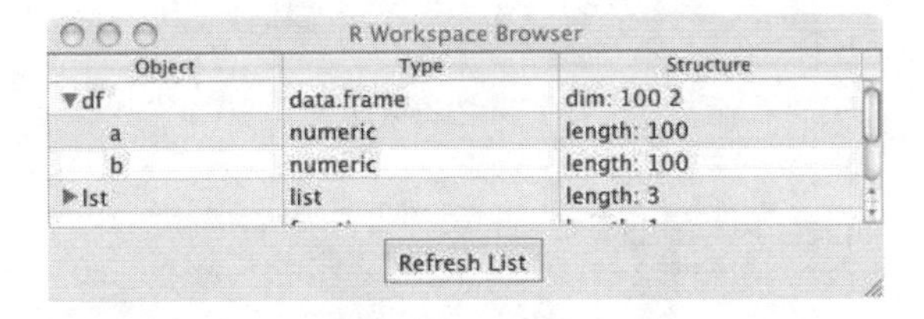

Figure B.5 Workspace browser for Cocoa GUI in Mac OS X

```
> library(Rcmdr)
```

This opens the main window, as in Figure B.6, with a menu bar and a log window. If you are using the Windows GUI, `Rcmdr` works better in the single-document interface (SDI). The default setting for the GUI is to use the multiple-document interface (MDI). To make the change is done by setting the option for SDI, not MDI, using the `File::Options` menu item.

Once running, the `R` session may be interacted with by means of the menu bar and the subsequent selection of commands.

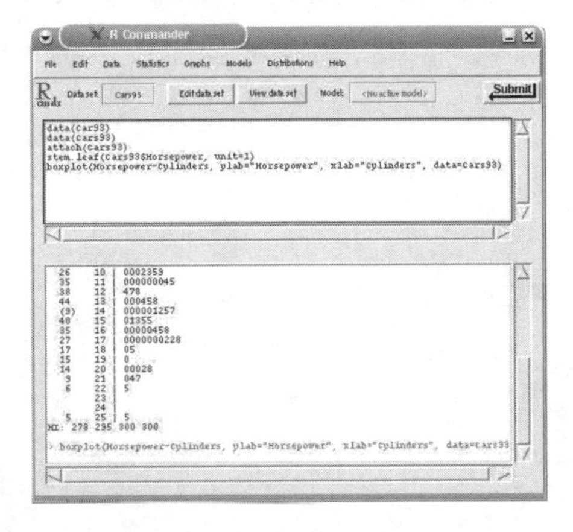

Figure B.6 Main `Rcmdr` **window**

In particular, first the user defines the active data set. This is done either with the `Data::Data in Packages...` menu item or the `Data::Active Data Set...` one. Once a data set is selected, `Rcmdr` ensures that the variable names become available to the other functions.

For example, creating a histogram of a univariate variable in the active data set is done from the `Graphs::Histogram...` menu (Figure B.7).

The desired variable is selected, as are some options, then `OK` is clicked to make the plot. In the console part of the `Rcmdr` window a command like the

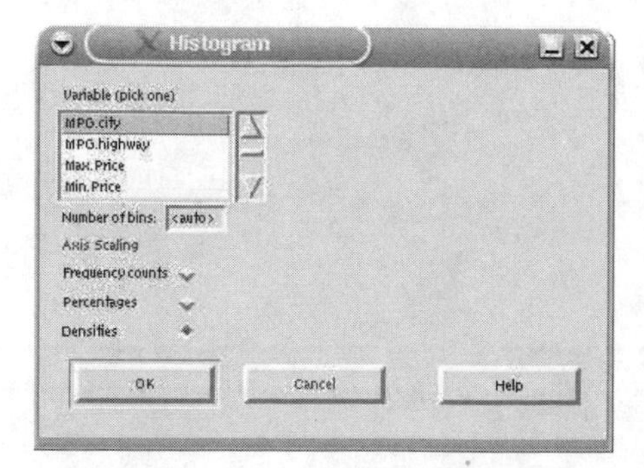

Figure B.7 `Rcmdr` **dialog for a histogram**

following is issued:

```
> Hist(Cars93$MPG.city, scale="density", breaks='Sturges', col="darkgray")
```

There are interfaces to many other graphical displays of data, as well as other common data-analysis features. For example, under the `Statistics` menu are dialogs to make summaries, to perform significance tests, and to fit statistical models.

Appendix C

Teaching with R

Using R in the classroom is not unlike using other software packages, but there are some features that are useful to know about that make things easier.

Getting students a copy of R for home use One of the great benefits of R is that it is free and it works on many different computing platforms. As such, students can be expected to use it for their homework without much concern about accessibility. Installing R on a computer is usually as easy as downloading the binary file by following the CRAN link at `http://www.r-project.org`.

However, R is a big download, around 20 megabytes. It may be better to have students burn R to CDs on campus, where a high-speed internet connection is likely available.

An alternative to installing R on a computer is running it from a CD. The Quantian distribution (`http://dirk.eddelbuettel.com/quantian.html`) is a version of Debian Linux on a single CD that includes R and a whole suite of other free software packages that are useful for scientific computing. Quantian is used by booting from the CD-ROM. R can be started from a terminal by the command R or from within an ESS session in XEmacs, a text editor that allows for interaction with the R process. This is done by starting XEmacs and then issuing the command `ALT-x R`.

Getting data to students Entering data can be a chore. One way to make data sets available to students is through an R package such as the one that accompanies this book. On CRAN, the packages `car`, `DAAG`, `Devore5`, `Devore6`, `ISwR`, and `MPV` contain many data sets that may be of interest to students of this book.

More informally, you can put data files or functions into a single file on a web site and then have the students `source()` these files in.

If the file is stored at the url `http://www.somewhere.com/rdata.txt`, then these commands will source it into the R session as though it were a local file:

```
> f = "http://www.somewhere.com/rdata.txt"
> source(url(f))
```

To make a file for reading in, use the `dump()` command. The syntax is

```
dump(list=..., file=...)
```

The `list=` argument should be a vector of names for the desired objects (data sets, functions, etc.) in the file. For example, `c("data.1","function.2")`. The `file=` argument should be a quoted file name. Typically, this is written to the current directory and then moved to a web site.

Making reports A report or handout in R that includes R commands and graphics can be made using a word processor or the LaTeX system. Using a word processor to present R materials is no different from creating other documents, except, perhaps, the inclusion of graphics.

Many newer word processors can import encapsulated PostScript files for figures. These are output by the `dev.copy2eps()` function. If your word processor imports pdf files, these can be made using the `pdf` driver with a command such as

```
> dev.print(file=filename, device=pdf)
```

To save a graphic in a format suitable for inclusion in a word processor in Windows, use the right mouse button to pop up the "Save as metafile..." dialog. The graphic can then be included in a document by inserting it as a figure.

The LaTeX text-processing system is freely available typesetting software that is well suited for technical writing. The `Sweave()` function of the `utils` package integrates with LaTeX to produce documents that automatically insert R commands and output into a document. One especially nice feature is the automatic inclusion of figures. For more information on LaTeX see `http://www.latex-project.org/`. More information on `Sweave()` is available in its help page: `?Sweave`.

Some projects for teaching with R At the time of writing, there are at least two projects under way that are producing materials for using R in a classroom setting.

The StatDocs project, by Deborah Nolan and coworkers, aims to create a framework for presenting statistics materials using R and some add-on packages from the Omegahat project, `http://www.omegahat.org`.

The Stem and Tendril project, `http://www.math.csi.cuny.edu/st`, by this author and colleagues, is producing freely available projects for statistics computer labs that can be used at various levels in the statistics curriculum.

Appendix D

More on graphics with R

This appendix revisits many of the plotting techniques presented in the text offering more detail and organization. In addition, functions are provided for two new graphics: a squareplot and an enhanced scatterplot.

This appendix provides a good deal of extra information but is by no means inclusive. More facts are available in the manual *An Introduction to R* that accompanies R, and in the help pages `?plot` and `?par`. The organization of this appendix was inspired by some excellent notes by R pioneer Ross Ihaka, which were found at `http://www.stat.auckland.ac.nz/~ihaka/120/`.

D.1 Low- and high-level graphic functions

Consider what the command `plot(x,y)` does when making a scatterplot. First it sets up a plot device if one isn't already present. Then areas for the figure and the margins are assigned, the limits on the x- and y-axes are calculated, the axes are drawn and labeled, a box is drawn around the graphic, and, finally, the points are plotted.

If we issue the command `plot(x,y,type="n")` instead of the command `plot(x,y)`, all but the last step of plotting the points is done. We refer to this as setting up the plot figure. High-level graphic functions will do this; low-level graphic functions add to the existing figure. The `plot()` function allows for many arguments to adjust this process. For example, we can plot connected line segments instead of points; we can add colors, labels, and titles; and we can adjust the axes. More details are in the help page `?plot`. What we focus on next is how to set up a figure step-by-step so that we can gain full control over the details. This is useful when we are designing our own graphics, or when we are not satisfied with the defaults.

D.1.1 Setting up a plot figure

The following steps are for the creation of any graphic, not just a scatterplot.

A plot device

A graphic is drawn to a plot device. Generally speaking, a default plot device is used. If we want, though, we can create a new device that allows us to control the type of device and the size of the figure. How we do so varies according to our system and our intentions. In Windows, the function `windows()` will create a device; on a UNIX machine running X Windows, the function `X11()` will create a device; and on Mac OS X, the function `quartz()` will make a new device using the default windowing system. The arguments `width=` and `height=` set the size of the graphic figure. An argument `pointsize=` will set the default size of any text. Other devices are available to make various graphic formats, such as PDF.

More than one device can be opened in this way. Each device is assigned a number. This is returned by `dev.list()`. Only one device is the "active" device. To switch between devices, use `dev.set()`, with the device number as the argument.

Once a device is set up, the `plot.new()` function tells the device to finish any current plot and prepare for a new plot.

The margins of a figure

A plot figure has space for the main graphic, and space in the margins for drawing axes and labels. Graphical parameters are consulted to allocate this space. We use `par()` to work with the parameters.

The `par()` function Graphical parameters are stored internally. To access and set them we use the `par()` function. All the current parameters are returned by calling `par()` without arguments. To retrieve a single value, use the name as the argument. Values can be set using the `name=value` style of R functions. For example, `par("mai")` returns information about the parameter `mai=`,* which controls the plot margins, and the value of `mfrow=` is set with `par(mfrow=c(2,2))`. The help page, `?par`, contains information about the numerous options.

Many of the graphical parameters may also be set as arguments to a high-level plot function. This can temporarily set the value for the graphic. Using `par()` will set them for the device.

Before changing the graphical parameters, we should save the current setup,

* We typeset graphical parameters with an extra = to emphasize that they are arguments to `par()`. This is not typed when the value of the argument is retrieved by `par()`.

in case we wish to return to the original choice. We can save the values to a list op with

```
> op = par(no.readonly=TRUE)
```

(Some of the arguments are read only and can't be changed. This command saves only those that are not read only.) Restoring the old parameters is done with

```
> par(op)
```

When changing values inside a function,

```
> on.exit(par(op))
```

will issue the command when the function exits, so the user doesn't have to worry about doing so.

Several parameters can be set by `par()`. For now we mention those used to make allocations in a figure for the graphic and the margins.

A device is set up with a certain size given by `din=`. A figure's size is given by the graphical parameter `fin=`. The figure is made up of its margins and the plot region. These are stored under `mai=` and `pin=`.

For example,

```
> par("din","fin","pin","mai")
$din                                # device size
[1] 6.995 6.996

$fin                                # figure size
[1] 6.995 6.996

$pin                                # plot size
[1] 6.133 5.717

$mai                                # margin sizes
[1] 0.7088 0.5698 0.5698 0.2919
```

The first three specify the sizes in inches for width and height. As margins may be nonsymmetric, there are four values specified: the bottom, left, top, and right.

The values add up as shown by

```
> .5698 + .2919 + 6.133             # widths
[1] 6.995
> .7088 + .5698 + 5.717             # heights
[1] 6.996
```

The margins can also be specified by the number of lines of text that can be printed using `mar=`. The actual size is relative to the current point size of the text. The argument `plt=` can specify the plot region in terms of the device size.

These areas are usually set up for us when the plot device is. At this point, the overall width and height are given, and R makes the necessary computations for the margins. If we are unsatisfied with the defaults, we can override the calculated values. If the margins or space for the main graphic are made too small, an error message will be given.

Multi-graphic figures

In a few examples in this text, more than one graphic is plotted on the same figure. For example, when we plotted the diagnostic plots after fitting a linear model, four graphics were returned. These can be seen one after another, or can be forced to show up all four at once if the graphic parameter `mfrow=` is set to `c(2,2)`.

The parameter `mfrow=` stores two values as a vector. A value of `c(3,2)` says to split the figure into six equal pieces, with three rows and two columns. The default is `c(1,1)`. Figures are filled in row by row. Use `mfcol=` to fill in column by column. To advance to the next area to draw a graphic, we can use a high-level plot command or the function `plot.new()`.

The `fig=` graphical parameter controls the amount of the current figure the next graphic will use. By default, this is set to `c(0,1,0,1)`, indicating the whole figure. The first two numbers are the *x* value, and the second two are *y*. They refer to the lower left of the figure. For example, a value like `c(0,1/2,1/2,1)` will use the upper-left part of the figure. The parameter `new=` should be set to `TRUE` to add to the existing figure with a high-level plot command.

In the example at the end of this appendix, the `layout()` function is used to break the figure into four regions of varying sizes. This function uses a matrix to set up a grid and an order to fill in the plots. The size of the grid is controlled with the arguments `widths=` and `heights=`. In the example, the function call is

```
layout(matrix(c(1,0,              # which order to place graphs
                3,2),
                2,2,byrow=TRUE),
       widths=c(3,1),             # 3/4 wide for col. 1
       heights=c(1,3),            # 3/4 wide for row 2
       respect=TRUE)              # make square
```

The matrix specifies how the figure is filled. In this case upper left, lower right, lower left. The value of `widths=` says to make the first column three times as wide as the second; for `heights=` the top row is one-quarter the height of the bottom one.

When there are multiple plot figures per overall figure, there is an outer margin that is controlled by `omi=`. Values are specified in inches. The arguments `oma=` and `omd=` allow you to specify values in lines of text and fractions of a whole respectively.

Setting up the coordinate system and axes of a graphic

When creating a plot figure, the *x*-and *y*-limits are specified, allowing locations to be specified in a Cartesian manner. This is done with `plot.window()`, using the arguments `xlim=` and `ylim=`. Values contain the minimum and maximum values of the desired range. Additionally, the parameter `asp=` can be set to give the aspect ratio in terms of width/height.

Once the coordinates are chosen, axes for a figure can be added with the `axis()` function. The common arguments are

```
axis(side=..., at=..., labels=...)
```

The value of `side=` is specified with 1, 2, 3, or 4, with 1 being the bottom of the figure and other values moving around clockwise. The value of `at=` allows us to override the default placement of the tick marks. The `labels=` argument can adjust the labels on the tick marks.

The `axis()` function can be used with high-level plotting functions to create custom axes if you explicitly ask for the axes not to be drawn when first plotting. This is done by specifying the arguments `xaxt="n"` and `yaxt="n"` to the high-level plot function.

Adding titles and labels to a figure

The `title()` function adds titles and labels the axes. The main arguments are

```
title(main=..., sub=..., xlab=..., ylab=...)
```

The value of `main=` is written in the top margin of the figure. The value of `sub=` is written in the bottom margin below the x-label, which is specified with `xlab=`. The y-label is specified with `ylab=`.

If more control is needed, the `mtext()` function will add text to the margins of a figure.

Adding a box around the graphic

Most of the high-level plot functions in R surround the main graphic with a box. This can be produced using the `box()` function. The argument `bty=` can be set to describe the type of box to draw. The default is a four-sided box. Many graphics in this book were produced with the value `bty="l"` to draw only two sides, like an "L". Other possible values are `"o"`, `"7"`, `"x"`, `"u"`, and `"]"`.

The value of `bty=` can be set globally using `par()`. If this is done, then just calling `box()` or a high-level plot command will draw the box the desired way.

■ **Example D.1: Setting up a plot window** To illustrate: if `x` and `y` are to be plotted in a scatterplot, we can mimic the work of `plot(x,y,type="n")` using the following sequence of commands:

```
> plot.new()
> plot.window(xlim=range(x), ylim = range(y))
> axis(1); axis(2)
> box()
> title(xlab="x",ylab="y")
```

■

D.1.2 Adding to a figure

Several functions were mentioned in the text to create or add to a figure. Examples would be `plot()` and the similar function `points()`.

For the plotting functions in R, several parameters can be set to change the defaults. When plotting points, some arguments refer to each point plotted, such as `col=` or `pch=`. This type of argument can be a single number or a vector of numbers. Usually the vector would be the same size as the data vectors we are plotting; if not, recycling will be done.

Adding points to a graphic

Both `plot()` and `points()` plot points by default. Points have several attributes that can be changed. The plot character is changed with the argument `pch=`. As of R version 1.9.0, the numbers 0 through 25 represent special characters to use for different styles of points. These are shown in Figure D.1. As well, we can use one-character strings, such as letters, or punctuation marks like "." and "+". To print no character, use a value of `NA` or a space character.

The size of the point is controlled by the `cex=` argument. This is a scale factor, with 1 being the current scale. Changing `cex=` using `par()` will apply the scale factor to the entire graphic.

The color of a point is adjusted with the `col=` argument. Again, trying to change this globally using `par()` will result in more than just the points having the specified color.

Specifying colors in R Though the default color for objects in a graphic is black, other colors are available. They are specified by name (e.g., `"red"`,`"white"`, or `"blue"`), by number using red-green-blue values (RGB) through `rgb()`, or by a number referring to the current palette. Over 600 names are available. See the output of the function `colors()` for the list. The value "transparent" is used when no color is implied. This is useful for producing graphic images that allow the background to come through.

The function `gray()` was used often (without comment) in the text to create gray scales. The argument is a vector of values between 0 and 1, with 0 being black and 1 white.

Some useful functions produce palettes of colors: for example, `rainbow()`, `heat.colors()`,`terrain.colors()`,`topo.colors()`, and `cm.colors()`. These functions have an argument `n=` that specifies the number of colors desired from the palette.

Additionally, the `brewer.pal()` function in the `RColorBrewer` package (not part of the base installation) will create nice-looking color palettes. Once the package is installed, running `example(brewer.pal)` will display several such palettes.

R stores a current palette that allows for reference of colors by a single number. This palette is returned and set by the function `palette()`. The default palette is:

```
> palette("default")              # clear out any changes
> palette()
[1] "black"   "red"     "green3"  "blue"    "cyan"    "magenta"
[7] "yellow"  "gray"
```

With this palette, asking for color 7 returns yellow.

The palette may be customized by specifying a vector of colors using their names or RGB values (or a function that returns such a vector). For example, a gray-scale palette can be made as follows:

```
> palette(gray((0:10)/10))
> palette()
 [1] "black"   "gray10"  "gray20"  "#4C4C4C" "gray40"  "#808080"
 [7] "gray60"  "#B2B2B2" "gray80"  "#E6E6E6" "white"
```

With this palette the color 7 is a shade of gray given the name `gray60`.

Adding lines to a graphic

The `plot()` and `points()` functions have an argument `type="l"` that will connect the points with lines instead of plotting each point separately. The `lines()` function is convenient for adding lines (really connected line segments) to a graph. If paired data is given in the vectors `x` and `y`, then `lines(x,y)` will connect the points with lines. Alternatively, a model formula may be used.

For these three functions, the values of `x` and `y` are connected with lines in the order in which they appear. If any of the values of `x` or `y` is `NA`, that line segment is not drawn. This can be used to break up the line segments.

It should be noted that although these functions draw straight lines between points, they are used to plot curves. This is done by taking the x values very close together, so that their straightness is smoothed out. For example, these commands will plot the function $f(x) = x^2$ over the interval $(-2,2)$:

```
> x = seq(-2, 2, length=200)     # 200 points
> y = x^2
> plot(x,y,type="l")
```

If we did this with `length=5` when defining `x`, the graph would look very clunky.

The `abline()` function is used to draw a single line across the figure.

The characteristics of the lines drawn can be varied. The color of the line segments are set with the `col=` argument. The line width is set with `lwd=`. Values bigger than 1 widen the line. The line type is set with `lty=`. A value of 1 through 6 produces some predefined types. These are shown in Figure D.1. If more control is sought, consult the `lty=` portion of the help page `?par`.

The changes can be made each time a line is drawn. Issuing the command `par(lty=2)` will cause all subsequent lines on the device to be drawn with style 2 (dashed) by default.

Adding a region to a graphic

The `rect()` function will plot rectangles on a figure, as is done with a histogram. A rectangle is specified by four numbers: the x- and y-coordinates of the lower-left corner and the upper-right corner. More general regions can be drawn using the `polygon()` function. This will plot a polygon specified by its x-and y-coordinates. For `polygon()`, unlike `lines()`, the first and last points are connected. This creates a figure containing an area.

The `col=` argument specifies the interior color of the regions; the line (or border) color is set using `border=`. An alternative to filling with color is filling with angled lines. These are specified by the arguments `angle=` (default is 45°) and `density=` (larger values produce more lines).

Adding text to a graphic

Adding text to a graphic can be done with `text()` or, in the special case of adding a legend, with `legend()`. The main arguments for `legend()` are the position, which can be specified interactively with `locator()`; the text to be added with `legend=`; and any of `pch=`, `col=`, and `lty=` as desired. These are usually vectors of the same length as the number of legend items we wish to add.

The `text()` function will add labels to a graph, with the option to format the text. The positions are specified with (x,y) values (or `locator()`). Text is centered at the (x,y) point, although the `at=` argument allows for adjustments. The text to add is given to the `labels=` argument.

Basic formatting can be done using the `font=` argument. A value of 1 will produce the default text, 2 bold text, 3 italic text, and 4 bold-italic text.

Math expressions can be printed as well. The full details are in the help page `?plotmath`. The basic idea is that `R` expressions are turned into mathematical expressions and then printed. For example, `expression(x==3)` will print as "$x = 3$." (The `expression()` function makes "expressions" that can subsequently be evaluated.)

■ **Example D.2: Showing values for `pch=` and `lty=`** The following commands produce Figure D.1, which illustrates the various plot characters and line types.

```
X11(width=5,height=2,pointsize=12)          # new UNIX device
par(mar=c(0,4,1,1))                         # small margins
plot.new()                                  # new plot
plot.window(xlim=c(-.5,26.5),ylim=c(0,8), asp=1)  # set up limits
k = 0:25                                    # pch values to plot
zero = 0*k                                  # same length as k
text(k, 8 + zero, labels=k)                 # add numbers
points(k,7 + zero, pch=k, cex=2)            # add plot characters
i = 6:1                                     # which line types
abline(h=7-i,lty=i)                         # where to plot line
axis(2,at=1:8,                              # at= for where
```

```
     labels=c(paste("lty =",i),"pch","k"), # labels for what
     las=2)                              # las=2 gives orientation
```

■

Figure D.1 Example of `pch=` and `lty=` values

D.1.3 Printing or saving a figure

R prints its graphics to a device. Usually this device is the plot window we see, but it need not be. It can also print to a file and store the output in different formats. For example, R can store the current graphic in Adobe's PDF format. This can be done by printing the current screen device to a pdf device:

```
## .. create a plot, and then...
> dev.print(file="test.pdf",device=pdf)
```

Adobe's PDF format is great for sharing graphs, but it isn't always the desired format for inserting into documents. By changing the argument `device=` to `png` or `jpeg` those file types will be saved. For some, PostScript is a convenient format. Encapsulated PostScript can be created with the function `dev.copy2eps()`. For Windows GUI users, the plot window has menus for saving the graphic in png, jpeg, bmp, postscript, PDF, and metafile formats. The Mac OS X GUI allows the user to save graphics in its native PDF format.

D.2 Creating new graphics in R

In this section we illustrate some of the aforementioned techniques by creating graphics for a squareplot and a scatterplot with additional information in the margins.

■ Example D.3: A squareplot alternative to barplots and pie charts

The *New York Times* does an excellent job with its statistical graphics. Its staff is both careful to use graphics in an appropriate manner and creative in making new graphics to showcase details. A case in point is a graphic the *Times* uses in place of a barplot, dotplot or pie chart that we will call a squareplot.

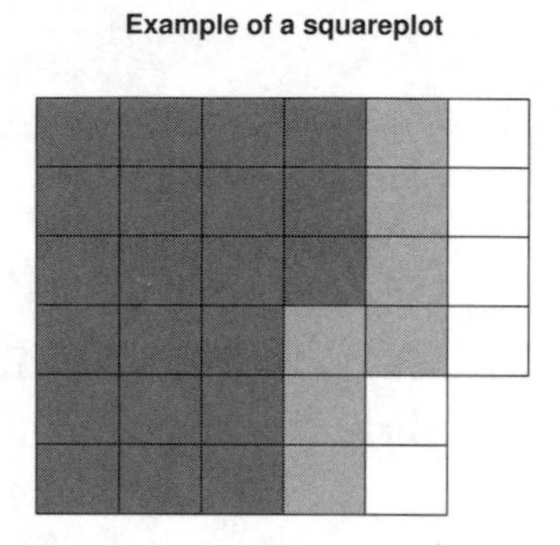

Figure D.2 Squareplot of `c(21,7,6)`

The squareplot shows counts of categorical data. Unlike the barplot, the squareplot makes it easy to see the exact counts. Unlike the dotplot, it is can be read without consulting the scales. Unlike the pie chart, the squareplot's relative areas are easy to discern.

The basic idea is to use squares of different colors to represent each count. The squares and colors are laid out in a way that facilitates counting values and comparing different plots. Figure D.2 shows an example, from which we can count that the categories have counts 21, 7, and 6.

The `UsingR` package contains the function `squareplot()`, which is reproduced below. Creating the graphic is pretty simple. A helper function to draw a square using `polygon()` is defined. Then the larger square is defined and laid out. An empty plot is made. Then a new vector, `cols`, is created to match the colors with the counts. This is done with `rep()` to repeat the colors. Finally, the squares are made and colored one-by-one using a for loop. The functions `floor` for the integer portion of a number and `%%` for the remainder after division are employed.

```
squareplot <- function(x,
                        col = gray(seq(.5,1,length=length(x))),
                        border=NULL,
                        nrows=ceiling(sqrt(sum(x))),
                        ncols=ceiling(sum(x)/nrows),
                        xlab = deparse(substitute(x)),
```

```
                        main = NULL,
                        ...
                        ) {
  ## create a squareplot ala the New York Times. Used as an
  ## alternative to a segmented barplot when the actual
  ## count is of interest.

  ## helper function
  draw.square <- function(x,y,w=1,...) {
    ## draw a square with lower left corner at (x,y)
    polygon(x+c(0,0,w,w,0),y+c(0,w,w,0,0),...)
  }

  ## size of big square
  square.size = max(nrows,ncols)

  ## setup window with plot.new() and plot.window()
  ## arguments to ... are passed along here
  plot.new()
  plot.window(xlim=c(0,square.size),ylim=c(-square.size,0),
       ...)
  title(main=main, xlab=xlab)

  ## vector with colors
  cols = rep(col,x)

  for(i in 1:sum(x)) {
    x.pos = floor((i-1)/nrows)          # adjust by 1
    y.pos = (i-1) %% nrows

    draw.square(x.pos,-y.pos -1,col=cols[i])
  }
}
```

■

■ Example D.4: A scatterplot with histograms

The scatterplot is excellent at showing relationships between two variables. However, the distributions of the individual variables are hard to see. If we add histograms along the axes of the scatterplot, the individual distributions become clearer. This example comes from the help page `?layout`. A similar graphic using boxplots is found using the `scatterplot()` function in the `car` package.

An example is illustrated in Figure D.3. This shows the per-capita gross domestic product (GDP) of several countries versus their CO_2 emissions. Notice the outlier that appears in the regression analysis, and the CO_2 variable. It does not appear as an outlier for per-capita GDP. It's a simple analysis but leaves us wondering which country this is and why.

```
> attach(emissions)
> names(emissions)
[1] "GDP"        "perCapita" "CO2"
> scatter.with.hist(perCapita,CO2)
```

A listing of the `scatter.with.hist()` function is given below. First, the

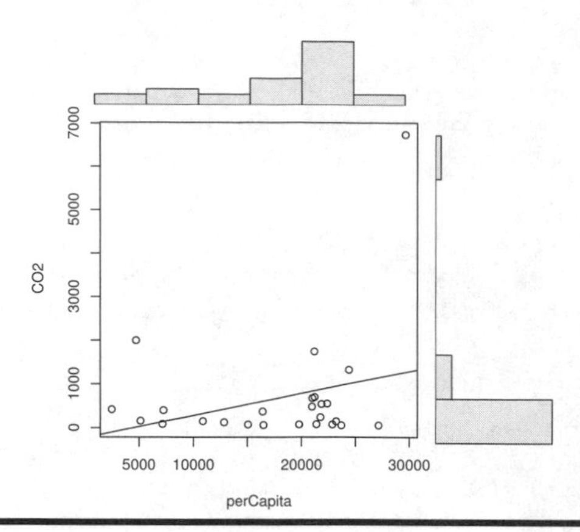

Figure D.3 Per-capita GDP versus emissions by country

old `par()` settings are saved. As these will be changed during the function, it's nice to return them as we found them. Then the layout of the graph is specified with the `layout()` function.

Then the histograms are drawn. Care is taken to get a similar number of breaks and a similar size. As well, the line `par(mar=c(0,3,1,1))` sets up the margins so that not too much white space appears.

Finally, the scatterplot is drawn. The expression `deparse(substitute(x))` finds the name of the variable that we called the function with for the label. We use `switch()` to add one of several trend lines: the regression line given by `lm()`, the fit given by `lowess()`, and the Friedman super-smoother fit given by `supsmu()` from the `stats` package. This logic could have been implemented with if-then-else lines, but that approach is more cluttered. The `invisible()` function is used to return quietly.

This function could be improved by adding an interface to the model formula notation. The techniques to do that are discussed in Appendix E.

```
scatter.with.hist <-
  function(x,y,hist.col=gray(.95),trend.line="lm",...) {

    ## Make a scatterplot with trendline and
    ## histograms of each distribution.

    on.par <- par(no.readonly = TRUE)
    on.exit(par(on.par))                # see ?par for details

    nf <- layout(matrix(c(1,0,          # which order to place graphs
                          3,2),
                        2,2,byrow=TRUE),
                 widths=c(3,1),         # 3/4 wide for col. 1
                 heights=c(1,3),        # 3/4 wide for row 2
                 respect=TRUE)          # make square
```

```
  layout.show(nf)

  n<-length(x)
  no.breaks = max(nclass.scott(x),nclass.scott(y))
  xhist <- hist(x,breaks=no.breaks, plot=FALSE)
  yhist <- hist(y,breaks=no.breaks, plot=FALSE)
  top <- max(c(xhist$counts, yhist$counts))

  ## adjust margins for better look
  par(mar=c(0,3,1,1))
  barplot(xhist$counts, axes=FALSE, ylim=c(0, top),
          space=0,col=hist.col)

  par(mar=c(3,0,1,1))
  barplot(yhist$counts, axes=FALSE, xlim=c(0, top),
          space=0,col=hist.col, horiz=TRUE)


  par(mar=c(4,4,1,1))
  x.name = deparse(substitute(x))
  y.name = deparse(substitute(y))
  plot(x,y,xlab=x.name,ylab=y.name,...)
  if(!is.null(trend.line) && !is.na(trend.line)) {
    switch(trend.line,
           "lm" = abline(lm(y~x)),
           "supsmu" = lines(supsmu(x,y)),
           "lowess" = lines(lowess(x,y)),
           NULL
           )
  }

  invisible()                          # restores par settings
}
```

■

Appendix E

Programming in R

One of R's advantages over many other statistical software packages is that at its core is a programming language with a consistent and relatively modern syntax. This allows us to write functions that simplify our work and extend the functionality of R to our problems at hand. The goal of this appendix is to introduce some of the key programming concepts and give enough examples of simpler stuff. The curious reader can find *much* more information in either *An Introduction to R* or *R Language Definition*, manuals accompanying R, or in the books *S Programming* by Venables and Ripley and *Programming with Data: A Guide to the S Language* by Chambers.

E.1 Editing functions

Programming can be repetitive: we write a function, test it, find errors, fix them, and repeat until we are happy with the result. Knowing how to make this process as painless as possible alleviates some of the tedium and lets us focus on what is important. A recommended approach is to use text files and an external editor (such as Notepad in Windows) to edit files. The `edit()` function can also manipulate functions from the command line.

E.1.1 Using `edit()`

Let's begin with the most studied of examples from computer programming: the "hello world" function. We can define such a function on the command line with

```
> hello = function() {cat("hello world\n")}
> hello()
[1] "hello world"
```

This is a basic function that prints (using `cat()`) the response “hello world” no matter what the input is. The “\n” tells R to print a new line. If we wanted to make a change to this function we could retype the definition with our desired changes. This is facilitated by using the arrow keys. However, using the `edit()` function will let us use an editor to make the changes, thereby providing more control over the editing process. We call `edit()` like this:

```
> hello = edit(hello)    # assign results of edit() back to hello
```

This command opens a text editor* to the function and allows us to edit. Make these changes, save, and exit.

```
function(x) {
  cat("hello",x,"\n")
}
```

(No prompts are given, as we are doing this in the editor and not on the command line.) Now our function can take an argument, such as:

```
> hello("kitty")
hello kitty
```

The function `fix()` is an alternative interface to `edit()`, which does not require us to assign the value back.

E.2 Using functions

A function in R is defined by the following pieces:

```
function ( arguments ) body
```

The body term is a block of commands. If there is more than one command in the body, they are put into a block using braces. A function declaration returns an object with class attribute “function.” A class attribute allows R to organize objects by type. The result of a function declaration is usually assigned to a variable or function name, although sometimes, such as with `sapply()`, functions are used anonymously. In the example above,

```
hello = function(x) {
  cat("hello",x,"\n")
}
```

the keyword `function` is used. The argument is simply the variable denoted x, and the body is the single command `cat("hello",x,"\n")`. In this simple case, where the body is a single command, the function could be written without the braces, as in

```
hello = function(x) cat("hello",x,"\n")
```

* The default editor varies depending on the operating system used. The editor can be changed using the `options()` function, as in `options(editor="jedit")`.

E.2.1 Function arguments

When we use a function, we typically pass it arguments, so that we can get different outputs. The arguments to a function are specified as a list of variable names separated by commas, such as

```
arg1, arg2, arg3
```

When a function is defined, default values may be specified using the `name=value` syntax, as in

```
arg1=default1, arg2=default2, arg3=default3
```

Arguments need not have defaults, but defaults are generally a good idea. The defaults can depend on other argument values, as R performs lazy evaluation. This will be illustrated in the upcoming example. A catch-all argument, `...`, can be used to pass unspecified, named arguments along to function calls made inside the body of the function. Once a function is defined, the `args()` function will return the argument list.

Since there can be many arguments to a function, there needs to be a convention for how arguments are handled when a function is called. R functions can be called with either named arguments or positional arguments. The named arguments are matched first. They appear in the function call as `name=value`. The `name` should match one of the function's arguments, or they will be passed along if a `...` argument is used. Named arguments may be truncated, as long as the truncated form uniquely identifies the argument. This will not work, though, if the argument appears in the function definition after a `...` argument. The use of named arguments is encouraged.

When a function is defined, the arguments have a natural position. If a function is called without named arguments, R will try to match the arguments in the function call by their position. This can make for less typing, but it is harder to debug.

Finally, if a function call does not include one of the function's arguments but a default is given, this default value will be used. If no default is specified, an error will occur.

We illustrate how R handles function arguments by an example.

■ **Example E.1: Our histogram function (how R handles function arguments)** The default `hist()` function in R is a bit lacking. First, as a histogram reminds us of the underlying density, it should look like a density. That is, it should be normalized so the area is 1. For similar reasons, it is nice to estimate the density and add it to the graphic. Finally, following the `truehist()` function of the MASS library, we use the "Scott" rule to select the bins.

Here is a first attempt at what we want:

```
ourhist = function(x) {
  hist(x,breaks="Scott",probability=TRUE)
  lines(density(x))
}
```

We can type this in on the command line, or define a function stub and then use `edit()`, as in

```
> ourhist = function(x) {}
> ourhist = edit(ourhist)          # now edit
```

Try it out.

```
> x = rnorm(100)
> ourhist(x)
```

It works fine. But what if we wanted to use a different rule for `breaks=`? It would be nice to be able to override our settings. One way would be to define a `breaks=` argument:

```
ourhist = function(x,breaks) {
  hist(x,breaks=breaks,probability=TRUE)
  lines(density(x))
}
```

A typical usage yields

```
> ourhist(x)
Error in hist.default(x, breaks = breaks, probability = TRUE) :
        Argument "breaks" is missing, with no default
```

Oops, we haven't set a default value for breaks and we didn't directly specify one, such as `ourhist(x,"Scott")`. Immediately after we list the argument for the function, we can supply a default value in the pattern `name=value`. Try this:

```
ourhist = function(x,breaks="Scott") {
  hist(x,breaks=breaks,probability=TRUE)
  lines(density(x))
}
```

Both `ourhist(x)` and `ourhist(x,breaks="Sturges")` will now work. The two commands show a difference in the number of bins, the second using the Sturges rule to compute the number instead of the default.

Still, the histogram drawn looks bland. Let's add some color to it—the color purple. The `hist()` function has a `col=` argument to set the color of the boxes. We can make the color be purple by default with this modification:

```
ourhist = function(x,breaks="Scott",col="purple") {
  hist(x,breaks=breaks,probability=TRUE,col=col)
  lines(density(x))
}
```

Trying it out gives

```
> ourhist(x)
> ourhist(x,"Sturges")          # use different bin rule
> ourhist(x,"Sturges","green")  # green before purple
> ourhist(x,"green")            # Oops
Error in match.arg(tolower(breaks), c("sturges", "fd",
"freedman-diaconis",   :  ARG should be one of sturges, fd,
freedman-diaconis, scott
```

We see that we can make changes to the defaults quite easily. The third line uses the Sturges rule and green as the color.

However, we also see that we can make an error. Look closely at the last line. We want to change the color to green but keep the default for the breaks rule. This didn't work. That is because `R` was expecting a breaks rule as the second argument to the function. To override this *positional matching of arguments* we use *named arguments*. That is,

```
> ourhist(x,col="green")
```

will work. Why? First R matches by named arguments, such as `col="green"`. Then it tries to match by partial matching of names. For example,

```
> ourhist(x,c="green")
```

will work, as no other arguments begin with the letter `c`. Finally R tries to match by position.

Default values can be written to depend on arguments given during the function call. The use of *lazy evaluation* of arguments allows this to happen. For example, it is common to have graphic labels depend on the name of the variable passed into a function. To supply a default title for the histogram can be done as follows

```
ourhist = function(x,breaks="Scott",col="purple",
                   main=deparse(substitute(x))
                   ) {
  hist(x,breaks=breaks,probability=TRUE,col=col,main=main)
  lines(density(x))
}
```

Now the default value for the main title is found by applying `substitute()` and `deparse()` to the argument `x`. This has the effect of making a character string out of an R expression. The term "lazy" refers to the fact that the value for `main=` isn't determined until it needs to be—when the function is called, not when the function is first defined.

There are many other things we might want to modify about our histogram function, but mostly these are already there in the `hist` function. For example, changing the x-axis label. It would be nice to be able to pass along arguments to our function `ourhist()` to the underlying `hist` function. R does so with the `...` *argument*. When our function contains three periods in a row, `...`, in the argument and in the body of the function, all extra arguments to our function are passed along. You may notice if you read the help page for `hist()` that it too has a `...` in its argument list.

Again, modify the function, this time to

```
ourhist = function(x,breaks="Scott",col="purple",...) {
  hist(x,breaks=breaks,probability=TRUE,col=col,...)
  lines(density(x))
}
```

Then we can do these things

```
> ourhist(x,xlab="histogram of x") # change the x label
> ourhist(x,xlab="histogram of x",col="green") # change both
```

■

E.2.2 Function body and return values

The function body is contained in braces if there is more than one command. The body consists of a collection of commands that are executed in the order given, although control statements can change this. The last command executed is the return value of the function. This can be forced by the `return()` function. If the return value should not print out, the `invisible()` function can be used for the return. This is used with many plotting functions.

Inside a block of commands, the `print()` or `cat()` functions are used to force printout to the console. Just evaluating the variable name will not force it to print as it does on the command line. In the Windows GUI, the printing may be buffered. The function `flush.console()` will print any buffered output.

Inside a function body, variable names are resolved by seeing if the name has been assigned inside the body or matches a variable name. If not, then values in the environment in which the function was called are used. If the variable name still can't be found an error will occur. Assignment inside a function body will not affect values outside the function body unless it's done with the `<<-` assignment operator or `assign()`.

This next example involves both return values and conditional evaluation.

■ **Example E.2: An EDA function (return values)** The `summary()` command is used to give textual summaries of a given data object. However, in many cases a graphical summary is also appreciated. We write a function that returns the summary but that also presents a graphical summary.

We name the function `eda()`. This first attempt will make a few plots and then return the summary command.

```
eda = function(x) {
  old.par = par(no.readonly = TRUE)     # See par examples
  on.exit(par(old.par))
  par(mfrow=c(1,3))                      # 3 graphs

  hist(x,breaks="Scott",probability=TRUE,col="purple")
  lines(density(x))

  boxplot(x,horizontal=TRUE)             # boxplot with points
  rug(x)                                 # marked by rug()

  qqnorm(x)                              # normal probability plot

  return(summary(x))                     # return summary
}
```

Looking at the body of the `eda()` function we see that the `par()` settings are saved into `old.par`. The `on.exit()` function executes commands when the function exits. In this case, it returns the original settings for `par()`. This usage is illustrated in the help page for `par()`. As this function changes the plot device by setting `mfrow=c(1,3)` to create three graphs on one screen, it is nice to return to the old settings when we leave. The three graphs are straightforward. Finally, the last line uses `return()` to return the value of the `summary()` function. In general, the last line evaluated is returned, but specifying the return value eliminates surprises.

Try it out a few times. For example, look at the output of `eda(rnorm(100))`.

Functions like this are pretty handy. It would be nice to improve it a bit. In particular, the function as written is good only for univariate, numeric data. What about eda for bivariate data? Categorical data? If we have categorical data, we might want to plot a barplot and return a summary. ■

E.2.3 Conditional evaluation

R has some standard ways to deal with conditional evaluation. Depending on a value of some condition, one of many things can be done.

if()

The `if()` function allows us to do something based on a condition. It takes two forms: an "if-then" form

```
if (condition) {
  statement(s) if condition is TRUE
}
```

and an "if-then-else" form

```
if (condition) {
  statement(s) if condition is TRUE
} else {
  statements(s) if condition is FALSE
}
```

The `condition` is a logical expression, such as `x > 0` or `x == y`. For example, the following is a really bad way of defining an absolute-value function:

```
abs = function(x) {
  if(x < 0) {
    return(-x)
  } else {
    return(x)
  }
}
```

The logic should be clear. If the value of `x` is less than 0, the return value is `-x`. Otherwise it is just `x`. This example will not work for a vector of numbers with more than one element as the built-in `abs()` function will.

Suppose we wanted to improve our `eda()` function defined previously by making it aware of different types of `x`. For example, if `x` is a factor, we want to present our summary differently. One simple way to do this is to have a conditional statement in the function, such as

```
if(is.factor(x)) {
  ## do factor summary
} else if(is.numeric(x)) {
  ## do numeric summary
} else {
  ## do a default summary
}
```

We could write such conditions for all the different types of data objects we have in mind. Sometimes the `switch()` function can help simplify the coding.

There are problems with this example. If we want to add a new type of variable, then the original `eda()` function needs to be available. This is fine for functions we write, but what about functions written by others for the system? We wouldn't want to modify those. Fortunately, there are styles in R for writing functions that eliminate this concern that are described in the section on object-oriented programming.

E.2.4 Looping

Computers, unlike humans, are very happy to do things again and again and again. Repeating something is called looping. There are three functions for looping: `for()`, `while()`, and `repeat()`. The latter is least common and will be skipped here.

for() loops

The standard `for()` loop has this basic structure:

```
for ( varname in seq ) {
  statement(s)
}
```

The `varname` is the name of a variable, the `seq` can be any vector or list, and the statements get executed for each value of `seq`. When `seq` is a vector, `varname` *loops* over each value. The statements are evaluated with the given value for `varname`. When `seq` is a list, `varname` refers to each successive component in the list (the values `seq[[i]]`, not `seq[i]`).

A simple example is a countdown program:

```
> for(i in 10:1) print(i)        # say blastoff
```

In this example `varname` is the variable `i`, and the `vector` is the set of numbers 10 through 1. We didn't use braces, as only one statement is called.

To illustrate further, let's consider a primitive way to compute a factorial. The definition of the factorial of a positive integer is $n! = n \cdot (n-1) \cdot \cdots \cdot 2 \cdot 1$.

```
fact = function(x) {
  ret = 1
  for(i in 1:x) {
    ret = ret*i
  }
  return(ret)
}
```

The loop runs over the values of the vector `1:x`. At each stage the running result is multiplied by the new value. We can verify that `fact(5)` returns `120`. (This function is already implemented in the `factorial()` function. Even so, it would be better written using the `prod()` function to avoid the loop.)

The statements `next` and `break` (no parentheses) can be used to skip to the next value in the loop (`next`) or to break out of the loop (`break`).

Using `while()`

The `for()` loop above is great to use when we know ahead of time what values we want to loop over. Sometimes we don't. We want to repeat something as long as a condition is met. For example, as long as a number is positive, or a number is larger than some tolerance. The function `while()` does just this. Its template is

```
while (condition) {
  statement(s)
}
```

Here is a simple example that counts how many tails there are before the first heads.

```
tosscoin = function() {
  coin = "tails"                # initialize condition
  count = -1                    # get counting right this way
  while(coin == "tails") {
    coin = sample(c("heads","tails"),1)
    count = count + 1
  }
  cat("There were",count,"tails before the first heads\n")
}
```

The usage of the functions `while()` and `for()` can often be interchanged. For example, we can rewrite the factorial example above with the `while()` function as follows:

```
fact1 = function(x) {
  ret = 1
  while(x > 0) {
    ret = ret*x
    x = x-1
  }
  return(ret)
}
```

There is no real savings here. In fact, neither function is a very good way to perform this task.

E.3 Using files and a better editor

If you plan on doing any significant programming in R, then it is worthwhile to familiarize yourself with using external files to store your work (rather than the default workspace) and using a different editor to facilitate interaction between the external files and an R process.

E.3.1 Using an external editor

The `edit()` function makes small changes easy, but the changes are not saved in a convenient way for future reference. They can be saved in the workspace, but this can become quite cumbersome. It is often better to keep functions in files in the working directory. Many functions can be kept in a single file, or each one in its own file. Commands can also be stored in a file for subsequent editing.

The basic idea is to edit the file and then have its contents parsed and evaluated in R line by line. This last step is done with the function `source()`. This process is repeated until you are satisfied.

For example, you can save the `hello` function in a file called "hello.R" using a text editor of your choice (e.g., Notepad) and then read the contents in using the `source()` function, as in

```
> source("hello.R")
```

Specifying the file name can be done conveniently with the `file.choose()` function. More details on specifying a file are given in Chapter 1.

Better text editors

If you are going to be programming a lot in R, it makes sense to use an editor well suited to the job. The default editor in Windows, Notepad, is quite primitive; the default editor (often vi) in UNIX may be too cryptic. The editor can be changed using `options()`, but to which one? Good editors should do some work for you. For writing programs in R what is desired are features such as on-the-fly code formatting, syntax highlighting, integration with an R process, and debugging help.

There are several choices. The most advanced and powerful is a combination of Emacs (either XEmacs, `http://www.xemacs.org`, or GNU Emacs, `http://www.gnu.org/`); and ESS, (`http://www.analytics.washington.edu/statcomp/projects/ess/`). Emacs is a text editor and ESS extends the editor so that it can interface with many statistical software packages. This setup works under Windows, Linux, and Mac OS X. This working environment, or integrated development environment (IDE), provides an interactive shell with TAB completion of command names and parentheses matching; a history mechanism; and integrated help pages. Additionally, functions and script files can be edited and evaluated in an R session with a simple command that can be typed or controlled by the mouse. The editing has built-in indenting, which allows you to identify quickly the block of a program being edited. The only real drawback is the complexity of learning Emacs.

Many people prefer other tools for editing. A list of editors that integrate with R appears on the R GUI web page at `http://www.r-project.org/` under the "IDE/Script Editors" link.

E.4 Object-oriented programming with R

Object-oriented programming (OOP) is a more advanced topic than those covered in the rest of this book. This section is included here as the freely available materials on OOP are not that accessible even for those with an understanding of

OOP principles. The goal is to give a brief introduction to R's implementation(s) for those who have some previous experience with OOP.

OOP is especially useful for large programming projects. It became popular in the 1990s and is now ubiquitous. Most languages taught (e.g., Java, C++) have an OOP component. OOP requires an initial investment in time to structure the data and methods, but this pays off in the long run. Often, for statistical exploration, it is programming overkill, but it should be considered when you are programming any larger projects.

The nature of object-oriented programming in R has changed with the introduction of the base `methods` package and the add-on `OOP` package. At the time of writing there are four types of OOP for R: the original S3 methods; the `R.oo` package, `http://www.maths.lth.se/help/R/`; the newer S4 methods; and the `OOP` package. The `OOP` package extends the S4 style; the `R.oo` package extends the S3 style.

The notion of a class and associated methods are integral to OOP, and their implementation will be discussed below. But first, for OOP in R, an understanding of method dispatch is helpful.

E.4.1 Method dispatch

Method dispatch comes in when R decides which function to call. For example, when the R `print()` function is invoked, say with the command `print(x)`, what happens? If the function invoked is a **generic function**, R first looks at what `x` (the first argument) is and then, based on this, uses an appropriate idea of "print." To determine what `x` is, R considers its class attribute. Many R objects have a class attribute; others have an implicit class attribute, such as data vectors. Their implicit class attribute is inherited from their mode: for example, "character" or "numeric." The `class()` function determines the class of an object. Once the class is determined, say it is "classname," R looks for the appropriate function to call, depending on the type of function. As `print()` is an S3 generic function, R looks first for the function `print.classname`. If it finds it, it uses that function. If not, it goes again to the next value of the class of `x` (it can have more than one) and tries again. Finally, if everything fails, it will use the function `print.default()`. This process of resolution is called *method dispatch*.

Users of R rely on this all the time. For example, we've seen that many different plots are produced by the workhorse function `plot()`: `plot(x,y)` will produce a scatterplot, `plot(y ~ x)` will also, `plot(y ~ f)` will produce boxplots (for a factor `f`), and `plot(lm(y~x))` will plot four diagnostic plots. There are many other usages we didn't show, such as `plot(sin)`, which will plot the sine function over $[0, 1]$. All use the function name `plot` (a "generic" function). This generic function uses the class of the first argument to call the appropriate idea of plot. The end result for the user is far fewer function names to remember. The end result for the programmer is that it is much easier to add new types of data objects.

To illustrate the notion of method dispatch, we create a simple function that tells us how "large" a data variable is.

■ **Example E.3: Defining a `size()` function (an example of method dispatch)** When exploring the built-in data sets that R provides or that are otherwise available to us, we may not know the size of the data set. If the data set is really large and undocumented, then just typing the variable name can be very annoying. Fortunately, there are functions to help us know the size of a variable. For example, `length()` will tell us the length of a vector or list and `dim()` will tell us the dimension of an array or data frame. If we want a single command to tell us the size, we can define one. Let's call it `size()`.

We want `size()` to adapt itself to the various types of variables it sees—just like the `summary()` function. In order for R to do this, we first need to define `size()` to be a generic function as follows:

```
> size = function(x,...) UseMethod("size")
```

This says that, when encountering the function `size()`, use the method based on the class of `x` (the first argument).

Now we need to define some methods. First we define a default method (called `size.default()`) and some others for different classes:

```
> size.default = function(x) length(x)
> size.data.frame = function(x) dim(x)
> size.list = function(x) lapply(x,size)
```

The default for `size()` is the `length()` of the object. This works well for vectors. For data frames the number of rows and columns are returned. For lists we define the size to be the dimension, or the size of each entry in the list. We could also define functions to be dispatched for matrices and arrays if desired.

Let's see what it does:

```
> size(1:5)                    # for integers
[1] 5
> size(letters)                # for characters
[1] 26
> size(factor(letters))        # for factors
[1] 26
> size(data.frame(a=1:5,b=letters[1:5])) # for data frames
[1] 5 2
> size(list(a=1:5,b=letters,c=list(d=1:4,e=letters))) # for lists
$a
[1] 5
$b
[1] 26
$c
$c$d
[1] 4
$c$e
[1] 26
```

We see that the list example recursively uses `size()`. ■

E.4.2 S3 methods, S4 methods, and the OOP package

In the previous example, we defined methods for lots of classes. Now we give an example of defining a class, methods for this new class, and creating a new instance of the class. For a concrete example, we will create a "string" class using S3 methods, S4 methods, and the OOP package.

R has many built-in functions for handling strings, but their syntax can be confusing. This example (influenced by the xtable package) defines a class for strings and gives it some syntax similar to the String class in the Ruby programming language (http://www.ruby-lang.org/).

This example covers the following: creating a new class, creating new instances of the class, and defining methods for instances of the class. Inheritance, another important part of OOP, is briefly mentioned. The code is available in the accompanying UsingR package.

Extending the usual syntax by overloading In the upcoming example, it makes good sense to define "slicing" and "adding" of strings. When doing these things with a data vector, the operators [and + are used. By overloading these operators, the same natural syntax can be used with strings.

When overloading an operator, we must take care with S4 methods. This is because the default arguments (the signature) of the new function must match those of the current implementation. For example, the [operator has this signature: i,j,...,drop. (See ?"[" for details.) Any overloading of "[" must contain this same set of arguments.

S3 methods

Creating a new class using S3 style is very simple. The "class" is an attribute of an object that can be set with the class() function, as in class(x) = "String".

We make a function, string(), to create new instances of our String class and a function, is.String(), to inform us if an object is in our String class.

```
string = function(x) {
  if (!is.String(x)) {
    x = as.character(x)
    class(x) = "String"
  }
  return(x)
}
is.String = function(x) return(class(x) == "String")
```

(We write these functions as though they are "sourced" in from a file. The good way to organize this would be to include all these new functions in a single file, as they are in a file in the UsingR package.)

Now when a generic function is called on an instance of this class, the function for the String class, if present, is used. For example, the String class should have some basic functions, such as length(), summary(), and print().

```
length.String  = function(x,...) return(nchar(x))
```

```
summary.String = function(x,...) return(table(strsplit(x,"")))
print.String   = function(x,...) cat(x,"\n")
```

At this point we can see if it works as expected. For example:

```
> bart = string("I will not skateboard in the halls")
> bart
I will not skateboard in the halls
> length(bart)
[1] 34
> summary(bart)
  I a b d e h i k l n o r s t w
6 1 3 1 1 2 2 2 1 4 2 2 1 2 3 1
```

The `summary()` function splits the string into characters and then makes a table. By modifying the `summary.string()` function other summaries could be given.

It would be nice to be able to access the letters in a `String` object as we do the elements in a vector. First, let's define a function `slice()` and then use this to define access via `"["`. To do so, `slice()` is made a "generic" function in order for method dispatch to work, and then defined as desired:

```
slice = function(x,index) UseMethod("slice")
slice.String = function(x,index) {
  tmp = unlist(strsplit(x,""))[index]
  return(string(paste(tmp,sep="",collapse="")))
}
```

To make `slice()` generic, we used `UseMethod()`. The definition of `slice()` uses `strsplit()` to split the string into characters.

To make the vector notation work on the `String` class, we overload the square-bracket meaning by defining a function as follows:

```
"[.String" = function(x,index) slice(x,index)
```

The quotation marks are necessary with this function definition as they prevent `[` from being interpolated. The `"["` function is already a generic function, so we don't need to use the `UseMethod()` call.

If we knew we didn't want to use `slice()` in any other context, we needn't have made it generic. For example, we might want to concatenate strings using the (overloaded) `"+"` symbol:

```
concat.String = function(x,y,sep="") {
  x = string(paste(x,string(y),sep=sep))
  return(x)
}
"+.String" = function(x,y) concat.String(x,y)
```

We imagine using just the + notation to add strings, not `concat()`. This is why we didn't bother making `concat()` a generic function. Due to this, the definition of `"+.String"` must use `concat.String()`, and not simply `concat()`.

To see this in action, we have

```
> Simpsons = string("Homer, Marge")
> Simpsons[1:5]
Homer
> Simpsons + ", " + " Bart, Lisa, Maggie"
Homer, Marge, Bart, Lisa, Maggie
```

S4 methods

We now create the String class using S4 methods. S4 methods are well documented in the help pages (see ?Methods). They are, unfortunately, a little difficult to read through. We will see in our simple example that certain aspects of S4 methods are much stricter than S3 methods, thereby forcing good programming habits.

This example defines the class and methods incrementally. Alternatively, these can be defined all at once.

First, we define a new class using the setClass() function. An S4 object has slots for storing the data that need to be specified. A slot stores a component, like a list element, allowing one to set and access the data.

For the simple String class, the one slot stores the value of the string using the class character.

```
setClass("String",representation(string = "character"))
```

The second argument to setClass() uses the representation() function to define what type of data will be stored in the slots. The name and class of each slot is specified, forcing the proper type of data in that slot. Only character data, or data that can be coerced to character data, can be stored in the string slot. The example shows only one slot being created; more than one can be created by separating the definitions by commas.

Creating an instance of a class is done with the new() function, as illustrated in this helper function:

```
string = function(x) {
  new("String",string=as.character(x))
}
```

The slot holding string is set with the value of as.character(x). Only slots that are already defined for the class can be filled (defined by setClass() or through inheritance). Otherwise an error message, "Invalid names for slots," will be returned.

Trying this out we have

```
> string("I will not eat things for money")
An object of class "String"
Slot "string":
[1] "I will not eat things for money"
```

The string() function works, but the printout is not as desired. Instead of print(), when the methods package is used, S4 methods use show() for their default print call. We can make a show() function tailored for our class using setMethod(). The setMethod() function needs a name, a class, and a function defining the method.

```
setMethod("show","String",
          function(object) {
            cat(object@string,"\n")
          })
```

Now objects of our String class will use this method, instead of the method showDefault(), to print. The function definition must have the proper formal arguments, or an error will be thrown. This enforces a uniformity among methods that share a name. That isn't the case with S3 methods. The new syntax, `object@string`, illustrates how we access a slot in an object. The argument to `show()` is called `object` and is an instance of the `String` class. The slot `string` is desired, which is accessed using the `@` syntax, as in `object@string`.

Now the default printing is more suited for a string.

```
> string("I will not waste chalk")
I will not waste chalk
```

The `show()` function is an S4 standard generic function, meaning for us that method dispatch will work. The `setMethod()` function will also work with S3 methods, such as `summary()`, and "primitives," such as `length()`. For example, we can create such methods for the `String` class as follows:

```
setMethod("length","String",function(x) nchar(x@string))
setMethod("summary","String",
          function(object,...) {
            return(table(strsplit(object@string,"")))
          })
```

We need to use the argument `x` in `length()` and `object` in `summary()` to match their formal arguments (signature). The signature of a function can be found from its help page (e.g., `?summary`), or with the function `args()` (e.g., `args(summary)`).

New methods can be created as well. For example, to create a `slice()` method and the `"["` syntax, we first create a generic function `slice()`, adapt it to the `String` class, and then link `"["` to this.

```
setGeneric("slice", function(object,index) standardGeneric("slice"))
```

This sets up `slice()` as a generic function for which method dispatch will apply. The definition of the function sets up the signature of the `slice()` function, namely an `object` and an `index`. The definition of `slice()` for the `String` class is

```
setMethod("slice","String",
          function(object,index) {
            tmp = paste(unlist(strsplit(object@string,""))[index],
              sep="",collapse="")
            return(string(tmp))
          })
```

To define the `"["` method, we again use `setMethod()`. It does not need to be made generic, but the signature must match.

```
setMethod("[","String",function(x,i,j,...,drop) slice(x,i))
```

There is a long list of formal arguments. The `"["` method works for vectors, data frames, arrays, and now strings, so it needs to have lots of flexibility in how it is called. We have little flexibility in how a new method is defined, though: at the minimum, we must include all the arguments given.

We can try this out:

```
> pets = string("Snowball, Santa's little helper")
> pets[-1]
nowball, Santa's little helper
> pets[length(pets)]
r
```

The OOP package

For those used to programming in an OOP language, the S3 and S4 methods might be called "object oriented," but they don't look object oriented. Usually, the syntax for calling a method is `object-separator-methodname`, such as `$object->print()` in PERL or `object.print()` in Ruby, C++, and Java. The separators `->` and `.` are not well suited for R, as `->` is reserved for assignment and `.` is used for punctuation. Rather, the natural `$` is used.

This object-method style forces the attention on the object and not the function. In addition, many methods can change the value of the object in place, meaning that to change the object, instead of requiring an assignment statement such as `object = object.method(newValue)`, a command such as `object.method(newValue)` can be used. Using S3 and S4 methods, this is not available. The add-on package OOP from `http://www.omegahat.org` allows for such programming style. This package is still in development and is not merged into the main R distribution. It may not even work with newer versions of R (it missed the 1.8.x series). What follows works with R version 1.9.1 and OOP version 0.5-1.

First, we must load the library with the command `library(OOP)`. It takes a bit of time to load.

To define a new class using OOP, the `defineClass()` function is used, as in

```
defineClass("String")
```

The "slots" are now called "fields." To set up the fields, the `defineFields` method is used on the `String` class.

```
String$defineFields(string = "character")
```

We have only a single field of type character. The `defineFields()` method is placed after the separator `$`, and the object here is `String`.

The `new()` function is called to make a new instance of a class. There is a default call, but we can override it by defining a method called `initialize`.

```
String$defineMethod("initialize",
                    function(val) {
                      set.string(as.character(val))
                    })
```

The `set.string()` function assigns to the `string` field. For each field, an assignment method `set.fieldname()` is created, as well as a variable `fieldname` containing the values. The `initialize()` method is called by the `new()` method. In this case, it sets the field value of `string` to the value of the argument after coercion to a character.

Now we can create objects or instances of the `String` class using `new` as follows:

```
> bart = String$new("I will not drive the principal's car")
> bart
Object of class "String":
Field "string":
[1] "I will not drive the principal's car"
```

Our printout needs work, but we can see that we have an object of class "`String`" and that the `string` field is as expected.

For our String class we want the `length()`, `summary()`, and `show()` methods as before.

```
String$defineMethod("length",function() return(nchar(string)))
String$defineMethod("summary",function() {
  return(table(strsplit(string,"")))
})
String$defineMethod("show", function() cat(string,"\n"))
```

The formal arguments needed for S4 methods are not necessary here.

We can try these out now:

```
> barfriends = String$new("Moe, Barney, Lenny, Carl")
> length(barfriends)
[1] 1
> barfriends$length()
[1] 24
> barfriends$summary()

  , B C L M a e l n o r y
3 3 1 1 1 1 2 3 1 3 1 2 2
> barfriends
Object of class "String":
Field "string":
[1] "Moe, Barney, Lenny, Carl"
> barfriends$show()
Moe, Barney, Lenny, Carl
```

The function `length()` must be called as `barfriends$length()` and not as `length(barfriends)`. The latter dispatches the wrong function. This also causes `show()` to work now as it did with S4 functions. A workaround is to define a `print()` method using either S4 or S3 methods,

```
setMethod("print","String", function(x,...) x$show()) #S4 style
```

or

```
print.String = function(self) self$show() # S3 style
```

Now the `show()` method will be called when the object is "printed."

```
> barfriends
Moe, Barney, Lenny, Carl
```

The same thing can be done to use the `"["`, `"+"`, etc., syntax. For example, defining a method to split the string by a pattern and use `"/"` for a shortcut can be done as follows:

```
String$defineMethod("split",function(by="") {
  unlist(strsplit(string,by))
})
setMethod("/","String",function(e1,e2) e1$split(e2))
```

Again `strsplit()` does the hard work. We defined a new method, `split`, but didn't have to worry about matching the formal arguments of the previously defined `split()` function (an S3 generic method). However, when we use the S4 style to use the `"/"` syntax, we need our function to match the formal arguments it uses (`e1,e2`).

We can now "divide" our strings as follows:

```
> flanders = String$new("Ned, Maude, Todd, Rod")
> flanders$split()              # into character by default
 [1] "N" "e" "d" "," " " "M" "a" "u" "d" "e" "," " " "T" "o" "d"
[16] "d" "," " " "R" "o" "d"
> flanders/" "
[1] "Ned,"   "Maude," "Todd,"  "Rod"
```

As mentioned, when using OOP we can modify the object in place. For example, we might want to make the string uppercase. This can be done by defining an `upcase` method as follows:

```
String$defineMethod("upcase",function() {
  set.string(toupper(string))
})
```

Applying the function gives

```
> flanders$upcase()
> flanders
NED, MAUDE, TODD, ROD
```

The `upcase()` method uses both the `string` variable and the `set.string()` function, which are created from the field names. Simple assignment to `string` will not work, although the `<<-` assignment operator will.

Inheritance

In OOP, inheritance allows us to define fields and methods for a class and have them available to all subclasses. Subclasses are used to extend a class by adding functionality that is desired in the specific case, keeping the core functionality in the parent class for other subclasses to share.

The OOP package allows for the fields and methods to be inherited in a simple manner. For example, to create a Sentence class to extend the String class is easy. When defining the class we need only say it "extends" the String class as follows:

```
defineClass("Sentence", extends = "String")
```

Now, instances of the Sentence class inherit all the methods of the String class.

```
> flanders = Sentence$new("the Flanders are Ned, Maude, Todd, and Rod")
> flanders
the Flanders are Ned, Maude, Todd, and Rod
```

This shows that the `print()` method was inherited. We can add more methods to the Sentence class that are specific to that class. For example, we might want to ensure that our sentences are punctuated with a capital letter for the first letter, and a period for the last character.

```
Sentence$defineMethod("punctuate",function() {
  ## add a period, and capitalize the first letter
  chars = split("")
  chars[1] = toupper(chars[1])
  if(chars[(length)(chars)] != ".")
    chars[(length)(chars)+1] = "."
  set.string(paste(chars,sep="",collapse=""))
})
```

We define punctuate() using defineMethod(). The split() method uses the method defined for the String class, whereas the length() functions are for the length of a character vector. OOP would try to use the length() method we defined, so we need to place the function in parentheses to use the one we want. Warning messages will remind us if we get this wrong.

We can now punctuate our sentences:

```
> flanders$punctuate()
> flanders
The Flanders are Ned, Maude, Todd, and Rod.
```

Inheritance with S4 methods is very similar. Again, the classes that the new class extends (the superclasses) are specified at the time a new class is made. For example:

```
setClass("Sentence",contains=c("String"))
```

The argument is contains= and not extends=.

New instances are made with new() as before:

```
sentence = function(x) {
  new("Sentence",string=x)
}
```

Methods for the new class are added in a similar manner:

```
setGeneric("punctuate", function(object) standardGeneric("punctuate"))
setMethod("punctuate","Sentence",function(object) {
  ## add a period, and capitalize the first letter
  chars = split(object,"")
  chars[1] = toupper(chars[1])
  if(chars[length(chars)] != ".")
    chars[length(chars)+1] = "."
  return(sentence(paste(chars,sep="",collapse="")))
})
```

In this case, the object is not modified. Rather, a new sentence is returned. Again split is used from the definition in String. Here the function length() is correctly dispatched, and there is no need to include it in parentheses.

For more information on S4 classes, the help page ?Methods gives pointers and links to other help pages. There is an informative article on S4 classes in Volume 3/1 (June 2003) of R News (http://www.r-project.org/). For the OOP package the help page ?defineClass provides useful information and links to the rest of the documentation. An out-of-date, but still informative description by the OOP authors is contained in the Volume 1/3 (September 2001) of R News.

Problems

E.1 Write your own sample standard deviation function, std, using the sample variance formula

$$s^2 = \frac{1}{n-1}\sum(x_i - \bar{x})^2.$$

E.2 Write two functions `push()` and `pop()`. The first should take a vector and an argument and add the argument to the vector to make it the last value of the vector. The function `pop()` should return the last value. How might you modify `pop` to return the value *and* the shortened vector? How might you change `pop` and `push` when `x` is an empty vector?

E.3 Write a short function to plot a simple lag plot (cf. `?lag.plot` in the `ts` package). That is, for a vector `x` of length `n`, make a scatterplot of `x[-n]` against `x[-1]`. Apply the function to random data (`rnorm(100)`) and regular data (`sin(1:100)`).

E.4 Write a function to find a confidence interval based on the t-statistic for summarized data. That is, the input should be $\bar{X}$, S, n, and a confidence level; the output should be an interval.

E.5 Newton's method is a simple algorithm to find the zeroes of a function. For example, to find the zeroes of $f(x) = x^2 - \sin(x)$, we need to iterate the equation

$$x_{n+1} = x_n - \frac{x_n^2 - \sin(x)}{2x_n - \cos(x_n)}$$

until the difference between x_n and x_{n+1} is small. In pseudo-code this would be

```
while( delta > .00001) {
  old.x = x
  x = x - (x^2 - sin(x))/(2*x - cos(x))
  delta = | x - old.x |
}
```

The answer you get depends on your starting value of x. This should be passed in as an argument. Implement this in `R` and find both roots of $f(x)$.

E.6 Type in the `size()` example above. Try the following:

```
> x = rnorm(100)
> size(x<0)
```

Does it work? If not, write a new method to handle this case. (Use `typeof()` to find the type of `x<0`.)

E.7 Add a new S3 method to the String class. We want to "divide" strings by breaking the string up into pieces based on the numerator. For example, dividing "now is the time" by " " (a blank) would return a vector of words split up by the blank spaces. This can be done with the function `strsplit()` as illustrated here:

```
> x = "now is the time for all good men"
> y = " "
```

```
> unlist(strsplit(x,y))          # unlist to get a vector
[1] "now"  "is"   "the"  "time" "for"  "all"  "good" "men"
```

Write an S3 method so that this could be done as

```
> x = string("now is the time for all good men")
> x/y
[1] "now"  "is"   "the"  "time" "for"  "all"  "good" "men"
```

E.8 Define an S3 "subtract" method for our String class that extracts all instances of the string y from x. Such as

```
> x = string("now is the time for all good men")
> y = string(" ")
> x - y                          # remove the blanks
nowisthetimeforallgoodmen
```

(You might want to use the "divide method" and then `paste()`.)

E.9 Write S4 methods `upcase()` and `downcase()` for the String class which interface with `toupper()` and `tolower()`.

E.10 Write an `OOP` method `strip()` that removes unnecessary spaces in the string. For example, `" this has too many spaces "` would become `"this has too many spaces"`.

Index

>, 5
... argument, 170, 389
:, 13
<-, 8
=, 8
?, 12
NA, 20
[[]], 120
#, 6
%in%, 20
if statement, 175
letters, 38
while loop, 175
AIC(), 309
IQR(), 51
I(), 96
TukeyHSD(), 328
UseMethod(), 400
abline(), 93
anova(), 308, 337
aov(), 317
apply(), 72, 123
apropos(), 12
attach(), 25, 116
barplot(), 33
binom.test(), 188
boxplot(), 65
  first argument a list, 119
  model formula, 126
box(), 377
browseEnv(), 21
bwplot(), 133
cat(), 388
cbind(), 70
chisq.test(), 252
choose(), 151
class(), 397, 399
colnames(), 71
colors(), 378
  cm.colors(), 378
  gray(), 378
  heat.colors(), 378
  rainbow(), 378
  terrain.colors(), 378
  topo.colors(), 378
cor(), 87
cut(), 60, 112
c(), 9
data.entry(), 12
data.frame(), 114
  access, 115
  list notation, 121
  matrix notation, 116
  size of, 115
  vector notation, 121
  with predict(), 94, 295
densityplot(), 132
detach(), 26
diff(), 15, 84
dimnames(), 115
dim(), 115
dotchart(), 37
dump(), 27, 372
ecdf(), 266
edit(), 169
example(), 12, 199

factor(), 38, 136
file.choose(), 28
fivenum(), 52
fix(), 169, 388
for() loop, 394
ftable(), 107, 129, 340
function(), 169
getwd(), 27, 28
glm(), 346
help.search(), 12
hist(), 57
identify(), 97
interaction.plot(), 338
is.na(), 20
jitter(), 90
kruskal.test(), 320
ks.test(), 268
lapply(), 124
layout(), 376, 384
length(), 115
levels(), 38
lines(), 60, 78, 296
    NA, 379
    col=, 379
    lty=, 379
    lwd=, 379
list(), 114
lm(), 92, 280
    data=, 95
    subset=, 95, 280
    extractor function, 282
        AIC(), 309
        anova(), 308
        predict(), 294
        residuals(), 282
        summary(), 293
locator(), 97
loess(), 101
ls(), 21
mad(), 55
margin.table(), 72
mean(), 44
    na.rm=, 44
    trim=, 46
median(), 44
new(), 401
nls(), 352
objects(), 21
on.exit(), 392
oneway.test(), 316
order(), 120
pairs(), 109
palette(), 378
par(), 374
    fig=, 155
    mfrow=c(1,3), 392
plot.window(), 376
plot(), 60, 82
    cex=, 43, 86, 378
    col=, 378
    lty=, 86
    pch=, 86, 378
    type=, 60, 86
    xlab=, 43
    usage
        boxplots, 127
        densityplots, 60
        model formula, 95
        scatterplot, 82
polygon(), 380
prop.test(), 187, 220
    alt=, 220
    p=, 220
qqline(), 80
qqnorm(), 80
qqplot(), 80
quantile(), 50
range(), 48
rbind(), 70, 261
read.csv(), 114
read.table(), 28, 114
recordPlot(), 78
replayPlot(), 78
rep(), 14, 318, 319, 349
rm(), 21
rownames(), 71
rug(), 60, 155
sample(), 143, 147, 182, 250
sapply(), 124, 177
scale(), 52, 125, 209
scan(), 27
scatter.smooth(), 101
sd(), 49
seq(), 14
setClass(), 401
setwd(), 28
shapiro.test(), 272
show(), 401

smooth.spline(), 101
source(), 27, 173, 396
split(), 129
stack(), 129, 317, 318
stem(), 42
stepAIC(), 309, 348
stripchart(), 42
subset(), 119, 304
  select=, 119
  subset=, 119, 348
summary(), 52, 293, 347
supsmu(), 101
switch(), 384
t.test(), 193, 223
  paired=TRUE, 242
table(), 32
text(), 65
title(), 59
t(), 75
unlist(), 140
unstack(), 129
update(), 304
url(), 29
var.test(), 248
var(), 49
which(), 18
while(), 174, 395
wilcox.test(), 208
with(), 25
xtabs(), 127, 261, 340
xyplot(), 133
command line
  +, 7
  ;, 6
density(), 60
help(), 11

adjusted R^2, 292
alternative hypothesis, 214
ANOVA, 313
assignment, 7

Bernoulli random variable, 150
binary variable, 343
binomial random variable, 150
bootstrap sample, 176
boxplot, 64
built-in data sets, 24

c.d.f., 146
central limit theorem, 161
class, 398
class attribute, 136, 388, 397
coefficient of determination, 292
coefficient of variation, 55
command line, 411
  >, 5
confidence ellipse, 300
confidence interval, 184, 185
  t-based, 190
  TukeyHSD(), 328
  difference of means, 201
  for proportion, 184
  nonparametric, 207
  regression coefficients, 292
contingency tables, 106
CRAN, 23
critical values, 215
cumulative distribution function, 266

data frame, 113
data recycling, 11, 122
data sets in UsingR
  BushApproval, 199
  MLBAttend, 330
  MLBattend, 40, 130, 283, 311
  OBP, 54, 58, 226
  age.universe, 181
  alaska.pipeline, 299, 300
  alltime.movies, 65
  babies, 109, 117, 206, 226, 232, 247, 275, 301, 326, 334, 347
  babyboom, 67, 273
  batting, 90, 103
  baycheck, 312
  best.times, 299
  breakdown, 104
  bright.stars, 256
  brightness, 68, 195, 275
  bumpers, 66
  bycatch, 176
  cabinet, 55, 211
  cancer, 113, 114, 129
  carsafety, 112, 135, 322, 331
  central.park.cloud, 32
  central.park, 35, 39
  cfb, 46, 63, 67, 124, 212, 335
  coins, 75
  deflection, 298, 300, 311

diamond, 103
dvdsales, 76, 125
emissions, 99, 287
ewr, 108, 115, 118, 123, 125, 329
exec.pay, 51, 55, 68, 232
father.son, 159
fat, 89, 102, 311
female.inc, 135, 322
firstchi, 66
five.yr.temperature, 101
florida, 76, 97
galileo, 304, 310
galton, 89, 90, 248, 284, 299
grades, 71
grip, 342
hall.fame, 67, 124, 322, 357
healthy, 357
homedata, 82, 89, 92, 195, 298
kid.weights, 42, 85, 95, 134, 135, 195, 334
last.tie, 62
lawsuits, 67, 68, 178
mandms, 30, 255
math, 66
maydow, 84
midsize, 358
mw.ages, 104
nba.draft, 148
normtemp, 66, 82, 89, 125, 195, 226, 247, 275, 335
npdb, 40, 55, 130, 331
nyc.2002, 199
nym.2002, 30, 54, 67, 89, 334
oral.lesion, 265
pi2000, 54, 55, 66, 68, 255
primes, 29
reaction.time, 80
reddrum, 358
salmon.rate, 231
samhda, 112, 221, 253, 261
scrabble, 256
stud.recs, 81, 195, 226, 269, 270, 272, 275, 302, 309
student.expenses, 105
tastesgreat, 356
too.young, 103
twins, 81, 89
u2, 125, 211
urchin.growth, 354
yellowfin, 353
data vector, 9, 138
degrees of freedom, 290
density, 145
density estimate, 60
distribution of a random variable, 141
dot chart, 37

empirical distribution, 266
error sum of squares, 314
error term, 90
ESS, 360, 371, 396
external packages, 23, 362
extra sum of squares, 307

five-number summary, 64
for loop, 167
for loops, 165
frequency polygon, 59

generalized linear model, 346
generic function, 95, 126, 397

H-spread, 53

i.i.d., 146
identically distributed, 146
independence
independent events, 143
independent random variables, 258
sequence of random variables, 146
independent, 258
influential observation, 99
inter-quartile range, 51

lag plot, 55
law of large numbers, 160
least-trimmed squares, 100
level of confidence, 185
link function, 345
log-odds ratio, 345
logical expressions, 18
logical operators, 19
logistic regression, 345
logistic-regression, 343
long-tailed distribution, 62

margin of error, 185
marginal t-tests, 289, 307
marginal distribution, 72

matrix, 70
  access entries, 116
  create with `dim()`, 140
mean sum of squares, 290
method of least squares, 91, 302
mode of a distribution, 61
model formula, 92, 127, 132
  `*`, 338, 339
  `+`, 278
  `-`, 278
  `:`, 338
  `^`, 339
  `I()`, 278
  simple linear regression, 92
multinomial distribution, 250

nested models, 307, 337
nonlinear models, 343
nonparametric, 207, 228
normal quantile plot, 79
null hypothesis, 213

observational study, 3
outlier, 63, 99

p-value, 214
  t-test, 223
  one-way ANOVA, 316
  sign test, 229
  test of proportions, 219
p.d.f., 146
partial F-test, 308
Pearson correlation coefficient, 292
Pearson's chi-squared statistic, 251
percentiles, 50
pie chart, 36
pivotal quantity, 197
plot device, 367, 374
  `X11()`, 374
  `quartz()`, 374
  `windows()`, 374
population mean, 144
population standard deviation, 144
prediction interval, 295
predictor variable, 90

quantile-quantile plot, 79
quantiles
  pth quantile, 50
  quartiles, 50
  quintiles, 50

random sample, 146
random variable, 141
range, 48
rank-sum statistic, 210
ranked data, 88
regression coefficients, 90
rejection region, 215
residual, 91, 280
residual sum of squares, 280
resistant measure, 44
response variable, 91
robust statistic, 193

sample, 146
sample mean, 43
sample median, 44
sample standard deviation, 49
sample variance, 49, 56
sampling distribution, 148
scatterplot, 82
scatterplot matrix, 109
short-tailed distribution, 62
significance level, 214
significance test, 214
  t-test, 223
  sign test, 229
  test of proportion, 219
  chi-square test, 252
  Kolmogorov-Smirnov test, 267
  partial F-test, 308
  rank-sum test, 245
  Shapiro-Wilk test, 271
  signed-rank test, 230
  two-sample t-test, 238
  two-sample test of proportion, 234
simple linear regression model, 90, 277
skewed distribution, 62
skewed left, 62
skewed right, 62
slicing, 17
slots, 401
Spearman rank correlation, 88
spike plot, 143
standard error, 185, 233
standard normal, 153
standardized residuals, 285
startup file, 132, 361

statistical inference, 2, 141
statistically significant, 214
stem-and-leaf plot, 41
strip chart, 42
summation notation, 47
symmetric distribution, 61

tails of a distribution, 62
test statistic, 214
    χ^2 test statistic, 251
    t-test, 222
    Kolmogorov-Smirnov, 267
    proportion, 218
    sign test, 228
    signed-rank test, 230
transform, 95
treatment coding, 326
treatment sum of squares, 315
trimmed mean, 46
type-I error, 215
type-II error, 215

unimodal distribution, 61

weighted average, 48
working directory, 28

z-score, 52